www.wadsworth.com

www.wadsworth.com is the World Wide Web site for Thomson Wadsworth and is your direct source to dozens of online resources.

At *www.wadsworth.com* you can find out about supplements, demonstration software, and student resources. You can also send email to many of our authors and preview new publications and exciting new technologies.

www.wadsworth.com
Changing the way the world learns®

INTERPERSONAL
PROCESS IN THERAPY

An Integrative Model

FIFTH EDITION

EDWARD TEYBER
California State University, San Bernardino

THOMSON

™

BROOKS/COLE

AUSTRALIA • CANADA • MEXICO • SINGAPORE • SPAIN • UNITED KINGDOM • UNITED STATES

THOMSON
BROOKS/COLE

Interpersonal Process in Therapy: An Integrative Model, Fifth Edition
Edward Teyber

Executive Editor: Lisa Gebo
Assistant Editor: Alma Dea Michelena
Editorial Assistant: Sheila Walsh
Technology Project Manager: Barry Connolly
Executive Marketing Manager:
  Caroline Concilla
Marketing Assistant: Rebecca Weisman
Senior Marketing Communications Manager:
  Tami Strang
Project Manager, Editorial Production:
  Lori Johnson

Art Director: Vernon Boes
Print Buyer: Doreen Suruki
Permissions Editor: Kiely Sisk
Production Service: G & S Book Services
Copy Editor: Laura Larson
Cover and interior illustration:
  Bill Stanton
Cover Printer: Phoenix Color Corporation
Compositor: G & S Book Services
Printer: R.R. Donnelley, Crawfordsville

Printed in the United States of America
1 2 3 4 5 6 7 09 08 07 06 05

For more information about our products, contact us at:
Thomson Learning Academic Resource Center
1-800-423-0563

For permission to use material from this text or product, submit a request online at **http://www.thomsonrights.com.** Any additional questions about permissions can be submitted by email to **thomsonrights@thomson.com.**

Library of Congress Control Number:

ISBN 0-534-51564-9

Thomson Higher Education
10 Davis Drive
Belmont, CA 94002-3098
USA

Asia (including India)
Thomson Learning
5 Shenton Way
#01-01 UIC Building
Singapore 068808

Australia/New Zealand
Thomson Learning Australia
102 Dodds Street
Southbank, Victoria 3006
Australia

Canada
Thomson Nelson
1120 Birchmount Road
Toronto, Ontario M1K 5G4
Canada

UK/Europe/Middle East/Africa
Thomson Learning
High Holborn House
50–51 Bedford Row
London WC1R 4LR
United Kingdom

Latin America
Thomson Learning
Seneca, 53
Colonia Polanco
11560 Mexico
D.F. Mexico

Spain (including Portugal)
Thomson Paraninfo
Calle Magallanes, 25
28015 Madrid, Spain

Dedicated to those who are struggling to change

CONTENTS

FOREWORD

I am deeply honored to have been asked to write the foreword to this book. It is a classic in the field. I have used previous versions of Ed Teyber's book for years in my graduate class in counseling theories and strategies. The students always recommend that I use it again the next year, whereas they routinely suggest that other books should not be used again. And many of the students go on to proclaim that interpersonal therapy is the best fit for their emerging theoretical orientation after reading it.

So what is it that students like so much about this book? First, the writing style is very clear and accessible, something that is quite rare in psychology books. Second, I think it strikes a chord for students. They can readily identify with the family dynamics presented in the book, and they can see that working through relationships in their own lives would be very important. Most student therapists are terrified when they first try to use process comments, but they also quickly note that the energy in the room changes and intensifies. And this active, authentic approach helps students feel the freedom of being able to talk more openly about interpersonal problems, albeit tentatively, and try to resolve problems with clients, classmates, and others in their own lives.

Students have often asked me where this theory fits in the range of theoretical orientations. My answer is that it fits perfectly within the niche of insight-oriented, psychodynamic theories, in what is known as interpersonal theory. Teyber fits well with other classic interpersonal theorists such as Cashdan, Kahn, Kiesler, Safran and Muran, Strupp and Binder, and Sullivan because of his focus on the centrality of interpersonal relationships. He adds to these theories, though, by including thinking from attachment theory, emotion theory, and family systems theory, thereby expanding and enriching the conceptualization of interpersonal functioning. This approach also draws on theories and techniques from client-centered, cognitive-behavioral and other approaches, but Teyber's real emphasis is on the therapeutic relationship and the treatment process. Hence, this integrative method allows for using interventions from a variety of theoretical orientations to reach specific clients, but the focus is on the interaction between the therapist and client as different interventions are applied, rather than the techniques per se.

I am also often asked whether Teyber's approach is compatible with my three-stage integrative model (Hill, 2004). I reply that it is very compatible, especially in its egalitarian, client-centered philosophical foundation. However, we emphasize different things in our approaches. I focus on content (the helping skills themselves) and teach it through specific examples. My approach is to help student therapists learn to modify their interventions by paying careful attention to the process (being aware of their intentions, carefully choosing their interventions, watching for the client reactions, checking in with clients about their reactions, and modifying interventions based on client need). Building on these same helping skills, Teyber focuses more on the process of how things occur (the metacommunication level) and suggests various ways of intervening in the therapeutic relationship and working with their interpersonal process.

In addition, I focus relatively equally on three stages of therapy (exploration, insight, and action), primarily using immediacy as one skill within the insight stage to help repair ruptures in the process and help clients gain insight into their interpersonal functioning. Teyber also addresses issues in the exploration and action stages but focuses much more intensively on immediacy (for example, using process comments, self-involving comments, and interpersonal feedback throughout every stage of treatment). He also includes other interventions (how to establish a working alliance, working on emotions, responding to resistance, helping the client generalize changes to relationships outside therapy), but his focus is on how these and other treatment issues are being played out or expressed in the current interaction between the therapist and client. The therapy process might not look very different if you watched Teyber and me doing therapy. We do, however, emphasize different things when we write about therapy. Teyber zooms in and magnifies the immediacy process and helps us understand what is going on in the therapeutic relationship, whereas I pull back and provide more of an overview of the entire therapy process.

That said, perhaps not surprisingly, I believe that when students are first learning counseling skills, a focus on the helping skills is most appropriate. Students need something to grab onto and take with them into sessions, so being able to know how to ask open questions and offer restatements and reflections of feelings gives them something concrete to do to help the client explore. Often they also need to alter communication styles that work with friends but don't work so well in a therapeutic situation (such as giving advice or reassuring). Once they master these basic skills, though, the focus shifts and they are ready to go on to conceptualize clients and look more deeply into the relationship, as Teyber does so effectively (and they typically quit focusing on the individual helping skills). Hence, at an advanced undergraduate level, I teach just the helping skills. But at the beginning master's level, I use my helping skills text (Hill, 2004) and supplement it with primary readings in each of the three stages so that students are exposed to lots of different theories and learn how to think for themselves about what fits for them. Hence, in the exploration stage, students read books written by humanistic, client-centered theorists. In the insight stage, they read psychodynamic, existential, and interpersonal theorists (here's where Teyber fits). And in the action stage, they read cognitive and behavioral theorists.

The final question that students always ask when they read this book is how to pronounce Teyber's name. I just learned that it is pronounced "Tiber," like the river in Rome.

In closing, I want to thank Ed Teyber for his innovative work and advancing the theory of psychotherapy. Although I have never met Ed personally, I have formed a relationship with him through reading his book. I feel that I know him because of his open, engaging, direct style. I am a great fan of his work and would love to sit around with him and discuss theories and brainstorm about how we could test these theories in a research setting.

Clara E. Hill
University of Maryland

PREFACE

The relationship between the therapist and the client is the foundation for therapeutic change. Across different theoretical approaches, and in short- or longer-term modalities, researchers find that the therapist's ability to establish a strong working alliance early in treatment may be the best predictor of treatment outcome. However, developing therapists need help conceptualizing the therapeutic relationship and learning how to intervene in the current interaction or interpersonal process that is occurring between the therapist and client. Thus, this text presents a comprehensive treatment approach that teaches student therapists how to use themselves, and the relationships they establish with their clients, as the most important way to help clients change.

Clinical training is often so stressful for new therapists because they are painfully uncertain of how to proceed with their clients. They take helpful courses on counseling theories and helping/micro skills, but student therapists need more specific help as they approach their first clients in session. Fully cognizant of their limited experience and knowledge, these trainees are often anxiously aware that they do not really know what to do or how to help their clients. Although bright and caring, many feel inadequate and worry about making "mistakes" and doing something wrong that would hurt their clients. As a result, new therapists need a conceptual framework to help them understand where they are trying to go in treatment, and why, in order to help their clients change. The interpersonal process approach replaces the ambiguity that students often have about treatment with a well-developed framework for understanding how change occurs, and the role of the therapist–client relationship in the change process.

This text helps new therapists understand therapeutic relationships by focusing on the current interaction or interpersonal process that is transpiring between the therapist and the client. Specifically, it helps developing therapists learn how to use "process comments" and other "immediacy interventions," such as metacommunication, self-involving statements, and interpersonal feedback, to intervene in the here-and-now, current interaction with clients. Speaking in an approachable, conversational tone, this text draws from a variety of theoretical perspectives in encouraging new therapists to work with the interpersonal process dimension to facilitate change. The model teaches readers to (1) identify significant cognitive patterns and relational themes in the client's behavior, (2) help the client recognize how these patterns function both for better and for worse in the client's life, (3) change how these faulty patterns and outdated coping strategies are played out in the session in the immediate transactions between the client and therapist, and (4) generalize the relearning from such shared experiences beyond the therapy setting. This treatment model demonstrates a way of being with clients in session that is genuine, empathic, and collaborative, and also challenges clients to explore the interpersonal process they are enacting with the therapist and others.

Describing the course of treatment from the initial session through termination, this text clarifies each of the major issues that arise in treatment, and shows how theory leads to practice. With clarity and immediacy, it highlights the challenging clinical situations that new therapists are facing, and captures the questions and concerns that are most salient for student therapists as they begin seeing clients. The interpersonal process model is an integrative approach that incorporates cognitive-behavioral, family systems, and interpersonal-dynamic theories. This integrative framework encourages student therapists to draw flexibly from varying theoretical perspectives and to develop their own personal styles, while incorporating components of the interpersonal process approach.

Complex clinical concepts are introduced throughout each chapter, illustrated with numerous clinical vignettes and sample therapist–client dialogues that student therapists will find informative and compelling. Clinically authentic and personally engaging, this text will help students understand the therapeutic process and how change occurs. It is written for graduate student therapists in practicum/internship courses who are seeing clients and for students in upper-division and prepracticum courses that provide an in-depth, applied introduction to counseling and therapy.

Over the years, I have appreciated the opportunity to return to this work in subsequent editions, trying to clarify the interpersonal process model further and make it more practical and accessible for new therapists. The goals for this Fifth Edition were to further develop certain core concepts, such as empathic understanding, working collaboratively, and "rupture and repair" in the working alliance. I also wished to elaborate how maladaptive relational patterns and cognitive schemas are expressed in the therapeutic relationship; and clarify more fully how the experiential or in vivo relearning that has occurred with the therapist is generalized to other relationships.

This edition also highlights how the cognitive, affective, and interpersonal domains interact and reciprocally influence each other. It draws more fully from varying theoretical orientations to make the process dimension and immediacy interventions more understandable and applicable. Preparing trainees to practice in a managed-care milieu, I have clarified how the interpersonal process approach is applied to other treatment modalities, including couples therapy, group therapy, family therapy and time-limited treatment.

This edition also includes structured guidelines to help student therapists write the case formulations, treatment plans, and process notes that are necessary in a managed-care environment. In addition, an attachment-informed perspective, including the importance of understanding the range of attachment-related affects, especially the taboo affect of shame, has been expanded.

Finally, it is challenging for new therapists to work "in the moment" with clients and try out these process-oriented interventions. To assist with these here-and-now interventions that bring intensity and a more authentic dialogue, I have provided many more sample therapist–client dialogues and case vignettes to illustrate effective and ineffective ways to intervene with the process dimension. I hope they are helpful as readers strive to become more effective therapists.

Regarding ancillaries, the Fifth Edition provides instructors and students with a Student Workbook and an Instructor's Manual. The Student Workbook accompanying this text is an especially important aid to help readers integrate the personally challenging and evocative material presented in the text. It provides a study guide with key concepts and terms, and sample multiple-choice and short essay questions. The Student Workbook also includes extensive case studies of characters (clients) from well-known films (such as *Ordinary People*; "Dorothy" in *The Wizard of Oz*) and plays (*Death of a Salesman*) from an interpersonal process perspective. There are also self-assessment measures that readers can utilize to explore their progress in developing process-oriented skills, and questionnaires to help identify their own countertransference propensities. In each chapter, there are self-reflection questions to help students integrate and apply the material in that chapter to their own lives and professional development.

An Instructor's Manual also accompanies the text and is available both in print and online. The manual contains an instructor's lecture outline, in-class student exercises, and sample essay questions and multiple-choice questions for every chapter. Also, clips from an instructional video are posted on the Book Companion website at: http://counseling.wadsworth.com/teyber. The clips portray myself demonstrating interpersonal process concepts from the text in different situations that are often challenging for beginning therapists. Critical thinking questions based on the clips are also posted.

ACKNOWLEDGMENTS

Many friends and colleagues have continued to help me learn more about this work and to present it more effectively. In particular, I wish to thank two long-time friends, Margaret Dodds-Schumacher and Faith McClure. I have appreciated their expertise as editors, teachers, and therapists and thank them for their generous contributions to the Fifth Edition.

Two former graduate students, now colleagues, have accompanied me on every step of this revision. Karen Dobbins and Dianne Swanson have helped greatly, and I appreciate all they have given. I also wish to thank Debra Douwenga for all of her assistance. Additionally, I would like to express my appreciation to the reviewers who made suggestions for this revision: Cynthia Glidden-Tracey, Arizona State University; Diane Harris-Wilson, San Francisco State University; Adina Smith, Montana State University; and Ellen Whipple, Michigan State University.

Finally, it is a pleasure to work with Alma Dea Michelena and so many good friends and colleagues at Brooks/Cole.

ABOUT THE AUTHOR

Edward Teyber is a professor of psychology and director of the Psychology Clinic at California State University, San Bernardino. He received his Ph.D. in clinical psychology from Michigan State University. Dr. Teyber is also the author of the popular press book, *Helping Children Cope with Divorce* (Jossey-Bass, 2001), and coauthor with Dr. Faith McClure of *Casebook in Child and Adolescent Treatment: Cultural and Familial Contexts* (Brooks/Cole, 2003). He has published research articles on the effects of marital and family relations on child adjustment and contributes articles on parenting and postdivorce family relations to newspapers and magazines. Dr. Teyber also enjoys supervision and clinical training, and he maintains a part-time private practice.

PART ONE

Introduction and Overview

CHAPTER ONE The Interpersonal Process Approach

The Interpersonal Process Approach

Claire, a 1st-year practicum student, was about to see her first client. She had long and eagerly awaited this event. Like many of her classmates, Claire had decided to become a therapist while working on her undergraduate degree. Counseling * had always been intrinsically interesting to her. For Claire, being a therapist meant far more than having a "good job"; it was the fulfillment of a dream. How meaningful, she thought, to make a living by helping people with the most important concerns in their lives.

At this moment, though, Claire felt the real test was at hand: Her first client would be arriving in a few minutes. Worries raced through her mind: What will we talk about for 50 minutes? How should I start? What if she doesn't show up? What if I do something wrong and she doesn't come back next time? Claire also wondered whether her client—a 45-year-old Hispanic woman—would have difficulty relating to her, a Caucasian woman in her mid-20s. Needless to say, Claire was intent on finding a way to help this client with her problems. She had learned something about therapy in her undergraduate psychology classes and a good deal more from her volunteer experience with callers on the local crisis hotline. But even with these experiences and a supervisor to guide her, Claire was painfully aware of her novice status and the fact that she didn't know very much about actually doing therapy.

Although some of Claire's classmates thought she could be a little idealistic, they did share her excitement about becoming a therapist. Many of them were

---

* The terms *counseling* and *therapy* are used interchangeably throughout this text.

older than Claire and far more experienced in life. Some had raised children; others had already had careers as teachers, nurses, and businesspeople. These new therapists were often coming from life roles in which they had already been successful and felt confident. A counseling career held new hopes for these more seasoned classmates as well, but it also evoked anxiety about their ability to become effective therapists. With each successive step—completing undergraduate requirements, being accepted into a graduate program, and now starting their first practicum—they were approaching their new career goals. Like Claire, however, they knew that realizing their dreams of a rewarding new career would also depend on their ability to help others. And with the arrival of their first clients, that ability was about to be tested.

Therapists Need a Conceptual Framework

Usually, beginning therapists' anxieties about their ability to help are not to be dismissed as just neurotic insecurity or obsessive worrying. It is realistic to be concerned about one's performance in a new, complex, and ambiguous arena. Sadly, however, Claire and her classmates often lose sight of the significant personal strengths that they already possess and bring to their first counseling experiences. Most students who are selected for clinical training already possess sensitivity, intelligence, and a genuine concern for others. Such personal assets, and all they have learned from their own life experiences, will prove helpful to their future clients. In this regard, treatment outcome studies repeatedly find that success rates in treatment have more to do with the individual therapist than the theoretical orientation or type of treatment (Hovarth & Bedi, 2002; Wampold, Mondin, Moody, Stich, Benson, & Ahn, 1997; Seligman, 1995; Luborsky et al., 1986).

Personal experience, common sense and good judgment, and intuition are useful, indeed. However, in order for a therapist to be effective with a wide range of people and problems, these valuable human qualities need to be wed to a conceptual framework. Therapists working within every theoretical tradition become more effective when they understand specifically what they are trying to do in treatment. When new therapists know where they are going and why, they can be more consistently helpful to a wide range of people and problems. Without a conceptual framework as a guide, however, decisions about intervention strategies and case management are too arbitrary to be trustworthy, and the therapist's confidence in the treatment process lessens.

In the current managed-care milieu, far fewer psychological resources are available for clients. Whereas effective therapy has always required thoughtful treatment planning, this undertaking now takes on additional importance. Practitioners will not succeed in obtaining treatment authorizations, or in negotiating additional sessions for clients from insurance caseworkers, if they cannot formulate specific treatment goals and intervention plans. The problem most caseworkers have with requests for continuing treatment is when therapists do not formulate clear treatment plans for the direction of therapy or clarify how they will intervene to reach these goals (McWilliams, 1999). To begin learning how to do this, new therapists need comprehensive case examples of how clients' problems can be conceptualized and translated into intervention strategies. In this book, basic concepts are progressively built upon in each succeeding chapter, and case studies that illustrate such effective case formulations will be provided. In addition, in-depth case studies that follow the guidelines for writing case formulations and treatment plans in Appendix B are provided in the Student Workbook that accompanies this text.

Many therapists become painfully anxious in their early counseling work because they lack a practical conceptual framework. Uncertainty over how to conceptualize client problems and ambiguous guidelines for how to proceed in the therapy session heighten their insecurity and preoccupation with their performance. As a result, both their effectiveness with clients and their ability to take pleasure in this rewarding work are diminished.

Finally, the lack of a conceptual framework hinders novice therapists as they try to find their own professional identities. Before therapists can feel secure in their abilities, they will need to integrate a theoretical framework that is applicable with diverse clients and congruent with their own values and life experiences. This is a complex, developmental process as therapists in training need to explore and actively try out different ways of working with clients. That is, beginning therapists cannot establish their own professional identity by simply adopting the same approach or taking on what their instructors and supervisors do. To be effective in whatever theoretical orientation they eventually adopt, and to enjoy being a therapist, therapists in training need to try out different approaches for several years and eventually *choose for themselves* how they are going to work with clients.

With the therapists' personal development in mind, the purpose of this book is to provide a conceptual framework for understanding the therapeutic relationship and for learning how therapists can use the "interpersonal process" or current interaction with their clients to intervene and help clients change. Throughout this text, readers are encouraged to modify this framework to better fit their own personalities and therapeutic styles. The interpersonal process approach can be applied by therapists who choose to work within different theoretical orientations. With support from supervisors and instructors, new therapists will be able to personalize this model, integrate it with other theoretical approaches, and modify it to make it their own. My goal will be met if cognitive-behavioral therapists, interpersonal/psychodynamic therapists, existential-humanistic therapists, family systems therapists, and others can use the interpersonal process approach to make their work with clients more effective.

What will this conceptual framework entail? This framework will help therapists recognize the relational patterns and cognitive schemas that are shaping clients' problems. Tracking these themes, therapists will use "process comments" and other "immediacy interventions" to identify and change these hurtful or problematic patterns as they are occurring with others and, especially, in the therapeutic relationship. That is, therapists will do this "in the moment" as these patterns are occurring in the current interaction with the client, as well as when they are activated with others. With this real-life or experiential relearning, which is a powerful way to expand clients' schemas, therapists are using the therapeutic relationship as a social learning laboratory. Then, therapists are actively helping clients transfer this relearning or generalize the change that has occurred in their relationship to others in the client's life. Thus, to provide a focus for treatment and direction for how to proceed, the process-oriented therapist aims to

- identify these maladaptive cognitive and interpersonal patterns that repeatedly occur for the client and cause problems with others;
- use "process comments" to talk together and explore how aspects of the same problems that occur with others may be occurring between the therapist and client as well;
- engage the client in a partnership or working alliance to find ways to change this familiar but problematic pattern in their current interaction; and
- transfer this experiential or in vivo relearning to others in the clients' lives.

As we will see, addressing and resolving these faulty patterns in the current interaction with the therapist, and generalizing this experiential relearning to other relationships where the same patterns and problems are occurring, comprise an effective way to intervene for therapists working within different theoretical traditions.

Theoretical and Historical Context

The therapeutic model presented here is termed the *interpersonal process approach*. This is not a new theory of therapy and it is not based in any single theoretical orientation. Rather, it is a synthesis of related concepts that share a common focus on the therapeutic relationship—and how therapists can use their current interaction to help clients change. Theorists and researchers from many different theoretical orientations have all contributed to our understanding of the therapeutic relationship and how therapists can use it to guide how we intervene with our clients. Varying theories focus on different aspects of people and their problems. In particular, the interpersonal, cognitive, and familial domains have been emphasized in different treatment approaches. This text highlights the "process dimension" in each of these approaches and links these concepts about the current interaction between the therapist and client to the practice of individual therapy. These three domains are briefly introduced here, so as to provide a theoretical context for the clinical approach outlined in the remainder of the book.

The Interpersonal Domain

Let's begin with a historical perspective. The interpersonal dimension in counseling and therapy was originally highlighted by Harry Stack Sullivan in the 1940s and is articulated and studied today by Hans Strupp, Michael Kahn,

Donald Kiesler, Gerald Klerman, Myrna Wasserman, Lorna Benjamin, and many others. Sullivan (1968) first brought the relational focus to psychotherapy in the 1940s; he remains an enormously influential but insufficiently recognized figure. A maverick, Sullivan radically broke away from Freud's biologically based libido theory and was one of the first major theorists to argue that the basic premises of Freud's drive theory (for example, sexual and aggressive instincts) were inaccurate. Also, rather than Freud's focus on fantasy and intrapsychic processes, Sullivan emphasized clients' current behavior with others and the child's actual experience with parents. He argued that every major tenet of Freud's theory could be understood better through interpersonal or behavioral experiences.

Sullivan developed an elaborate developmental theory of psychopathology that emphasized what people do to avoid or manage anxiety in interpersonal relations. He viewed anxiety, and elaborate attempts to avoid or minimize it, as the central motivating force in human behavior. He described a painful core anxiety that was rooted in dreaded expectations of derogation and rejection by parents and others, and later by oneself. Following from this view, Sullivan conceptualized personality as the collection of interpersonal strategies that the individual employs to avoid or minimize anxiety, ward off disapproval, and maintain self-esteem.

According to Sullivan, the child develops this personality, or self-system, through repetitive interactions with her parents. More specifically, the child organizes a self around certain basic patterns of parent–child interaction. These repetitive interactions with parents, which may be anxiety arousing or rewarding, are structured into the child's personality through a collection of complementary self–other relational configurations. For example, children may develop internal images of themselves as helpless or insignificant and expectations of parents and others as demanding or critical. Alternatively, more fortunate children may evolve images of themselves as love-worthy and expectations that others can be trustworthy.

Having learned these self–other relational patterns, people systematically behave in ways that avoid or minimize the experience of anxiety. For example, suppose that particular aspects of the child, such as feeling sad or crying, consistently result in parental rejection or ridicule. The child learns that these parts of herself are unacceptable and represent a "bad self," and these anxiety-arousing aspects of the self are split off or disowned. The child also develops interpersonal coping styles (for example, a pleasing/placating manner or a macho/dominating manner) that preclude being subjected to anxiety-arousing rejection again. These coping styles are interpersonal defenses that originally were necessary to protect the self in the early parent–child relationships. Unfortunately, these interpersonal defenses are overgeneralized to other relationships in adulthood and become habitual behavior patterns as the child, now grown to an adult, anticipates that new experiences with others will repeat the relational patterns of the past.

Sullivan's early formulations have been further developed now by attachment researchers (using the terms *internal working models* and *relational templates*) and by cognitive-behavioral therapists (using terms such as *cognitive schemas, faulty expectations,* and *core beliefs*). In a historical perspective, however, this was pioneering work.

Sullivan developed his rich, if sometimes elusive, interpersonal theory through his clinical work with seriously troubled patients who were often hospitalized. He was a gifted therapist and had great respect for the personal dignity of his clients. His sensitivity to clients, compassion, and genius for understanding the intricate workings of anxiety inspired a new direction in American psychiatry. Sullivan influenced seminal thinkers in the interpersonal field, such as Carl Rogers (1951) and Erik Erikson (1968), and made far-reaching contributions to the theory and practice of counseling and therapy. He even prefigured family systems theory in his study of close relational systems. In fact, two of the founding fathers in the family therapy field, Don Jackson and Murray Bowen, trained with Sullivan. We will also be drawing on a host of short-term and time-limited, contemporary psychodynamic approaches to treatment that were shaped by this interpersonal orientation.

In this regard, Sullivan led the transition from "psychoanalytic" to "psychodynamic" therapy. That is, he helped bring about the more active therapeutic stance we will be adopting here—with an interested, responsive therapist who is highly engaged and involved, far from a neutral, blank screen. In sum, many of the basic concepts in the interpersonal process approach can be traced back in some form to Sullivan and other interpersonal theorists.

The Cognitive Domain

Approaches as seemingly diverse as object relations theory, attachment theory, and cognitive-behavioral therapy each help us understand and intervene with the faulty thinking that is central to clients' symptoms and problems. Let's begin with object relations—with its alienating jargon, it is usually the most difficult and remote theory for graduate student therapists to integrate. Following this overview, we will link the attachment researchers' core concept of "internal working models" to the "schema" oriented approaches of contemporary cognitive therapists.

Object Relations and Attachment Theory: Internal Working Models   In the present context, *objects* are people or, more precisely, internal representations of them. Object relations theory is about interpersonal relationships, especially those between parents and young children, and how those relationships are internalized as cognitive schemas or internal working models. In particular, object relations theorists try to account for the meaning of attachment and the emotional need for relationships with others. Sullivan's theory centers on relationships, but the question that drove his work was very pragmatic: What do people do to avoid anxiety in interpersonal relations? In contrast, the question for object relations theorists is more abstract and inferential. Object relations theorists are interested in understanding how formative interactions between parents and children become internalized by the child and serve as cognitive schemas that shape how children establish subsequent relationships with others. These cognitive models provide basic expectations or road maps for what will transpire in relationships. Although these early schemas for relationships will become more complex and change over time, object relations theorists believe that these formative parent–child interaction schemas provide the basic structure for developing a sense of self, organizing the interpersonal world, and shaping subsequent relational patterns.

Similar to many cognitive-behavioral and attachment-oriented therapists, object relations therapists use these schemas (they call them *internal working models* or *relational templates*) to conceptualize clients' problems and understand the problematic relational patterns that clients are enacting with the therapist and others. For example, in order to identify these schemas or clarify faulty relational expectations, therapists working from an object relations, attachment, and cognitive-behavioral framework all will ask themselves (and their clients) questions like these:

- What does the client want from others? (For example, clients who repeatedly were rejected might wish to be responded to emotionally, reached out to when they have a problem, or be taken seriously when they express a concern or preference.)
- What does the client expect from others? (Some clients might expect others to diminish them or take advantage and try to exploit them again.)
- What is the client's experience of self in relationship to others? (For example, they might think of themselves as being unwanted or burdensome.)
- What are the emotional reactions that keep recurring? (The recurring reactions might be, for instance, shame or affection.)
- As a result of these core beliefs, what are the client's interpersonal strategies for coping with his relational problems? (Possible strategies include pleasing or always going along with others, emotionally disengaging or physically withdrawing from others, or trying to dominate and control others.)
- Finally, what kind of reactions do these interpersonal styles tend to *elicit* from the therapist and others (such as disinterest, sympathy, or irritation)?

Object relations theorists believe that the primary motive of human development is to establish and maintain emotional ties to parental caregivers (object relatedness). From their point of view, the greatest conflicts in life are threats to, or disruptions of, these basic ties (separation anxieties or abandonment fears). In this regard, anxiety is a signal that emotional ties to caregivers are being threatened. If parents are emotionally available and reliably responsive to a child's attachment needs, the child is secure in his ties. Through the course of development, such children can gradually internalize their parents' availability and come to hold the same constant loving affect toward themselves that their parents originally demonstrated toward them. In other words, cognitive development increasingly allows young children to internally hold a mental representation of their parents when they are away. As these children progress toward maturity, they become increasingly able to soothe themselves, function for increasingly longer periods without emotional refueling, and effectively elicit support when necessary. *Object constancy* and more independent functioning develops as their ability to comfort themselves becomes the source of their own self-esteem and secure identity as capable, love-worthy persons. Furthermore, they possess the cognitive schemas or internal working models necessary to establish new relationships with others that hold this same affirming affective valence (Farber & Geller, 1994). For example, these secure children are more likely as adults to be able to successfully choose an emotionally responsive and supportive partner and enjoy a better marriage.

But what if the child cannot maintain secure emotional ties to the primary caregivers? In some families, parents are preoccupied, depressed, or even abusive. Children of such parents are biologically organized to have their attachment needs fulfilled, but their emotional security is compromised by inconsistent or unresponsive parents. These children are trapped in an unsolvable dilemma. They cannot succeed in eliciting security from their parents, nor can they forsake their need for attachment. In a word, they feel anguish. What do they do? Object relations theory tries to account for the different solutions and adaptations the child must make to this poignant and all-too-human dilemma (for example, cognitively distorting reality by denying that certain needs of their own, or reality-based problems with others, really exist; and idealizing others in unrealistic, problem-free ways).

Object relations theorists propose complex developmental theories to account for the cognitive and interpersonal mechanisms that children employ to protect against the painful separation anxieties aroused when they cannot maintain secure emotional ties. These theorists propose that all children mature through stages of development that progressively allow for a more integrated, stable view of self and others. However, if development is problematic, children remain "stuck" or regress under stress to less integrated stages. In particular, when parents are excessively inconsistent or unresponsive, the child must resort to *splitting defenses* to maintain ties to an idealized loving parent akin to the polarized or dichotomous thinking some cognitive behaviorists describe. That is, the child internalizes the "bad" (threatening or rejecting) aspects of the parent separately from the "good" (loving or responsive) aspects of the parent. In object relations theory, in contrast to classical psychoanalytic theories, symptoms and problems do not develop from unacceptable sexual and aggressive impulses that have been repressed, but from what has been painful or threatening in children's relationships with their caretakers. In healthier parent–child interactions that provide secure attachments, however, the young child can tolerate increasing levels of frustration and disappointment as she matures. Gradually, the child will be able to increasingly integrate her ambivalent feelings toward the sometimes frustrating and sometimes responsive parent into a stable and affirming relational schema of self and other, which becomes the basis of self-esteem and identity.

Under greater attachment threats, however, the child cannot do this without resorting to splitting defenses. These splitting defenses preserve the necessary image of an idealized, "all good," responsive parent with whom the child is internally connected. But the price is high: Reality is distorted, the self is fragmented, and the child becomes the one who is "bad." The frustrating parent is no longer "bad," which allows the child to view the external world as safe. The price, however, is inner conflict. The child believes that, if only he were different, parental love would be forthcoming. This situation is illustrated most poignantly by the physically abused child who is bruised and battered but, nevertheless, continues to idealize and defend the abusive parent. Typically, the child does this by maintaining that he is bad and actually deserves the punishment. This self-negating distortion enables the child to cope with his insecure attachment and believe that he has some control over events and is not helpless. In their initial caseloads, most beginning therapists will have clients who were physically or sexually mistreated and blame themselves or unrealistically assume responsibility for what went wrong.

Why is this complex and abstract theory relevant to interpersonally oriented therapists? Realizing that the child must find some way to maintain ties with a good, loving parent helps therapists to understand why clients persist in maladaptive behavior and self-defeating relationships. When clients relinquish symptoms, succeed, or become healthier, they often feel anxious, guilty, or depressed. Loyalty and allegiance to symptoms—maladaptive behaviors originally developed to manage the "bad" or painfully frustrating aspects of parents—are not maladaptive to insecurely attached children. Such loyalty preserves "object ties," or the connection to the "good" or loving aspects of the parent. Attachment fears of being left alone, helpless or unwanted, are activated if clients disengage from the symptoms that represent these internalized "bad" objects (for example, the client resolves an eating disorder or terminates a hurtful relationship with an abusive partner). The interpersonal process approach is designed to modify these cognitive schemas by expanding the client's expectations of what can occur in close personal relationships. We will see how therapists can do this by providing clients with experiential or in vivo relearning—the real-life experience that, at least sometimes, some relationships can be different.

Unfortunately, object relations theorists address the most sensitive and personal dimensions of human experience in impersonal and alienating terms. However, we will be able to use certain aspects of this theory to understand clients better, relate more empathically to them, and guide our therapeutic interventions. We will be exploring object ties and attachment disruptions further, in order to better understand why clients may resist change. That is, we will see how anxiety and guilt are evoked for some clients as their attachment ties to internalized parental figures are threatened by change. In particular, we will be returning to the concept of schemas (*relational templates* or *internal working models*) in order to understand and intervene with the cognitive component of clients' problems. Just as object relations and attachment theory help us understand how clients mentally represent themselves and the social world, cognitive behaviorists have been skillful in helping therapists intervene and change these faulty schemas that engender so many problems. In particular, let's explore the cognitive-behavioral schema approaches that clarify the central role of "early maladaptive schemas" in understanding symptoms and problems.

Cognitive-Behavioral Therapy: Schemas   In his early work, Beck (1967) emphasized the importance of *schemas* in depression, referring to schemas as "a cognitive structure for screening, coding, and evaluating the stimuli that impinge on the organism." Beck emphasized that repetitive themes occur in clients' thinking through "schema bias"—a consistent and selective bias in organizing information and interpreting events that results in "the typical misconceptions, distorted attitudes, invalid premises, and unrealistic goals and expectations" that clients present (p. 284). Whereas cognitive therapy has focused primarily on three aspects of cognition—automatic thoughts, cognitive distortions, and underlying assumptions—more recent theorists propose a primary emphasis on "early maladaptive schemas" (Young, 1999).

In his rich and substantive approach, Jeffrey Young integrates cognitive-behavioral therapy with object relations and psychodynamic approaches, and he places more emphasis on the therapeutic relationship, affective experience, and early life experiences. Young, Klosko, and Weishaar (2003) define early maladaptive schemas (EMSs) as stable and enduring themes that develop from ongoing patterns of parent–child interaction during childhood that are significantly dysfunctional. Young has identified 18 different schema types or core constructs that fundamentally shape an individual. For example, expectations that one's needs for security, safety, or acceptance will not be met, based on the client's actual experience in a detached, cold, or withholding family of origin; or expectations about oneself and the environment that interfere with the ability to function independently, based on an overprotective or enmeshed family of origin; and so forth. These schemas serve as templates for the processing or filtering of later experience and are taken for granted by individuals as a priori truths. They feel familiar and are not questioned by the individual—they are just the givens of one's existence and form the core of a client's self-concept and conception of the environment. EMSs are self-perpetuating and resistant to change as the individual will distort information to maintain the validity of these schemas (by selecting, for instance, a mate or spouse who repeatedly confirms the individual's faulty self-perception as inadequate and shame-worthy). Strong feelings accompany these EMSs, as intense anxiety (or other feelings) are activated when life situations activate the EMS. For example, imagine a graduate student who has a Failure schema, which he developed from repeated interactions with hypercritical parents who were impossible to please. He may experience intense anxiety or other distressing feelings as comprehensive exams, his dissertation defense, or other evaluation/performance situations that activate his Failure schema loom dreadfully on the horizon.

In schema-focused cognitive therapy, therapists identify primary or core schemas by paying close attention to

- events that trigger high levels of affect,
- the client's most serious and enduring life problems, and
- the early origins of the client's emotional distress.

The therapist intervenes with "empathic confrontation" and tries to find a balance between empathizing with the client's pain brought on by the schema and challenging the client with evidence that disconfirms the schema. The therapist is alert for EMSs and questions them whenever they come up—in the client's relationships with others and, especially, in the therapeutic relationship.

The cardinal issue here, which will be emphasized throughout the interpersonal process approach, is that discussing schemas or misperceptions in the abstract is far less powerful than questioning them as they are occurring—that is, when they are accompanied by strong affect. For example, the client is furious with the therapist and feeling humiliated because she (mistakenly) believes that the therapist thinks she is "fat and ugly" just after she disclosed a binge-eating episode. The opportunity for emotional relearning and expanding the client's schemas is far more powerful when the therapist challenges or questions the maladaptive schema *as it is occurring*. Later in this chapter, we are going to call this experiential or in vivo relearning a "corrective emotional experience." Said differently, the client who grew up in a foster home and has an Abandonment schema is apt to feel anxious, depressed, or experience other familiar symptoms that disrupt their life at the times when separation experiences are occurring in the therapeutic relationship (such as at the end of the session, when the therapist is preoccupied, just before the therapist leaves for vacation, when they discuss termination, and so forth).

Working in the moment, the key here is to *address the schema when it is activated or in play between the therapist and client*. Experientially, this approach is far more meaningful to the client than talking about the same issue in the abstract—for example, relaying how this same theme played out with someone somewhere else a few days ago. Drawing flexibly from a wide range of techniques, the therapist may address the EMS by reviewing contradictory evidence whenever it comes up in the session. The therapist can also intervene by helping clients write flash cards that describe the schema, carry the flashcard with them, and read it back to themselves when the schema is activated. This counters the EMS with a more rational response in the real-life moment when the schema is "triggered," and it feels to the client like it really is true and "It's my fault; I am to blame" or "I will always be left; I can't ever count on anybody." Additionally, experiential techniques of role play and rehearsal with the therapist, and holding imaginary dialogues with parent figures who originally instilled the schemas, and other techniques are employed to resolve the EMS. Most important, however, *the therapist tries to invalidate the client's schemas by providing a therapeutic relationship that counteracts what the client experienced in the past*. For example, within the appropriate limits and boundaries of the therapeutic relationship, the therapist tries to be responsive toward a client who was ignored, flexible with a client who grew up with rigidity, consistent with a client who grew up without discipline, and so forth. We will return to this concept later in this chapter and discuss "client response specificity"—how therapists can tailor their interventions flexibly to meet the differing needs of different clients. For now, we can see that theoretical approaches as disparate as attachment theory and cognitive therapy can both teach us much about the schemas that shape clients' lives.

The Familial/Contextual Domain

Children must adapt to their attachment figures, and they do so within a familial and cultural context. Therapists can glean a wealth of information about their adult clients' strengths and problems from family systems theory, which clarifies the familial rules and roles that shaped their clients' development. Cultural norms and values are imparted through family interaction, and, in this and many others ways, we develop our identity and become who we are in our families of origin. In particular, family process researchers have learned much about family interaction and relationships by adopting a multigenerational perspective.

In the late 1950s, Gregory Bateson, Virginia Satir, Jay Haley, and other pioneers began studying communication patterns in families (Goldenberg & Goldenberg, 2004). They found that intrafamilial communication followed definitive but often unspoken rules that determined who spoke to whom, about what, and when. Disrupting current relationships, adult clients often continue faulty communication patterns that they learned years before in their families of origin. For example, they

- address conflicts through a third party rather than speak directly to the person involved,
- allow others to speak for them and define what they are thinking or feeling, and
- never make "I" statements and communicate directly what they want or set limits with others and say no.

Early family researchers also learned that children are often scripted into familial roles, such as the responsible or good child; the problem or bad child; and the hero, rescuer, or invisible child. Following object relations theory, the most common configuration is the good child/bad child role split that occurs in so many dysfunctional families. Some therapists were the "good child" in their families of origin and had a sibling who fulfilled the "bad child" role and seemingly couldn't do anything right or was always the problem. Additionally, many therapists, and others who enter the helping professions such as nurses or ministers, were "parentified." In this role reversal, they fulfilled a caretaking or confidant role for their parent. Most adult clients are still filling these limiting childhood roles, which distort and deny many important aspects of who they really are. In the process of adapting in their families, these children "internalized" these roles as self-schemas. That is, these familial roles became their identity, and even though their parents now may live far away or even be deceased, they continue to place their families' expectations on themselves and to re-create these problematic relational patterns with others. For example, many clients who have played the good-child role in their families are still perfectionistically demanding of themselves, guilty about doing things for themselves, worried about the needs of others, afraid of their anger, and confused by their sad and empty feelings. Whereas the bad child may act out personal conflicts externally (for example, through substance abuse or promiscuity), the good child is more likely to seek treatment and enter therapy, presenting with symptoms of anxiety or depression.

There is great cultural variation in the balance of "separateness/relatedness" or degree of continuing responsibility that young adult offspring are expected to hold for their parents. However, within the parameters of these varying cultural contexts, many families also have unspoken rules about how, in late adolescence, offspring can leave home or "individuate" and further establish their own beliefs and values, work and careers, and marriage and families. For example, the oldest daughter may not be allowed to grow up and leave home successfully on her own. Perhaps she will seek emancipation through pregnancy, only to find herself even more dependent on her parents than before and forced to live at home again. A son may only be able to leave home through rejecting confrontations with parents, and then he may live in another part of the country and have little or no continuing contact with the parents. In better-functioning families, young adults find support and guidance for launching their own lives, while continuing to maintain strong family ties and connections. Family rules, roles, and faulty communication patterns often serve to maintain family myths that, in turn, function to avoid anxiety-arousing issues in the family. For example, common family myths include the following:

- Dad doesn't have a drinking problem.
- We are a happy family and nobody is ever sad.
- Mom and Dad never fight and are very happily married.

In sum, family myths and these other family characteristics are rule-bound *homeostatic mechanisms* that govern family relations and establish repetitive, predictable patterns of family interaction. They are homeostatic in the sense that they tend to maintain the stability of the family system, often making many changes uncomfortable or threatening.

Salvadore Minuchin (1984) and other family researchers have also explored the alliances, coalitions, and subgroups that make up the structure of family relations. In some families, for example, the maternal grandmother, mother, and eldest daughter are allied together, and the father is the outsider. This structural road map for reading family relations becomes even more illuminating when these family dynamics are examined in a three-generational perspective (Bowen, 1978; Boszormenyi-Nagy & Spark, 1973). It is fascinating—and sometimes disturbing—to draw "genograms" and see how these same family rules, roles, myths, and structural relationships can be reenacted across three or four generations in a highly patterned, rule-governed system (McGoldrick, Gerson, & Shellenberger, 1999).

Adding to the complexity of this tapestry, these family rules, roles, and communication patterns operate within a broader cultural context. For example, the oldest son in a traditional Chinese family or the oldest daughter in a traditional Latino family may be expected to assume significant, ongoing responsibilities for elders or other family members. It will be necessary for therapists to establish the extent to which these roles and behavior patterns are culturally sanctioned or determined.

The interpersonal process approach draws on these basic family systems concepts and considers the effects of parental relations and familial experience on clients' personality strengths and problems. Therapists want to help clients make realistic assessments of the resiliencies and vulnerabilities that coexisted in their families of origin. However, *for many clients and for many therapists, it is culturally taboo to speak critically of parents or talk about problems with someone outside of the family*. Alice Miller (1984) observes how, for some, this even breaks the Fourth Commandment: Honor thy parents. Herein, though, is one of the great strengths of family systems theory. The family systems therapist is trying to understand hurtful familial interactions, not to place blame or scapegoat any family member. The family therapist is concerned about the well-being of every family member, parents as well as children. In parallel, only as clients realize that their individual therapist wants to understand—rather than to blame their caregiver or make them bad in some way—do clients have the safety they need to explore threatening material and make significant gains in treatment. At times, however, therapists' own splitting defenses lead them to blame or reject parental figures. Consider the following example:

INEFFECTIVE THERAPIST: I can't believe your mother did that to you! She's so mean!

Out of guilt and loyalty to their caregivers, it will be far more difficult for clients with this therapist to make progress in treatment. In contrast, it will be far more effective if therapists eschew blame and try only to understand and empathize with their clients' experience.

EFFECTIVE THERAPIST: I'm sorry you were hurt so much when she did that.

There are some good things in almost every family, even in very troubled or abusive families, just as there are limitations and problems in the healthiest families. When clients continue to idealize parents and deny real problems that existed, as many do when they enter treatment, they are complying with binding family rules and protecting their caregivers at their own expense. On the other hand, however, clients cannot simply reject hurtful parents and emotionally cut off from them, or they will tend to reenact these same problematic relations with others. This means that therapists need to help clients find ways to keep parents (or the healthier, more benevolent aspects of them) alive inside themselves as "partial identifications." How can clients accomplish this?

In order to resolve problems and make progress in counseling, clients want to be able to come to terms with both the good news and the bad news in their families of origin. As already noted, however, clients cannot achieve this integration if the therapist also employs his own splitting defenses or dichotomous thinking. In other words, the therapist does not want to identify with or idealize the wounded child that exists within some clients and to reject the hurtful parent as "bad." Neither does the therapist want to support clients' continuing denial of familial problems and idealization of parents. Instead, the therapist's appropriate role is to try to understand what actually occurred in clients' development and help them realistically come to terms with both the good news and the bad in their experience.

Family systems concepts and other contextual approaches inform the interpersonal process approach. In particular, they will help us identify the problematic interaction patterns that clients are enacting with the therapist and others, and help both therapists and clients understand the familial context that shaped many current problems. With this awareness, clients are better able to stop reenacting problematic familial roles with others in their current lives, but without breaking off important but conflicted familial relationships.

Let's pause here for a moment to appreciate two reasons why familial experience has such a profound, lifelong impact on the individual. The first is the sheer repetition of family transactional patterns. The same types of affect-laden interchanges are reenacted thousands of times in daily family life. To illustrate, suppose a parent has difficulty responding positively to his child's success experiences because of his own depression, narcissistic envy, or competitiveness. When the young child enthusiastically seeks this parent's approval for an accomplishment, the parent might ignore the child or change the topic, compare the child unfavorably with a sibling's greater accomplishment, turn away and look vaguely sad or hurt, or take the success away from the child by making it his own. The same type of parental response usually occurs when the child shows the parent a favorite drawing, makes a new friend, wins a race at school, or earns a star from her teacher. This transactional pattern becomes a powerful source of learning when it continues over a period of years and even decades. As an adult, this child is likely to feel conflicted about completing her educational degree, taking pleasure in a promotion earned at work, or even enjoying her marriage or good friends. In this way, the most significant and enduring problems in people's lives develop from these habitual response patterns. Contrary to popular belief and portrayals in Hollywood's movies, long-standing symptoms and problems are shaped far more by repetitive family transactional patterns (*strain* trauma) than by isolated traumatic events (*shock* trauma) (Wenar, 2000). This represents a paradigm shift for many therapists in training, who now are beginning to listen for patterns and themes in daily interaction, rather than keying so exclusively on crisis events. In this way, enduring problems do not arise so much from crisis or wounding events per se but from the lack of validation or empathy from caregivers for the painful events that occurred.

Second, the impact of these repetitive transactions is magnified because of the intensity of the feelings involved. Parents are the pillars of the child's universe, and children depend on them with a life-and-death intensity. These repetitive transactional patterns have been reenacted in highly charged, affective relationships with the most important people in one's life. The child in the preceding example may well feel a desperate need somehow to win the parent's approval yet simultaneously feel increasing anxiety about trying to succeed or approaching the unresponsive parent. Thus, our sense of self in relation to others is learned in the family of origin and, in many ways, will carry over to adulthood.

In sum, family interaction patterns may be hurtful and frustrating for the individual, validating and encouraging, or, most commonly, a mixture of the two. For better or worse, these repetitive patterns of family interaction, roles, and relationships are internalized and become the foundation of our sense of self and the social world. Of course, other factors are influential as well.

Familial relationships are molded within a cultural context that can affirm or re-
pudiate parental behavior, and familial experience will not be relevant to many
of the situational problems that clients present. However, familial experience
provides our first and most long-lasting model for what goes on in close relationships. It will figure significantly in the individual's choice of marital partner
and career, in how adult offspring will, in turn, parent their own children, and
in many of the other enduring problems and satisfactions of adult life. Although
the influence of the family of origin is profound, problematic transactional patterns can be relearned. Change is indeed possible, and it occurs in part through
a relational process of experiential relearning, which we consider next.

Core Concepts

In the chapters ahead, we are going to apply three core concepts to understand
clients and guide our interventions: the process dimension, a corrective emotional experience, and client response specificity. In this section, these three
orienting constructs are introduced and then illustrated with a case example.

The Process Dimension

The relationship between the therapist and the client is the foundation of the
therapeutic enterprise. The nature of this relationship is the therapist's most important means of effecting client change; it shapes whether, and how much,
clients can change in treatment. However, in order to utilize the therapeutic relationship as a vehicle for change, therapists need to understand the meaning of
their interactions with their clients. Thus, the interpersonal process approach
focuses on understanding and intervening with what is going on between the
therapist and client right now in terms of their current interaction or *process*.

The therapist–client relationship is complex and multifaceted; different levels of communication occur simultaneously. For example, there is a subtle but
important distinction between the overtly spoken *content* of what is discussed
and the *process* dimension of how the therapist and client interact. In order to
work with the process dimension, the therapist is making a perceptual shift
away from the overt content of what is discussed and beginning to track the relational process of how two people are interacting as well. This means that, at
times, therapists need to step beyond the usual social norms and talk more directly with the client about their current interaction. For example:

> THERAPIST: Since I asked you about your canceling our session last week, and
> then coming late to our appointment today, you've become quiet. Maybe
> something about that didn't feel right to you. Can we talk about that?

OR

> THERAPIST: Right now, John, I'm having a little trouble keeping up. You're speak-
> ing fast, and we seem to keep jumping from topic to topic. What do you see
> going on between us here?

OR

THERAPIST: I'm honored that you choose to share such a sensitive part of yourself with me. What's it like for you to talk with me like this?

We are going to call these types of here-and-now, present-focused interventions *process comments*. Working in different theoretical contexts, therapists have used varying terms for these interventions that focus on what is going on between the therapist and client right now: *metacommunications, immediacy interventions, interpersonal feedback, working "in the moment," therapeutic impact disclosure, self-involving comments*, and others. Because this approach may conflict with familial rules or cultural prescriptions for new therapists to speak directly in these ways, it may feel awkward at first for some. For other therapists, it may seem impolite or disrespectful to use process comments and speak forthrightly about what may be going on between "you and me." However, new therapists soon will see how powerful it is when they can link the problem that the client is talking about with others to their current interaction in the therapist–client relationship (that is, highlighting how the client is utilizing the same thought processes or behavior patterns with the therapist, right now, that are causing problems with others). Or, similarly, the therapist is *nondefensive* and willing to explore and sort through with the client potential misunderstandings, inaccurate perceptions, or other interpersonal conflicts that may be going on between them in their real-life relationship. To help new therapists learn to apply these powerful but challenging interventions, each successive chapter will explore further how therapists can make these process comments effectively. That is, how therapists can metacommunicate, make process comments, and use other immediacy interventions in an empathic, respectful manner that clients will welcome as a *collaborative invitation for genuine understanding and honest communication*—something that does not make them feel awkward or uncomfortable (Crits-Cristoph & Gibbons, 2002; Kiesler, 1988). Consider the following example.

During their first session together, the client tells the therapist that he resents his wife because she is "bossy" and "always telling me what to do." He explains that he has always had trouble making decisions on his own, and, as a result, his wife has come to simply make decisions for him. Even though he felt that she was just trying to help him with his indecisiveness, he resents her "pushiness" and "know-it-all attitude." After describing the presenting problem in this way, the client asked the therapist, "What should I do?"

Let us look at the process dimension of their interaction. Suppose the therapist complied with the client's request and said, "I think the next time your wife tells you what to do, you should. . . ." If treatment with this particular client continues in this prescriptive vein, the therapist and client will quickly begin to reenact in their relationship the same type of conflict that originally led the client to seek treatment: the therapist will be telling the client what to do, just as his wife has done. The client has certainly invited this advice and will probably welcome the therapist's suggestions at first. In the long run, however, he will probably come to resent the therapist's directives just as much as he resents his wife's and will ultimately find the therapist's suggestions to be of as little help as hers.

Alternatively, the therapist might respond to the other side of the client's conflict and say, "I don't think I would really be helping you if I just told you what to do. I believe that clients need to find their own solutions to their problems." Frustrated by this nondirective response that he just finds evasive, the client responds, "But I told you I don't know what to do! It's hard for me to make decisions. Aren't you the expert who is supposed to know what to do about these things?"

This response throws the client back on the other side of his conflict and leaves him stuck in his own inability to make decisions. If this mode of interaction continues and comes to characterize their relationship, their process will reenact (recapitulate) the other side of the client's problem. His inability to initiate, make his own decisions, and be responsible for his own actions will immobilize him in therapy, just as it has in other areas of his life.

What we see here is that clients do not just talk with therapists about their problems in an abstract manner. Rather, they often re-create in their relationship with the therapist the same type of problematic interactions that originally led them to seek treatment. That is, clients also convey their problems in *how* they interact with the therapist (the process), bringing critical aspects of their problems with others into their current interaction. *This repetition or replaying of the client's problem is a regular and predictable phenomenon that will occur in most therapeutic relationships.* We will return to this scenario later in the chapter and explore more effective responses that take the process dimension into consideration.

Corrective Emotional Experience

The concept of the "corrective emotional experience" originally was introduced in the 1940s by two psychoanalytically oriented therapists, Alexander and French (1980), and it has been the cornerstone of many short-term psychodynamic and interpersonally oriented therapies. Although it was not appreciated by the psychoanalytic community of its day (they disdained it as being superficial), their more active, direct approach provided clients with a real-life experience of change in the here-and-now relationship with the therapist. Accurately anticipating the future of counseling and therapy, Alexander and French originally advocated the corrective emotional experience, in part, to shorten the length of treatment. This and other short-term, psychodynamic therapies that we will be exploring use different terms to describe the same direct, immediate experience of change in the real-life relationship between the therapist and client (Crits-Christoph & Barber, 1991; Strupp & Binder, 1984).

Therapists working within different theoretical orientations can all help their clients change by providing a new and more effective response to their old relationship patterns than they have usually found with others. That is, the therapist recognizes the maladaptive relational pattern or theme that usually occurs with others, and she works together with the client to alter or change this problematic pattern in their relationship. We are going to call this core concept a *corrective emotional experience* (some interpersonal and attachment-oriented therapists use the term *reparative relational experience*, whereas some behavior therapists describe it as *exposure trials*). When clients terminate prematurely or therapy reaches an impasse, the therapist and client are usually reenacting in their relationship subtle aspects of the same conflict that the client has been struggling with in other relationships, although neither of them may be aware of this reenactment. Based on their schemas and expectations, clients soon come to hold the same misperception or faulty expectation toward the therapist, or respond in the same problematic ways, that they are doing with others in their lives. For example, even though the therapist has not behaved accordingly, clients soon may believe the following:

- They are being controlled by the therapist and have to do everything his way, just as they have always done with others.
- They have to take care of the therapist and meet her needs, just as they have been doing with others in their life.
- They must please the therapist and win his approval, just as they keep striving to do with significant others.

When this phenomenon occurs, *the therapeutic process is metaphorically repeating the same type of conflicted interaction that clients have not been able to resolve in other relationships and have often experienced in earlier familial relationships.* As clients begin to play out with the therapist the same relational patterns that originally brought them to treatment, the therapist's goal is to respond in a new and more effective way that allows clients to resolve the problem and change the pattern within their relationship. As clients have this in vivo experience of change with the therapist, their schemas expand and become more flexible or realistic, and it becomes much easier to begin changing this problematic pattern with others in their lives. Providing a corrective emotional experience by responding in a new or better way to the old relational pattern is easy to say, but it is often challenging to do in affect-laden relationships with clients.

Let's emphasize this core construct again—it's easy to say, but hard to do. In successful therapy, clients experience a new kind of relationship that resolves rather than repeats the old problematic scenario. That is, clients participate in a relationship in which their maladaptive relational patterns are triggered or brought into play with the therapist in their interpersonal process. However, this time, the therapist does not continue to respond in the same problematic way that others have in the past. In order to achieve this, throughout each session, therapists are asking themselves the same process-oriented question:

Am I co-creating a new and reparative relationship,
or am I being drawn into a repetitive pattern that
is familiar but problematic for the client?

When the therapist and client are able to resolve these expectable reenactments, which occur in most therapeutic relationships, change has begun to occur. Clients learn that they no longer have to respond in their old ways (such as always having to be in control, be the responsible one, or take care of others) or always receive the same unwanted responses from others (for example, being ignored, not being taken seriously, or being competed with). Thus, they begin to develop a wider range of expectations and can respond in more flexible and adaptive ways in their relationship with the therapist. As their behavior changes with the therapist, it is often relatively easy to take the next step and help clients transfer this experience of change with the therapist and adopt similar, more adaptive responses with others outside the therapy setting.

We introduced this sequence earlier; let's return and examine it more closely. In order to provide a corrective emotional experience and help clients change, the therapist's intentions are to

1. identify the faulty beliefs and expectations, problematic relational scenarios, and ineffective coping styles that keep occurring and causing problems in the client's life,
2. anticipate how these patterns or themes that are disrupting relationships with others may be expressed or come into play in the current interaction with the therapist,
3. provide new or corrective responses that help to resolve rather than repeat this familiar but problematic pattern in the relationship with the therapist, and
4. help the client generalize or transfer this new way of interacting and begin to respond in these more effective ways with others.

In this way, new therapists will find that their clients often change in a two-step sequence. First, clients have an experience of change with the therapist when their old relational patterns and problematic expectations do not recur along the process dimension. The second phase of treatment often occurs as the therapist helps clients generalize this experience of change to other relationships in their lives. How does the therapist help clients transfer this experiential relearning that has occurred in their relationship? The therapist does this by

1. clarifying when and how these same cognitive and interpersonal patterns are triggered and disrupt other relationships as well;
2. formulating and rehearsing new, more adaptive responses that change how the client is participating in most current interpersonal problems; and
3. helping clients anticipate future situations that are likely to evoke their old problematic response patterns, so they can more actively choose how they wish to respond instead.

In this way, clients learn to respond in more flexible ways, rather than rigidly repeating old familiar response patterns that are not succeeding in many current relationships.

The basic idea here is that clients believe actions, not words. Long ago, Frieda Fromm-Reichmann (1960) captured this central tenet best by saying that the therapist must provide the client with an *experience* rather than an *explanation*. Interpretations, empathic understanding, cognitive restructuring, self-monitoring techniques, education, skill development, and other interventions are utilized in this approach and are helpful with most clients. However, they are not the primary seat of action in the interpersonal process approach. This is a performance-based or experiential learning model. Following Strupp (1980), clients change when they live through emotionally painful and ingrained relational scenarios with the therapist, and the therapeutic relationship gives rise to outcomes different from those anticipated and feared. That is, when the client re-experiences important aspects of her primary problem with the therapist, and the therapist's response does not fit the old schemas, the client has the real-life experience that relationships can be another way. When clients *experience* this new or reparative response, a response that differs from previous relationships and that does not fit the client's negative expectations or cognitive schemas, it is a powerful type of experiential relearning that readily can be generalized to other relationships (Bandura, 1997). This behavioral experience of change with the therapist is far more compelling than words alone can provide.

We are going to be exploring many different ways that therapists can use this corrective emotional experience to facilitate change. In particular, the new or corrective response from the therapist helps clients change by creating *greater interpersonal safety* for the client. This is more than the general safety that comes from being taken seriously and treated with respect in the therapeutic relationship. Instead, it is a more significant sense of safety or deep security that results when the client is not hurt again by receiving certain interpersonal responses that are familiar and expected, yet unwanted or even dreaded. As we have noted, *this pivotal moment is not the end point in treatment but a window of opportunity for a range of important new behaviors to emerge.* For example, in the next few moments immediately following a reparative experience with the therapist, clients may be empowered to

- resolve their ambivalence and be able to make important personal decisions;
- risk trying out new ways of responding with the therapist or significant others in their lives;
- make significant self disclosures or bring up relevant new issues or concerns;
- feel bolder and risk addressing unspoken misunderstandings with the therapist or talk directly about concerns that the client has with the therapist;
- feel better about themselves or forgive themselves for things they have felt unrealistically shameful or guilty about;
- have the safety to experience more fully the pain of how much this old pattern or theme has hurt them in other relationships;
- make meaningful connections between current behavior and developmental relationships where these schemas and patterns originally were learned.

In these ways, corrective emotional experiences with the therapist begin to expand maladaptive early schemas and repetitive relational patterns, and a wider range of responses becomes available to the client—often in the next few moments.

A single corrective emotional experience is not sufficient for sustained change, but it is often the pivotal experience that initiates change. Based on the interpersonal safety that arises when clients find that they consistently or dependably receive this new, reparative type of response from the therapist, the client becomes more willing to try out interventions from educational, behavioral, cognitive, and other theoretical approaches, and it facilitates clients' commitment to follow through on mutually set goals and interventions. Thus, when clients experience a different type of relationship with the therapist that does not go down the familiar but problematic pathways they have come to expect from others, the therapist is seen as someone who can help, *earning achieved credibility* (Sue & Zane, 1987). Clients invest further in the treatment process and become more willing to try out new coping strategies and ways of responding with others. In contrast, when clients successfully elicit the same type of reactions from the therapist that they tend to find with others (for example, the therapist begins to feel frustrated, controlled, discouraged, or disappointed, as significant others in the client's life often feel), intervention techniques from every theoretical perspective will falter. In this way, the interaction between the therapist and the client provides a "meta" perspective for understanding what is occurring and what to do in therapeutic relationships, and it can readily be integrated with other therapeutic modalities.

Using the Process Dimension to Provide a Corrective Emotional Experience
Linking our first two core constructs, the therapist needs to be able to work with the process dimension in order to provide this corrective emotional experience. Let's return to our earlier example of the client who complains that his wife tells him what to do. When the client asked his therapist what he should do, one option would have been for the therapist to offer a process comment that described their current interaction and made it overt as a topic for discussion:

THERAPIST: Right now, it seems that you're asking me to tell you what to do. But I'm wondering whether that will only bring up the same problem for us here in therapy that you are having at home with your wife. Let's see if you and I can figure out a way to do something different in our relationship. Rather than having me tell you what to do, let's try to work together to understand what's going on for you when you are feeling indecisive. Where do you think is the best place to begin?

CLIENT: I'm not sure.

THERAPIST: Take your time, and let's just see what comes to you.

CLIENT: (*pause*) I'll get criticized. No matter what I decide, she'll find something wrong with it.

THERAPIST: Good work—that's a great place for us to start. It feels like you can't get it right in her eyes—that you'll always do it wrong. It sounds like there's a lot of feeling in that for you. Tell me more about how that is for you.

CLIENT: Well, I guess I hate it so much because it makes me feel like a child. . . .

In this instance, the therapist's intention is to ensure that the same problematic but familiar pattern and consequences do not occur in their relationship. The therapist is trying not to take the bait and repeat the pattern by telling the client what to do. Instead, the therapist is trying to offer the client a new and different type of relationship—a collaborative partnership in which they can work together on the client's problems. If this collaborative effort continues to develop over the course of treatment, the client is having the real-life experience that he can make decisions and successfully initiate—at least in his relationship with the therapist. Then, with the therapist's support for this new behavior, he will rehearse, role-play, and try out these new behaviors with his wife and others. He will have successes and failures in his attempts to adopt this stronger way of relating to others, and, as we will explore later, the therapist will help him learn from both. However, the therapist's initial attempt to offer the client a corrective response will need to be repeated in many different ways for change to occur. If the therapist can maintain this type of collaborative relationship, this new interpersonal process will be corrective and facilitate other types of therapeutic interventions as well. Educational inputs, behavioral alternatives, psychodynamic interpretations, and interpersonal feedback will all become more effective with this corrective interpersonal process. This client's life has been ruled by his Subjugation/Control schema, but now he is a collaborator in the treatment process with his own voice and is not being told what to do again.

Countertransference Issues and Therapists' Fears about Making Mistakes As we have seen, the guiding principle in the interpersonal process approach is to provide clients with an experience of change. With this in vivo relearning, clients live out a new relationship with the therapist that disconfirms their faulty expectations and expands their schemas for what can occur for them in relationships. For example, clients learn that, at least sometimes, with the therapist and then with some others in their lives, they can ask for help, set limits and say no, give priority to their own needs, succeed without evoking envy, and so forth. Providing this type of corrective emotional experience is an engaging and rewarding way to work with clients, but it demands much on the part of the therapist. In particular, countertransference issues, and new therapists' fears of making "mistakes" with their clients, make it harder to provide a corrective emotional experience. Let's examine both.

First, it requires personal involvement to work with clients in this way—the therapist must be willing to engage in an authentic or genuine relationship and risk being affected by the client (Rogers, 1980). That is, in order to have the emotional impact necessary to provide a corrective emotional experience and propel change, the relationship must hold real meaning for both participants (Gelso & Hayes, 1998; Gelso & Carter, 1994). In this simple human way, it is the relationship that heals. If the therapist is merely an objective technician, psychologically removed and safely distant from the client, the relationship will not hold real meaning for either person, and will be too insignificant to effect change.

On the other hand, therapy will also falter if the therapist becomes inappropriately close. If therapists overidentify with the client or become too invested in the client's choices or ability to change, perhaps in order to shore up their own feelings of adequacy as a new helper, therapy will not progress. In this situation, therapists often lose sight of the process that they are enacting with the client and typically begin to respond in problematic ways that reenact aspects of the client's maladaptive relational patterns. Thus, one of the most common problems for beginning therapists is to experience the client's situation or concerns as identical to their own (with the therapist thinking, for instance, "She's just like me!"). The solution to this overidentification is a supportive yet honest supervisory relationship that helps therapists recognize their own countertransference issues and begin to see more realistically the ways in which their own experience actually differs from the client's. There may be important similarities, yet, in reality, the client's problems are never the same as the therapist's. Once able to differentiate their own issues from those of the client, therapists will often be able to see how the therapeutic process has been reenacting aspects of the client's conflict. This understanding often enables therapists to alter the way they have been interacting and establish a more productive interpersonal process.

Second, new therapists should expect to make many "mistakes" with their clients. For example, to become overinvolved or underinvolved with certain kinds of clients, to reenact the client's maladaptive relational patterns at times, and so on. Such mistakes are an inevitable part of the therapeutic process for beginning and experienced therapists alike. Few clients are fragile, however, and therapeutic relationships are often remarkably resilient. Unfortunately, most student therapists do not know this, and concern about making mistakes remains one of their biggest anxieties. New therapists will do better therapy—and enjoy it more—once they discover that *mistakes can be undone*. In fact, mistakes provide important therapeutic opportunities when therapists are willing to work with the process dimension—to talk with clients about problems or misunderstandings that may be occurring between them. For example:

> THERAPIST: I'm wondering whether I might have misunderstood something you said. I'm asking about that because it seems as if you have become more distant from me in the last few minutes, and you've mentioned that other people often don't understand you. What do you see happening between us here?
>
> CLIENT: Yeah, I guess I do feel a little bit distant. But I don't feel "misunderstood"; I just don't really like it when you don't say very much—you're pretty quiet, you know.
>
> THERAPIST: That's helpful to learn—I like your honesty. So let's change things. I'll start speaking up more and share more of my thoughts with you. And let's check back in on this later and see how it's going. How does that sound?
>
> CLIENT: Thanks, I think that would make me more comfortable.
>
> THERAPIST: What's it like for you when I'm quiet or don't say very much?
>
> CLIENT: I get really uncomfortable. I guess I start worrying that you're judging me or something . . . you know, that you won't respect me . . . .

As their performance anxiety declines, beginning therapists will find that they are better able to identify and work with the process dimension. By doing so, therapists can readily undo mistakes and realign the therapeutic relationship when it has gone awry. Following Hill (2004), beginning therapists are encouraged to recognize their mistakes, apologize if necessary, and *talk through the event with the client*. This restores the therapeutic relationship, and it provides clients with an effective model about how to deal with problems in relationships. New therapists are encouraged to let go of unrealistic performance demands they may place on themselves to be perfect and never make mistakes— it's just not human. Instead, try to relax and let yourself enjoy this meaningful work; just do the best you can, and try to learn from your successes and disappointments.

Client Response Specificity

We have already begun to see that beginning therapists need a theoretical framework to guide their therapeutic interventions. Let's look further at what a theory needs to provide if it is to be of help to therapists and their clients.

One feature of an effective clinical theory is that it must have the flexibility and breadth to encompass the diversity of clients who seek treatment. Although unifying patterns in personality certainly exist, every client is different. Each client has been genetically endowed with a unique set of features and each has been raised differently in her or his family. Socialization is different for women and men, members of different cultures have different experiences, and economic class shapes opportunity and expectations. To be helpful, a clinical theory must be able to help therapists work effectively with all of these highly diverse clients. Moreover, each therapist is a different person. Like clients, therapists differ in age, gender, ethnicity, sexual orientation, and developmental background. Therapists also bring diverse values, worldviews, and personal styles to their clinical work. How can any theory help such a diversity of therapists respond to the extraordinary range of human experience that clients present?

Our third core construct, *client response specificity*, will be one of our best tools. Client response specificity means that therapists need to tailor their responses to fit the specific needs of each individual client— one size does not fit all! There are no cookbook formulas or generic techniques for responding to the complex problems, diverse developmental experiences, and multicultural backgrounds that clients present. In this regard, we are going to see that the same therapeutic response or intervention that helps one client make progress in treatment will only serve to hinder another. For example, some clients respond well to warmth in the counselor, whereas others want objectivity (Bachelor, 1995). More specifically, a warm and expressive therapist can put off a distrustful or avoidant client; a businesslike therapist can fail to engage an anxious client in crisis. The key point is that therapists need to have the flexibility to listen to the cues, assess clients' responses, and search for the best way to respond to this particular client (Lazarus, 1993). As we are going to see, however, *therapists often fail to hear and accommodate to clients' feedback about what they want and don't want from them.*

In regard to client response specificity, the operative word is *flexibility*. This approach requires therapists to be flexible, so that they can respond on the basis of each client's own personal history and ways of viewing the world. It is certainly helpful to be familiar with the experiences of particular groups (for example, to know that African Americans' history includes slavery and racial discrimination) and the characteristics that accompany certain diagnostic categories (such as knowing that bipolar clients in a manic phase are at risk to act out with aggression toward others, sexual promiscuity, or suicide). Additionally, however, with client response specificity, our abiding intention is to attempt to understand and respond to each client as a unique individual. Thus, we will not seek treatment rules or intervention guidelines that apply to all clients or even to specific diagnostic categories or groups, such as men, Latinos, or Christians. Rather, we will be trying to find case-specific recommendations in which we explore what it means to this unique individual to be depressed, lesbian, or biracial. Let's examine further what this point means.

Interpersonal process therapy is a highly "idiographic" approach; it emphasizes the personal experience or subjective worldview of each individual client. Diagnostic categories will tell us something about the client, but they are only the map and not the territory. Personality typologies also highlight potential directions that therapy might take, but the actual therapeutic process and issues covered will be client-specific. Throughout each session, *therapists are encouraged to join the client in a process of mutual exploration to try to find the subjective meaning that each particular experience holds for this client.* For example:

> THERAPIST: Let me check this out with you and make sure I'm understanding what you're really meaning here. When I hear you say that, it sounds to me more like you are feeling sadness, or maybe even grief, than depression. Am I getting that right—can you help me say it better?

From the first session through the last, the therapist is taking on the personal challenge to be *accurately* empathic or, in different terms, to have the cognitive and emotional flexibility to decenter and enter the client's *subjective worldview*. The meaning that a particular experience holds for the client often differs greatly from the meaning that this same experience holds for the therapist or for other clients that the therapist might see. Therapists are encouraged to listen carefully to clients' language and choice of words, and to collaboratively explore clients' own metaphors.

> THERAPIST: Help me understand what you mean, or what you might be saying about us and our relationship, when you tell me that you just want to "sail right out of here."

When someone cares enough about you to listen seriously, and works hard to understand just exactly what it is that you are trying to say, meaning is created.

With client response specificity, there is no standardized treatment protocol. Each individual client's developmental history, cultural context, and current life circumstances serve to guide the therapist's treatment plans and intervention strategy. Thus, the interpersonal process approach does not advocate a particular therapeutic stance toward clients in general, such as being directive or nondirective, prescriptive or exploratory, active or neutral, supportive or challenging, and so forth. Each of these modalities will be helpful with a particular client at times and ineffective with another. Thus, this approach asks something different and far more demanding of therapists:

- to conceptualize the specific relational experiences that this client needs in order to change, and
- to be flexible enough to modify their interventions and respond in the ways that this particular client utilizes most productively.

This recommendation may sound obvious or easy to do, but, unfortunately, researchers tell us that most therapists do not do this very well. Therapists continue to respond to clients in the same manner, based on their theoretical orientation—even when their approach is not working! Psychotherapy process studies find that most therapists do not demonstrate flexibility in their technical approach to the client, or have the interpersonal range to modify their interventions, and *provide the responses that each client could utilize best* (Strupp, 1980; Najavits & Strupp, 1994; Mash & Hunsley, 1993). With client response specificity, our aim is to assess how clients are responding to each of our interventions in an ongoing, moment-by-moment way, and flexibly adapting to respond in the ways they find most useful. To help with this, the counseling literature on *intentions* (Hill & O'Grady, 1985) suggests that therapists ask themselves the following questions:

Where am I right now, and what do I want to accomplish?
[For example, recognize the client's hurt.] OK, what is the
best way to do that? [Reflect how much that hurt when she
did that.]

In this way, therapists are encouraged to think about their intentions—what they are trying to accomplish *with each response or turn of the conversation* (Ivey & Ivey, 1999; Hill, 2004). Therapists will be better prepared to respond to the specific needs of this particular client, at this particular moment, when they ask themselves in an ongoing way, "What is needed right now, and how can I provide that?"

Let's use an example to illustrate client response specificity more concretely, and then examine more closely how therapists can assess clients' positive or negative responses to their varying interventions. Consider an often debated question: Should therapists self-disclose to their clients? Client response specificity emphasizes that self-disclosure (or any other intervention) *will hold very different meanings for different clients.* In fact, the same response may even have the opposite effect on two different clients with contrasting developmental histories and cultural contexts. For example, if a client's parent was distant or aloof, the therapist's judicious self-disclosure may be helpful for the client. In contrast, the same type of self-disclosure is likely to be anxiety arousing for a client who had been the confidant of a depressed parent. Greater sharing with the therapist may help the first client learn that, contrary to his deeply held beliefs, he does matter and can be of interest to other people. In contrast, for the second client, the same type of self-disclosure may inadvertently impose the unwanted needs of others and set this client back in treatment, as the therapeutic relationship is now paralleling the same, problematic relational theme that this client struggled with while growing up. Considering client response specificity in this way, therapists working within every theoretical orientation will be more effective if they first consider how the client's cognitive schemas or relational patterns are likely to shape the impact of their interventions on this particular client.

Like our other two core constructs, client response specificity is a complex and multifaceted clinical concept that will take time to learn and apply with clients. Let's illustrate it further by posing another common question: Should therapists give opinions when clients ask for advice? As before, the therapist's response is informed by each client's particular circumstances. For example, offering advice may be counterproductive for a compliant, intimidated female client who is trying to become more independent and assertive with her dominating husband, who undermines her confidence and fosters her dependency on him. For such a client, it may be better for the therapist to ask, "What do you think would be best to do?" In contrast, giving advice and directives may work well with clients whose caregivers were unable to help them solve their problems, perhaps because the caregivers were too self-absorbed or preoccupied with their own problems to be capable of or interested in helping their children. To withhold suggestions or problem-solving advice from these clients may only reenact developmental conflicts and impede their progress. Once again, the same response from the therapist will have very different effects on different clients. From this metaperspective, techniques or interventions from any theoretical approach may help or hinder the client; it depends on whether the response serves to reenact or resolve maladaptive schemas and relational patterns for this particular client.

Finally, let's highlight another important aspect of client response specificity: assessing clients' responses to therapists' interventions. Therapists can learn how to assess the effectiveness of their interventions by paying attention to how the client utilizes or responds to what they have just done. For example, if the therapist observes that the client becomes anxious, irritated, or distant in response to the therapist's self-disclosure, the therapist is learning how to respond most effectively to this particular client. It's not that self-disclosure is ill advised for another client, or an ineffective way to respond in general, but it is not helpful for this particular client. As we noted earlier, when the therapist provides a corrective response, many clients will make progress right away by feeling safer and acting more boldly, bringing forth relevant new material, becoming more engaged with the therapist, and so forth. In contrast, when the therapist responds in a way that recapitulates problematic old patterns, clients will behaviorally inform the therapist of this by acting "weaker" in the next few minutes — for example, by being more distant, indecisive, compliant, and so forth (Weiss, 1993; Hill, Thompson, & Corbett, 1992). Thus, if therapists track clients' immediate responses to the varying interventions they employ (providing interpersonal feedback, making process comments, cognitive reframing, and so forth), clients will behaviorally show therapists how best to respond However, as emphasized earlier, therapists then must be flexible enough to adjust their interventions to provide clients with the responses they can utilize best.

To sum up, working in this highly individualized way adds complexity to the therapeutic process and places more demands on the therapist. However, such an approach gives therapists the flexibility to respond to the specific needs and unique experiences of the diverse clients that seek help. Client response specificity is a core construct in the interpersonal process approach. We will be developing it further in subsequent chapters and, in particular, we will use the guidelines provided for keeping process notes (see Appendix A) and for writing case conceptualizations (see Appendix B) to help therapists formulate the specific relational experiences their clients need.

Teresa: Case Illustration of Core Concepts

The following vignette illustrates our three core concepts: the process dimension, a corrective emotional experience, and client response specificity. In particular, the vignette highlights how the process dimension and the way the therapist and client interact together can either resolve or reenact the client's problem. Usually, therapists do not literally reenact with clients the same hurtful responses that they have received from others (in this case, the problem is sexual abuse). However, unintentionally or without being aware of it, the way in which they interact often thematically evokes the same issues that the client is struggling with in other relationships (in this case, compliance and having to go along with the counselor and others). As you read this case example, think about what you would do similarly or differently if she were your client.

A 1st-year practicum student is talking for the first time with her 17-year-old client, Teresa, about Teresa's sexual contact with her stepfather. The content of what they are talking about is sexual molestation. Depending on the process they enact, however, the effectiveness of this discussion will vary greatly. On the one hand, suppose the counselor is initiating this discussion and pressing Teresa for disclosure about what occurred. The therapist is a graduate student who is genuinely concerned about Teresa's safety. However, her "need to know" is intensified by her concerns about her legal responsibilities as a mandated reporter, and her concern that her supervisor will want to know more details or facts about Teresa's molestation.

In response to the counselor's continuing press for disclosure, Teresa complies with the counselor's authority and reluctantly speaks about what happened. Useful information may be gained under these circumstances, but the opportunity for therapeutic progress is lost because aspects of Teresa's problem are being reenacted with the counselor in the way they are interacting. How is this a problematic reenactment along the process dimension? Teresa is again being pressured to obey an adult, comply with authority, and do something she doesn't want to do. Of course, being pressured to talk about something she doesn't want to disclose in no way retraumatizes her as the original abuse did. However, their interpersonal process of demand/comply is awry and will reevoke in Teresa similar types of feelings and concerns that the abuse initially engendered. That is, her helplessness will lead to depression and her compliance (having to go along) to feelings of shame. Because the therapeutic process is thematically or metaphorically reevoking the original problem, her disclosure is likely to hinder her progress in treatment and actually slow the process of reempowerment.

The situation may be further complicated if Teresa belongs to an ethnic group where family loyalty is highly prized or if she is a member of a religious community where obedience to authority and hierarchical relationships are emphasized. In this cultural context, Teresa is being asked to violate rules sanctioned by her family's ethnic or religious group. Other family members and friends may not approve of the stepfather's behavior. Even so, they may not provide Teresa with the validation and support she needs because she took this information to someone outside the family (the therapist) or because she violated their religious proscriptions that emphasize obedience and forgiveness. Thus, although disclosure violates family rules and loyalties for most victims, this approach may take on additional significance and add more distress for Teresa if she were a Hispanic adolescent who belonged to a conservative religious community.

What should the counselor in this example do instead? Wait nondirectively for Teresa to volunteer this information—while she may continue to suffer ongoing abuse at home? Of course not. But by attending to the process dimension, the therapist may be able to begin providing Teresa a reparative or corrective experience while gathering the same information. That is, instead of pressing for disclosure, what if the therapist *honored* Teresa's "resistance" or cultural prescriptions. For example, instead of pressing for more disclosure, the therapist could metacommunicate and make a process comment by inquiring supportively about Teresa's reluctance to speak:

THERAPIST: All of this is so difficult right now. What's the hardest thing about talking with me?

OR

THERAPIST: It seems hard for you to talk with me right now—maybe something doesn't feel safe. I'm wondering what might happen if you share this and let me try to help you. What could go wrong?

OR

THERAPIST: Perhaps talking with me about this violates family rules, and you're concerned what your mother or stepfather might think. Would you feel OK talking about your family's rules and what's OK or not OK to talk about outside your family?

OR

THERAPIST: Let's work together and try to find a better way to talk about this. Can you tell me one thing that we might be able to change or do differently that might make this a little bit easier for you?

The purpose of these process comments is to create a different interpersonal process and ensure that Teresa does not feel that she has to comply with the therapist—as she has with her stepfather and others. In response to these invitations, Teresa is likely to present a number of reality-based concerns that make it harder for her to disclose. For example, Teresa might reply that

- her mother won't believe her;
- her stepfather will be sent away;
- she will be told at church that it is her fault;
- others will tell her that she should stop causing problems, be forgiving, and not talk;
- she will be chastised for not keeping it in the family and trying to resolve it there;
- the counselor may not believe her or take her parents' side;
- the counselor may want to remove her from her home immediately;
- she fears that the counselor or others will regard her as "dirty," "ruined," or shame-worthy in some way;
- the therapist may feel uncomfortable talking about this sensitive issue and will not want to explore it fully, and so forth.

If the therapist can help Teresa identify and resolve her concerns about disclosing, then she can find it safe and empowering to begin talking about what happened—which does not occur if the therapist simply directs her to disclose. Although this process difference may seem subtle, its effect is powerful and will have far-reaching impact on the course and outcome of treatment. Teresa will either begin the empowerment process or confirm her problematic expectations that she must always go along with what others want, by the way she shares her trauma with the therapist. With the process comments and invitations suggested here, Teresa is able to *participate* in the decision process with an authority figure, have her concerns expressed and taken seriously, and, to the extent possible, accommodated as best they can. For example:

THERAPIST: Yes, Teresa, you don't have to do this alone. Your aunt Norma brought you to see us today, and she is waiting in the lobby. If you like, we can invite her to sit in with us while we talk.

In a way that is new and different, Teresa can at least have some shared control over what she says, and to whom, and still remain supported. This corrective interpersonal process with the therapist will allow Teresa to be able to begin acting in similarly empowered and self-affirming ways in some of her other relationships. The therapist can then help Teresa to systematically begin discerning other people in her life with whom it is safe to be similarly assertive (her aunt Norma, her minister, and certain friends and family members) and those who will punish this or demand that she merely comply again (her stepfather, her soccer coach, and other friends and family members).

But what if the therapist's efforts to create a different interpersonal process don't work and Teresa still does not want to speak or discuss the secret? The therapist is a mandated reporter and still will need to contact Child Protective Services. In some cases, the therapist's attempts to empower the client by giving her more choice or participation in the reporting/treatment process clearly will have helped, and other times it will seem as if it hasn't made a difference. Even if not, however, Teresa has found that the therapist is sincerely trying to find ways to include and empower her rather than merely demand that she comply again and do what the therapist wants. In this way, a small but significant difference is occurring that will facilitate her recovery, as Teresa sees that the therapist's intentions are different than what she has come to expect from others.

In sum, concepts and techniques from differing theoretical orientations all can be helpful, but they are not likely to be effective unless the interaction or process that transpires between the therapist and client is enacting a solution to the client's problems. In the chapters ahead, a variety of therapeutic interventions will be utilized, ranging from providing empathy and a sense of being understood, to helping the client recognize thought processes or behavior patterns that are maladaptive, to role-playing new behaviors. Per client response specificity, however, the effectiveness of these interventions will depend on whether the client's maladaptive schemas and patterns are reenacted or resolved along the process dimension.

Treatment Approach

Drawing on some of the concepts introduced in this chapter, we now provide a brief overview of the therapeutic model developed in Chapters 2–10. For most clients, certain basic developmental tasks were left unfinished in their family of origin. Although enduring problems are often complex and have multiple causes, this unfinished business usually includes the faulty beliefs and ineffective coping strategies that clients have adopted to cope with insecure attachments and other developmental conflicts. These developmental deficits in parenting often have to do with a lack of empathy or being understood, support for their autonomy and self-efficacy, and realistic limits on their behavior so they can learn both self-control and concern for others. Therapists help clients resolve the problems that result in two stages: (1) by providing a corrective emotional experience with the therapist and (2) by generalizing this experience of change to outside relationships. The therapist provides a corrective emotional experience by providing a relationship that enacts a resolution of the client's maladaptive relational patterns, rather than going down the same problematic lines again. Providing such reparative experiences is often easier in theory than in practice. As we will explore, clients are adept at engaging the therapist in aspects of the same scenarios that are disrupting other relationships.

Simultaneously, the client is seeking to avoid reexperiencing painful and unwanted but expected conflicts in the therapeutic relationships (schemas) and, at the same time, to find a new and more satisfying response to them (resolution). To this end, the client tries to assess safety and/or danger in the therapeutic relationship. Let's highlight this ambivalence: The client's relational problems with others are activated with the therapist and are especially likely to be played out along the process dimension. Throughout treatment, the client continues to assess whether the therapist responds in new and safer ways than others have in the past or whether the therapist responds in similarly problematic ways. For example, suppose the client starts to improve in therapy, risks acting stronger in her life, and begins successfully pursuing her own interests and goals (for example, actively seeks and successfully competes for a significant promotion at work). The client will be intensely interested in determining whether the therapist sincerely takes pleasure in the client's success or feels threatened and competitive, as the client's parent used to feel. Or, suppose that a different client reveals vulnerabilities and has emotional needs of the therapist. Will the therapist respond with genuine acceptance or, instead, feel threatened or overwhelmed by these needs, as the client's parent used to feel?

Clients begin to resolve their problems when the therapist's response repeatedly disconfirms such problematic schemas and expectations. Through this corrective emotional experience, clients find that it is safe to act in new and more adaptive ways—at least in some relationships. Clients can then begin exploring new aspects of themselves and better ways to address interpersonal conflicts or establish meaningful relationships with others. This begins the next phase of treatment: *generalizing their experience of change with the therapist to other relationships in their lives.* We will see that clients can do this, in part, by learning how they can change their own responses to others in current relationships—even when significant others continue to respond in familiar but unwanted ways.

As we have emphasized, therapists cannot simply tell clients that relationships can be different from those that the clients have experienced in the past; they must show them. Deeply ingrained schemas and relational patterns, which may involve themes of distrust or compliance, have fundamentally organized the client's life for decades. Such overlearned responses will change as the therapist and client address how similar relational themes are being activated in their relationship and, together, work out a different type of relationship that does not fit the old pattern. Based on this real-life experience of change, clients' schemas are disconfirmed and they can begin to transfer this relearning and apply it to other arenas in their lives where similar patterns are causing problems. For example, in this sequence of change, the client changes and becomes more assertive first in the immediate relationship with the therapist; then intermediately with certain others in her life who are safer or more accepting of this new behavior; and finally with others throughout her life (Egan, 2002).

However, finding this interpersonal solution, disconfirming the early maladaptive schema, and providing the real-life experience that relationships can be another way often entail conflict and anxiety for *both* the therapist and the client. In this regard, the power to effect enduring change does not come primarily from insight or new behavioral alternatives. Rather, the impetus for change often occurs when clients find that they are experiencing the same types of issues with the therapist (for example, control, power, mistrust) that cause problems in other relationships. This time, however, they are able to resolve the old problem. The problematic pattern is named or identified, and the therapist and client work together collaboratively to ensure that it is not repeated in they way they inter-act together. In this way, clients find that, at least sometimes, relationships can develop in better ways and that the therapist can help them apply the changes that have occurred in their relationships to others so they can begin to respond more flexibly and adaptively than they have in the past. This experiential re-learning is the basis of interpersonal process therapy. It places real demands on the personhood of the therapist but holds the great reward of helping people change.

Limitations

In this text, developing therapists are introduced to the practice of interperson-ally oriented therapy with individual adult clients. However, it is important for student therapists to learn how to work in different treatment modalities and in varying treatment lengths. Sadly, it has become more challenging for practi-tioners to earn a living in the contemporary managed-care milieu. Clinicians to-day must market themselves strategically (business courses and special training will be necessary for some). Trainees are encouraged to develop skills in a vari-ety of treatment modalities, including child, marital, group, and family. Al-though the treatment model presented here emphasizes individual adult treat-ment, the process orientation provides basic skills that are readily applied to each of these modalities. (For instance, the first journal published in the new field of family therapy was titled *Family Process*.)

The treatment fundamentals presented here also provide many of the basic skills necessary for brief therapy, including initial intakes, crisis inter-vention, and short-term therapy. Specific guidelines for linking the interpersonal process approach to time-limited, marital, group, and family therapy will be provided at various points throughout the text. In particular, the interpersonal process approach is readily adapted to short-term treatment modalities of 10 to 16 sessions. In recent years, interpersonal/dynamic theorists have developed effective, time-limited treatment approaches that are grounded in a relational or process approach (Davanloo, 1980; Levenson, 1995; Luborsky & Marks, 1991; Malan, 1976; Mann & Goldman, 1982; Safran & Muran, 1998; Sifneos, 1987).

However, before they can succeed in these short-term models, beginning therapists need to acquire certain basic clinical skills. For example, short-term work with this model requires therapists to "bring the conflict into the relation-ship" and to make overt how the therapeutic process may be reenacting aspects of the client's conflicts with others. Most new therapists are not ready to be so forthright and to address these relational reenactments with the client as quickly as time-limited approaches demand. Also, short-term treatment approaches require therapists to establish a treatment focus by identifying maladaptive schemas and relational patterns more quickly and accurately than most begin-ning therapists can do. Furthermore, in the closing session(s) of time-limited treatments, beginning therapists cannot be expected to manage the client's strong grief reactions and feelings of loss, which regularly are evoked as previous losses from death, divorce, and disappointments in other primary relationships are activated. It is also not realistic for beginning therapists to be able to work effectively with clients' negative transference reactions, particularly angry and blaming accusations that the therapist is betraying or abandoning them (just as others have done before), which are commonly evoked in time-limited therapy during the termination phase (Mann, 1973).

Beginning therapists will need more experience and specific training in brief psychotherapy before they can succeed in this essential treatment mode. However, relevant guidelines for working in this important modality will be introduced at many points. As they progress in their training, readers are encouraged to examine some of the excellent, short-term approaches available (Benjamin, 2003; Levenson, 1995; Strupp & Binder, 1984).

Aims

The interpersonal process approach emphasizes three broad aims that are essential to helping clients change. First, therapists want to learn how to establish a strong working alliance with a wide diversity of clients and how to restore it when the expectable misunderstandings and misperceptions that occur in every therapeutic relationship disrupt it at times. Change is predicated on the nature of the therapist–client relationship, and specific guidelines to provide a collaborative or working alliance are provided in the next chapter and throughout.

Second, new therapists need assistance to help them respond effectively to their clients' feelings. Therapy is a personal and private sharing in which therapists often respond to clients' vulnerability and help them with painful feelings. Making contact with clients in these very personal ways is an essential aspect of treatment, yet it is the dimension along which new therapists are most likely to stumble. Most therapists, like other people reared in our culture, have been socialized to avoid rather than approach strong feelings such as anger, fear, and shame. However, one of the most effective ways for therapists to facilitate change is to help clients come to terms with emotional reactions they have not been able to resolve on their own. Specific guidelines will be provided to help beginning therapists understand and respond effectively to their clients' emotions.

The third component of effective therapy is to conceptualize the client's problems and formulate what needs to occur in treatment for this client to change. This conceptualization should enable the therapist to identify the client's faulty beliefs and expectations, maladaptive relational patterns, and problematic coping strategies. Focusing on these issues will provide direction for the ongoing course of treatment. Because the most effective therapists are as adept conceptually as in the affective and relational domains, this text aims to facilitate the therapist's effectiveness in all three arenas.

This text will also provide an overview of the nature and course of a therapeutic relationship. Therapeutic relationships usually have a coherent life course and follow a predictable developmental pattern from beginning to end. Although variations occur with every client, therapists can often identify an ordered sequence of stages. At each successive stage of counseling, the therapist negotiates certain therapeutic tasks with the client. If the therapist achieves this aim, the next set of issues and concerns will often emerge from the client. An awareness of this developmental structure will help the therapist to formulate intervention goals and strategies for each successive stage. The chapters that follow are organized to reflect this developmental schema and to parallel the course of therapy from beginning to end.

Finally, readers of this book need to have realistic expectations for themselves. Novice therapists who are seeing their first clients will find much to help them in their initial work with clients. However, more information is presented here than a new clinician can fully integrate and apply. The concepts introduced in the interpersonal process approach are complex and challenging. Beginning therapists are advised to incorporate these concepts into their practice gradually, at their own pace. Especially during their 1st year of seeing clients, some may begin to feel that the more they learn, the less they know. With more experience, however, 2nd-year students will often begin to find that they are effectively employing many of these concepts with their clients. Typically, trainees are able to make these concepts their own and utilize them easily in two or three years. Because it will take some time to learn how to work in this way, the best approach is to be patient with yourself and enjoy the learning.

Closing

Throughout this text, beginning therapists will be encouraged to be themselves with clients rather than trying to fulfill the role of a therapist. Perhaps Kahn (1997) says it best:

When all is said and done, nothing in our work may be more important than our willingness to bring as much of ourselves as possible to the therapeutic session. . . . One of the great satisfactions of this work comes at the moment students realize that when they enter the consulting room, they don't need to don a therapist mask, a therapist voice, a therapist posture, and a therapist vocabulary. They can discard those accouterments because they have much, much more than that to give their clients. (p. 163)

With this point in mind, sample dialogues are provided throughout this book to illustrate the concepts presented. New therapists do not want to use these exact words, or try to say what a supervisor might say in a similar situation, but find their own ways to express these concepts. Therapists lose their own creativity and self-efficacy when they try to imitate someone else. Our own authenticity is essential. Let's turn now to the first stage of therapy: establishing a working alliance.

Suggestions for Further Reading

1. Attachment theory and its core concept of "internal working models" helps therapists identify the patterns and themes that link together clients' symptoms and problems, and helps therapists find a treatment focus. Readers are encouraged to read the excerpt "Representational Models" in Chapter 1 of the Student Workbook that accompanies this text. We will be working with the related concepts of cognitive schemas, relational templates, and internal working models throughout the chapters ahead.

2. Perhaps the single most helpful book for less experienced therapists to read is Clara Hill's (2004) practical and informative text, *Helping Skills: Facilitating Exploration, Insight and Action*. This highly readable text provides new therapists with the basic helping skills they need in their initial sessions with clients. Student therapists with a little more experience may benefit from Irvin Yalom's (2003) book *The Gift of Therapy: An Open Letter to a New Generation of Therapists and Their Patients*. In the beginning, it may not be easy to differentiate the process dimension from the overt content of what clients are discussing, and Yalom's book will help therapists understand and intervene with the process dimension.

3. One useful book for trainees interested in the interpersonal process approach is Michael Kahn's (1997) *Between Therapist and Client: The New Relationship*. This engaging book provides a historical overview of the major concepts and theorists of the therapist–client relationship. In particular, Kahn articulates the cardinal issue of *nondefensiveness* in the face of clients' attempts to "pull" or "hook" the therapist into their own relational reenactments. This elegant book deserves to be read and reread by beginning and experienced therapists alike.

4. Chapters 3 and 5 of Salvadore Minuchin's (1974) *Families and Family Therapy* provide a succinct presentation of structural family relations and family developmental processes. This discussion will help therapists who work with individual clients understand the genesis and familial context of their clients' problems.

5. A highly readable application of object relations and attachment theory to psychotherapy can be found in Chapters 6 through 9 of John Bowlby's (1988) book *A Secure Base*. An excellent conceptual overview of object relations theory may be found in *Freud and Beyond* by S. Mitchell and M. Black (1995).

6. The great novelists best bring to life the profound impact of childhood and familial experience on adult life. See, for example, John Steinbeck's (1952) *East of Eden* and Franz Kafka's (1966) *Letter to His Father*.

Responding to Clients

CHAPTER TWO

Establishing a Working Alliance

Conceptual Overview

Therapy is a profession based on trust. Clients enter treatment with a need: They are often in pain and asking for help with something they have not been able to resolve on their own. A trustworthy response is to respect the clients' requests for help and respond compassionately to their concerns. Yet to be most effective, the therapist also wants to respond in a way that helps clients achieve a greater sense of their own *self-efficacy*. That is, the goal is not only to resolve specific situational problems but to do so in a way that leaves clients empowered with a greater sense of their own ability to cope with the situations and stressors that are likely to cause problems or relapses in the future. Erik Erikson (1968) clarified that, in crisis, there is the opportunity for growth. In this way, the interpersonal process approach aims to address clients' presenting problems in a way that utilizes this opportunity to help clients live more fully.

Clients cannot attain these twin goals of resolving problems and achieving a greater sense of their own self-efficacy in a hierarchical or one-up/one-down therapeutic relationship. To do so, *clients need to share ownership of the change process* and be active participants who are working collaboratively with the therapist, rather than being passively "cured" or told what to do. Thus, this chapter presents a model for a collaborative relationship or "working alliance" that accepts the client's need for understanding and guidance yet equally encourages the client's own initiative and responsibility.

Although researchers have tried long and hard, they have not been able to find consistent empirical support for the long-term superiority of any one treatment approach over another. In contrast, they have found strong evidence of great variability in the effectiveness of individual therapists within each treatment approach (Teyber & McClure, 2000). That is, far greater differences in treatment effectiveness are found between therapists of the same theoretical orientation (that is, within-group differences) than for between-group differences (for example, interpersonal versus cognitive orientations). Arguably, the strongest finding in the psychotherapy outcome literature is that the most common feature of effective therapists, across different theoretical approaches, is the therapist's ability to establish a strong *working alliance* early in treatment. This chapter explores how therapists can use empathic understanding to establish a strong working alliance and use process comments to restore it when the misunderstandings and problems that inevitably occur temporarily rupture the working alliance.

The Working Alliance Is a Collaborative Relationship

At each successive stage of treatment, therapists have a different overarching goal to guide their interventions. In the first stage, the therapist's principal goal is to establish a working alliance with the client. A working alliance is established when clients perceive the therapist as a capable and trustworthy ally in their personal struggles—someone who is interested in, and capable of, helping them with their problems. To achieve this credibility with clients and become someone who matters to them, the therapist is able to successfully communicate that she

- grasps their predicament and recognizes their distress;
- feels with them and is empathic to their pain;
- is an ally on their side who has their best interests at heart; and
- has an abiding commitment to help them through this predicament.

The concept of the working alliance (also referred to as the *therapeutic alliance* and the *collaborative alliance*) was originally developed by the psychoanalyst Ralph Greenson (1967). He clarified that there are three separate but interrelated components in all therapeutic relationships, which he termed the real relationship; the transference component; and, most importantly, the working alliance. Found in many other theoretical contexts, the working alliance is akin to Bowlby's *holding environment* in attachment terms, where the client's distress is emotionally "held" or contained in the safe relational envelope of the therapist's understanding (Karen, 1998). In existential therapy, Rollo May (1977) elucidates the therapist's emotional *presence* with the client. He emphasizes how the therapist can be fully present with the client in an authentic manner throughout each session. Similarly, within a cognitive-behavioral framework, Safran and Segal (1990) underlined the importance of empathy and the therapeutic alliance. Although present in some form in almost every theoretical orientation, the working alliance is most closely linked to Carl Rogers's (1981) core conditions of genuineness, warmth, and especially, *accurate empathy*.

Researchers have defined the working alliance as a collaborative process whereby both client and therapist agree on shared therapeutic goals; collaborate on tasks designed to bring about successful outcomes; and establish a relationship based on trust, acceptance, and competence (Gaston, 1990). The therapist's ability to establish a successful working alliance in the initial sessions has emerged as perhaps the most important variable in predicting effective treatment outcomes in both short-term and longer-term treatment (Hovarth & Greenberg, 1994; Henry, Strupp, Schacht, & Gaston, 1994). A robust ingredient common to all psychotherapies, the working alliance is correlated with successful outcomes across a wide range of clients' symptoms and problems, and across differing approaches and theoretical orientations (Beutler, 1997; Gelso & Carter, 1994; Hovarth, Gaston, & Luborsky, 1993).

Collaboration: An Alternative to Directive and Nondirective Styles

One of the most important ways to establish a strong working alliance with clients is to work together collaboratively—as partners. In the initial sessions, one of the therapist's primary aims is to articulate clear expectations for working in this collaborative manner and, more importantly, to enact behaviorally these spoken expectations by giving clients the *experience* of working together on their problems. If a collaborative relationship is maintained throughout treatment, this interpersonal process will go a long way toward our goals of helping clients resolve their presenting problems and achieve a greater sense of self-efficacy. Thinking of the working alliance as a collaborative partnership, counselors may come to realize that therapy is not something therapists "do" to clients; it is a shared interaction that requires the participation of both parties in order to succeed.

At the outset, the therapist should ask clients what they know about counseling, explore their expectations of the therapist and the treatment process, and invite their thoughts and suggestions about how the therapist and client could best work together. When clients have been in treatment before, it is important to ask about what was helpful and what wasn't. It is especially important to do this if clients previously dropped out of treatment prematurely, felt they didn't change, or experienced problems or difficulties with the therapist that were not discussed or resolved. If you ask these clients why they stopped prematurely or what they didn't like about treatment, they often give one of these two types of responses:

CLIENT: The therapist didn't say very much—she was pretty quiet, and I never really knew what she was thinking.

OR

CLIENT: The therapist didn't really listen to me very well. He wanted to tell me what to do before he really heard what was wrong.

Let's look at how both of these nondirective and directive styles often fail.

A widely held misconception among clients is that the therapist is a doctor who will prescribe their route to mental health. Often, the therapist is perceived as the sole agent responsible for change—via advice, explanations, interpretations, or simply telling the client what to do. Therapists need to explain to some clients that the therapeutic process is different than seeing their medical doctor. Although the therapist is trained to work with problems such as those the client is presenting, therapy will be more effective if the client can be an active participant in treatment—for example, by

- orienting the therapist to the issues and concerns that are most important right now;
- identifying what the client and others have done in the past that has been helpful and what has not been helpful; and
- sorting through with the therapist shared treatment goals—brainstorming together what they would like to be able to change and how they can best work toward those goals.

Therapists may also explain that, even though the client has not been able to change on his own and the therapist will be an active partner, the client's willingness to work with the therapist to understand problems and explore solutions will be important. Unless the therapist works with this process dimension, a hierarchical doctor–patient or teacher–student relationship is likely to develop and undermine the client's self-efficacy.

Such a one-up/one-down relationship brings with it a number of problems. As we would expect from client response specificity, a doctor–patient relationship will work well with some clients in the short run. Many clients who enter treatment believe that they have to be compliant in close relationships. Regardless of what the therapist says about collaborative relationships, these clients still believe that the therapist, like others, truly wants them to merely follow his lead. Such clients will passively follow the therapist early in treatment and even actively seek direction from the therapist. However, this compliance will ultimately evoke shame and anger toward the therapist, just as it does in other relationships. Although most clients will not recognize or understand what is transpiring, this faulty interpersonal process will prevent them from utilizing the therapist's help, making progress, or even remaining in treatment. Even if clients don't have compliance issues, clients remain dependent on the therapist if this hierarchical mode is enacted. They will not be able to gain a greater sense of their own personal power as long as they hold the dysfunctional belief that the source of adequacy or responsibility for success resides with the therapist, rather than in themselves. Let's explore this point further.

If the interpersonal process approach isn't directive, is it a nondirective approach to treatment instead? No. Some clients bring cognitive schemas to the treatment setting that do not encompass collaborative relationships or mutuality, and these clients may insist that the therapist assume a directive or leadership role in the beginning. If clients bring to therapy an authoritarian upbringing and hierarchical schemas for relationships, then it would be fruitless to insist that they go along with a more egalitarian process. They have not experienced this in past relationships and cannot yet encompass it. Instead, the therapist can be flexible, meet the client on his terms, and accept the client's request to provide more direction or advice. At the same time, however, the therapist can start the change process by giving this client overtly spoken permission to initiate more whenever he might like. For example:

> THERAPIST: That sounds like an important issue for us to work with, Sue. Where do you think is the best place for us to begin?

The therapist can also find other ways to actively engage the client in a more collaborative treatment process.

> THERAPIST: Tell me about your strengths and successes, John. When you've had this problem in the past, what have you done, or how have others responded, that was helpful?

Finally, the therapist can also change this hierarchical pattern by finding ways to begin talking with the client about how they are interacting together. For example, if the client seems to be rejecting the advice and direction he has just elicited from the therapist—as often occurs—the therapist can begin to talk with the client about their current interaction or interpersonal process:

> THERAPIST: I'm a little confused here, John. You keep asking me what to do, but whenever I suggest something, you say "Yes, but. . . ." I'm wondering if one part of you feels like you are supposed to go along with what I say, but another healthier part of you doesn't want to be told what to do. What do you think might be going on between us here?

Whereas a directive stance may work in the short run, a purely nondirective approach will often sputter right from the start. Understandably, most clients feel frustrated when their requests for help or direction are merely reflected back (the "hot potato" game). A negative cycle may ensue: The client becomes increasingly frustrated and seeks some direction from the therapist, who may further eschew this role and talk about inner direction and finding one's own answers. This, in turn, further frustrates the client, who may not feel that she has the answer to anything at that moment and only sees the therapist's attempts to be nondirective as evasive. Many clients drop out of therapy prematurely because the therapist was too quiet or unwilling to accept their appropriate requests to be more actively involved and responsive. Therapists need to be an engaging presence in the relationship—actively involved and responsive, so clients are secure in knowing they have a helpful collaborator with them in the room. Therapists want to provide feedback about the relational and cognitive patterns they observe, help clients reframe problems and consider situations from new perspectives, offer an empathic understanding of the client's feelings and validate their experiences, provide interpersonal feedback, and use process comments to make their current interaction a social learning laboratory. Throughout the chapters ahead, this text will be illustrating this balanced therapeutic stance that is active, responsive, and engaging yet cannot accurately be characterized as directive or nondirective.

Beginning therapists who lose clients because of their own inactivity or lack of responsiveness often report feeling afraid of "making a mistake" or feeling insecure about "what to do." More subtly, new therapists' self-consciousness or inhibition may be prompted by their own concerns about acting more strongly and expressing their own thoughts and observations—having their own voice in the hour. Others hold themselves back because they feel apprehension about becoming someone important to the client and accepting the responsibility of trying to have an impact on the client's life. For many reasons that we will be exploring further, new therapists often are painfully concerned about "making mistakes" and hold themselves back from having an impact on their clients. For now, clients do need direction from therapists—especially when they are in crisis (James & Guilliland, 2000). However, the therapist can provide guidelines for how to proceed without falling back on a hierarchical or one-up/one-down relationship. How can this be achieved? In the initial interview, the therapist can make an overt bid to establish a collaborative relationship in which the therapist and client work together to resolve the client's problems. For example, this attitude may be conveyed by the following overture.

THERAPIST: Let's work together to figure out what's been going wrong. Tell me your thoughts about what the problem has been, and I'll follow along and join in.

CLIENT: I'm not sure where to begin with all of this.

THERAPIST: It's sounds like there's been a lot going on. What's been the hardest thing for you?

CLIENT: Well, it's about my wife and I; we're not getting along anymore. . . .

The therapist is trying to communicate, in words and in actions, that she will be an active and responsive ally. However, it is clear from this invitation that the client is not going to be a passive recipient, but an active participant. The therapist needs to follow through on this invitation by trying to create a collaborative partnership where clients are encouraged to contribute their ideas and find that their input is listened to, valued, and incorporated. Creating a working partnership is an intervention that establishes a new middle ground of shared control in the therapist–client relationship—a different therapeutic stance in between the polarized positions of prescriptively taking charge versus nondirectively following the client's lead.

For many clients, this collaborative partnership, in itself, provides a corrective emotional experience. Their collaborative interpersonal process is a new way of interacting that they may not have experienced in other close relationships. This will alter the client's early maladaptive schemas and it will provide an essential model for helping them change in their relationships with others. *Regardless of the therapist's theoretical orientation or techniques, one of the most important determinants of treatment outcome is whether the therapist and client can continue this process of mutual collaboration.* Let's look more specifically at ways to begin working collaboratively and establish a working alliance in the initial session.

Collaboration Begins with the Initial Interview

One of the most important concerns for new therapists is whether their clients will return after the first session and remain in treatment. Many clients do indeed drop out after the first session; such experiences painfully exacerbate new therapists' concerns about their own adequacy and performance. One of the most helpful guidelines for conducting a successful first session is for therapists to establish a collaborative relationship in their initial contact with the client. As we will illustrate, the therapist structures the session by providing the client with guidelines and direction for what is going to occur in the interview. However, it is essential that the treatment focus reflect the client's own wishes and that the treatment goals be experienced by the client as her own. The therapist's intention is to attend to the issues that the client views as most important. In other words, *the therapist is trying to help clients articulate their own goals and agenda, and then collaboratively join clients in addressing their concerns—as they see and experience them.* Therapists can help clients feel ownership of the change process by simply and briefly offering clients an open-ended invitation to talk about what brings them to therapy.

> THERAPIST: What's the difficulty that brings you to treatment? Help me understand what's wrong.
>
> CLIENT: There's so many different things going on, it's not easy to explain.
>
> THERAPIST: Things are complicated. Maybe you can start by telling me about the concerns you feel are most pressing right now.

This type of inquiry communicates several important things to the client. At the simplest level, it ends the opening phase of social interaction that occurs as the therapist and client are introduced and walk to the interview room. More important, it tells clients that the therapist is someone who is willing to talk directly about their personal problems—as they experience or perceive them—and is ready to respond to their need for help by listening. However, it does so in a way that still gives clients the freedom to choose where they want to start and leaves them in charge of how much they want to disclose. From the outset, clients are sharing control of the interview by choosing what they want to talk about, yet the therapist is an active participant who has offered some direction for where they are heading. This type of collaborative alliance does not occur if the therapist gives the client a more specific cue.

> THERAPIST: When we talked on the telephone, you said you were having trouble with your boss. What is the problem there?

OR

> THERAPIST: You've been having anxiety symptoms. How long has this been going on, and how severe have they been?

The difference between these two openings may seem insignificant. However, it is important to communicate that clients should talk about what they want to talk about and not feel that they have to follow the therapist's agenda.

From the start, we want the client to take an active role in directing the course of treatment, while still feeling that the therapist is participating as a supportive ally. *Enacting this interpersonal process in the initial session is more important than the content of what is discussed* and will shape the course and outcome of treatment.

In most cases, clients will readily accept the therapist's open-ended invitation and begin to share their concerns. The therapist can then follow the client's lead and begin to learn more about this person and his difficulties. Therapy is under way when this occurs. Before we go on to the next step, however, we need to examine two exceptions in which the client does not accept the therapist's offer to begin: when another therapist has conducted an initial screening interview and when the client has conflicts over initiating.

Previous Screening Interview   If another therapist has previously conducted an intake interview, the client may not be so ready to begin.

> CLIENT: (*impatiently*) I've already been through all of this in the intake with Dr. Smith. Do I have to go over it all again just for you?

> OR

> CLIENT: I don't know how much you already know about me. What has Dr. Smith told you about me?

Our initial goal is to develop a working alliance between the therapist and the client. In these two examples, the previous intake therapist is a third party who is psychologically still in the room with them. This is especially problematic for those clients who have come from families in which a third person was *triangulated* into every two-person relationship (Bowen, 1978). That is, whenever two people were close, or in conflict, a third family member would be drawn in and would disrupt the dyad (Haley, 1996; Minuchin, 1984). To keep therapy from recapitulating this dysfunctional family pattern, the therapist and client want to begin their own relationship as a stable dyad. As the family therapists teach us, it is important to maintain a dyadic therapeutic relationship that does not allow others to disrupt the therapeutic alliance.

> THERAPIST: I know that you have already spoken with Dr. Smith, and I have learned a little bit about you from his intake notes. But just you and I are going to work together from now on, and I'd like to hear about you in your own words. It may be a little repetitious for you, but this way we can begin together at the same point.

Third-party interference or triangulation in the therapist–client dyad may also occur in the initial therapy sessions because of the unseen but felt presence of the clinical supervisor. As we will see, this can occur from either the therapist's concerns or from the client's. Through insecurity, compliance issues, or other factors, beginning therapists may triangulate their supervisors into the therapeutic dyad. For example, the therapist might tell the client, "I'll have to ask my supervisor about that." More commonly, beginning therapists may silently invoke their supervisors—by wondering what they would say or do at a particular point in the session or how they would evaluate the therapist's performance at that moment. This self-critical monitoring does not facilitate treatment—and it takes the fun out of seeing clients! It diminishes both the therapists' self-confidence and their ability to be emotionally present with the client. Additionally, it disempowers therapists by inhibiting them from being themselves in the session—from finding their own words, acting on their own perceptions, listening to their own intuitions, and developing their own therapeutic styles. New therapists are encouraged to trust themselves enough to simply be a person sharing the life story of another person. Supervisory input and evaluation, though necessary, usually will be more productive when it is processed outside of the therapy hour. Again, in accord with our theme: The therapeutic process is awry when the dyadic relationship between the therapist and the client is broken by third-party interference.

Although we don't want the therapist to triangulate their supervisors into the dyad, clients may be curious or concerned about what the supervisor thinks of them or the progress they are making. This is a different situation and requires a therapeutic response. This is especially likely when a videotape or audiotape is being used during sessions. It is often informative for the therapist to explore clients' concerns or thoughts about the supervisor—it often reveals the cognitive schemas or transference distortions they bring to relationships. For example, if the therapist asks clients what they imagine the supervisor might be thinking about them, clients often reveal important concerns:

- I'm worried that he might be talking about me to other people.
- He probably thinks I'm a weak person because I'm so anxious about this— you know, doesn't respect me for being so worried about everything all the time.
- She's probably bored with me—like everybody else always is.
- I think she's probably critical of what I've been doing—you know, judging me.

By the client revealing these problematic projections in this way, the therapist has the opportunity to resolve them in treatment—and keep them from driving the client out of treatment. For example:

> THERAPIST: No, I don't think he's judging you at all. In fact, I think he probably feels compassion for what you've been coping with and respects your willingness to come here and work so hard on your problems.

In addition, the therapist has the opportunity to begin making connections and working on how these faulty schemas or problematic projections may be contributing to problems with the therapist and with others in their current lives. For example:

> THERAPIST: You feel he's judging you critically. Ouch, that wouldn't feel very good! Are there others in your life right now who may be judging you as well, or maybe it feels like I'm being critical, too?

Finally, some clients may feel uncomfortable about working with a trainee or beginning therapist:

CLIENT: Does your supervisor tell you what to say to me, or do they let you say what you think?

THERAPIST: I consult with a supervisor, but I have my own mind and will say to you what I think and believe.

Other clients may feel protective of the beginning therapist and try to improve in treatment or respond in ways that make the therapist look good to the supervisor. Although well intended, it is a highly problematic reenactment that will prevent the client from changing this problematic caretaking role with others in her life. In all of these ways, it helps clients make progress in treatment when they can talk through and resolve with the therapist such concerns and distortions. In sum, whenever therapists sense the presence of a third party psychologically in the room, they are encouraged to make this presence overt and explore the impact this third party is having on the therapeutic relationship.

Conflicts over Initiating   There is a second common circumstance in which the client does not respond to the therapist's initial request to begin. The problems that some clients bring to treatment involve conflicts over initiating. This type of client cannot begin at the therapist's request. By asking these clients to begin talking about whatever is most pressing or important for them, the therapist has inadvertently presented them with their central or presenting problem. It is difficult for this type of client to decide what to talk about, take the first step and initiate any activity, or assume any responsibility for the course of treatment. Thus, a therapist who nondirectively waits for the client to lead places an impossible demand on this client.

On the other hand, a directive therapist who takes charge and begins the session by telling this client what to talk about is often replaying the same unwanted scenario the client has found with others. Whether communicated in a friendly tone or impatiently, the therapist—like others in their lives, is again telling them what to do. Thus, treatment stalls right from the start if the client has problems with initiative and the therapist responds in either a directive or nondirective manner. What can the therapist do instead with this type of client? One of the best ways to find this more effective middle ground between directive and nondirective extremes is to make a process comment. That is, an effective intervention is simply to wonder aloud or ask about what may be occurring between them at that moment.

THERAPIST: It seems to be hard for you to get started. Maybe we can begin right there. Is it often hard for you to begin, or is there something about this situation in particular that is difficult for you?

By first identifying this as a problem, and then encouraging the client to explore it, the therapist has offered the client a focus and assisted the client in moving forward. However, the therapist has provided this focus without taking over and telling the client what to do, which would only reenact the familiar but unwanted relational pattern. The therapist's open-ended inquiry is supportive, in that it responds to the client's immediate concern. Yet the client can take this issue of initiating wherever he wants, and he is sharing responsibility for the direction of treatment. This type of response provides a new opportunity for clients to explore their problems in a supportive context. Our first goal is met as the client *experiences* a collaborative interaction with the therapist, rather than merely having a conversation about the need to work together—which usually doesn't take the client very far. In contrast, such a new, collaborative experience from the outset engenders hope that change is possible and sets positive expectations for the future course of treatment.

Finally, age, class, gender, culture, religion, and other issues of diversity are important to consider when initiating therapy. For example, new clinicians want to be sensitive to the fact that, in some cultures, it may be seen as disrespectful to presume to lead with one's elders, an educated person, or an authority figure such as a therapist. As we will explore further, therapists can best respond to such concerns by acknowledging differences and inviting clients to express or explore the concerns together.

Empathic Understanding: The Foundation for a Working Alliance

The therapist has begun the session by giving the client an open-ended invitation to talk about whatever feels most important to her. Responding to this bid, the client begins to share her concerns with the therapist and clarifies the background and context of her problems. The therapist's intention now is to try and find the subjective meaning that each successive vignette holds for the client and grasp what is most significant to the client *from the client's point of view*. In other words, therapists are striving to have the cognitive flexibility to decenter, enter into the client's subjective experience, and capture the core meaning that this particular issue holds for the client (McClure & Teyber, 2003). In particular, the client will feel deeply understood when

1. the therapist can reflect the most basic feeling or capture the key issue in what the client just said; and
2. the therapist can identify a common meaning, theme, or pattern that links together the client's varying concerns.

This type of accurate empathy earns credibility with the client and, perhaps more than any other component, is the basis for establishing the working alliance: Clients become more engaged with the therapist, invest further in the treatment process, and risk exploring their problems more fully. Let's explore this pivotal sequence more closely.

As used here, *empathic understanding* connotes a genuine feeling of warmth and concern for the client—it is not a technique but rather a respectful attitude or nonjudgmental stance toward the client. It tells clients that the therapist sees their situation, and that their distress matters to the therapist. When therapists can grasp their dilemma and articulate or express this understanding, they are providing the secure base to "contain" the client's distress that the attachment researchers describe (Bowlby, 1988). Although this is not sufficient to resolve clients' problems, such empathic understanding will often ease clients' initial distress; for some, their presenting symptoms may even abate. A secure attachment configuration is established, and the therapist's attuned responsiveness often begins to relieve the entire family of attachment affects:

- Anxiety may diminish from this reassuring emotional contact.
- Depression may lessen; the experience of being seen and accepted, rather than dismissed or judged, engenders hope in the client.
- Anger from frustration over not being seen or feeling invalidated by others may be assuaged by the therapist's affirmation.
- Shame over having emotional needs revealed by entering therapy and asking for help may be lessened by the therapist's respect.

Empathic understanding is the basis for establishing a successful working alliance, and a key concept for change in the interpersonal process approach. For example, Lafferty, Beutler, and Crago (1991) summed results across groups of more and less effective therapists from 11 different studies based on the extent of clients' symptom change. They found that clients of less effective therapists felt less understood by their therapists, whereas more effective therapists were seen as more empathic by their clients. The four sections that follow examine different aspects of this key concept of empathic understanding.

Clients Do Not Feel Understood

Most clients are concerned that others do not really listen to them, take them seriously, or understand what they are saying. Clients often describe themselves as feeling alone, unseen, different, or unimportant. Many clients feel this way because their subjective experience was not *validated* in their family of origin. While growing up, most clients repeatedly received messages that denied their feelings and invalidated their experience:

- You shouldn't feel that way.
- Why would a silly thing like that make you mad?
- You can't possibly be hungry now.
- I'm cold. Put your sweater on.
- How can you be tired? You've hardly done anything.
- You shouldn't be upset at your mother. She loves you very much.
- You don't really want to do that.
- We don't talk about those things.
- What's the matter with you? How could you feel that way!

Furthermore, family members often changed the topic or simply did not respond when the client expressed a certain feeling or concern. As Carl Rogers, Virginia Satir, and Irvin Yalom model so effectively, one of the most effective ways therapists can help their clients change is to affirm their subjective experience. R. D. Laing goes so far as to suggest that people stop feeling "crazy" when their subjective experience is validated (Laing & Esterson,1970).

Consistent invalidation in the client's family of origin has profound, long-lasting consequences. In its severe forms, some authors describe it as "soul murder" because clients lose themselves—their sense of self or their own voice—when they lose the validity of their own experience (Schatzman, 1973). Whereas disconfirmation has occurred to some extent in most clients' families, survivors of child abuse have experienced pervasive invalidation of some of their most important experiences in their lives (for example, "I don't know what you're talking about"; or, most poignantly, simply continuing as if nothing was said). One of the most serious consequences of such systematic invalidation is inefficacy or disempowerment. When reality based feelings and perceptions are repeatedly denied, offspring become incapable of setting limits, saying no, and refusing to go along with what does not feel right to them. More significantly, consistent denial of their experience leaves clients unsure of what has actually happened to them and of the subjective meaning that events hold—that is, they lose the ability to "have their own mind" and know what they know. They no longer trust their own perceptions of what may be occurring or making them uncomfortable. Typically, they won't register that something someone said or did bothered them until well after the problematic interaction is over, if at all. Denying the validity of their own experience, these clients characteristically say to themselves, "Oh, nothing really happened;" "It wasn't that bad;" or "It doesn't really matter" when it actually was significant to them. Routinely, they cannot find words to communicate their own experience or point of view with any clarity or specificity and, even if they can, do not expect others to understand or be interested in what they say.

When their subjective experience has been denied repeatedly, clients do not know what they are feeling, what they like or value, or what they want to do. In alcoholic, highly authoritarian, and abusive families, such invalidation is a pervasive, everyday experience that continues throughout their childhood and adolescence. In place of clear feelings and confident perceptions, a vague, painful feeling of internal dissonance results. Fortunately, this undifferentiated feeling state can be replaced with emotional clarity and a stronger sense of personal identity if the therapist consistently listens to clients with respect and validates their experience. To affirm clients and increase their self-efficacy, therapists listen intently, take seriously whatever matters to the clients, and reflect their understanding of what the clients have said. Although such simple empathy and validation may sound like common or ordinary human responses, they are not. Many clients have not had their most important feelings and perceptions validated in their most significant relationships—and in turn, they do not expect to be seen or understood by the therapist either. Thus, again and again, throughout each session, the therapist's abiding intention is to validate the client's experience by identifying and articulating the central meaning that this particular experience seems to hold for this client right now. For example, in the following questions and comments, *the therapist is not necessarily agreeing or disagreeing with the client but trying to capture the emotional meaning or key issue in what the client has just said.*

- It didn't seem fair to you.
- I'm wondering if you were frightened when he did that?
- It's been too much for you, more than you can stand.
- It felt great to be so effective and in charge!
- It was disappointing; you wanted more than that.
- Here again, are you having to take care of everyone else?

The therapist does not have to be an exceptionally perceptive person to understand the client's experience, and new therapists need to be reassured that, of course, they will not be accurate all the time. For example, when discussing parenting, Winnicott (1965) has a wonderfully reassuring phrase to diminish our performance anxieties: children need only "good enough mothering," just as clients, too, need their therapists only to be "good enough." Clients do not need their therapists to be "right" or accurate all the time, as long as it is clear that the therapist is sincerely striving to better understand the client and "get" what this experience means to them. The therapist's effort to understand conveys more empathy, and fosters more of a working alliance, than being "right." Thus, when therapists do not understand what the client is saying, they should not feign understanding by saying "Yes," "OK," or "I know what you mean" when they really don't understand. Instead, the therapist can acknowledge his confusion or ask the client to say something in a different way, so that he can understand better and they can get back on track together. By inviting this clarification or dialogue, *the therapist is intervening with a more collaborative approach to empathy.* For example:

> THERAPIST: That sounded important, Susie, but I didn't understand it as well as I wanted. Can you say that again, or put it differently?

Furthermore, if therapists feel they have been inaccurate, they can simply check out their perceptions and invite the client to clarify them. For example:

> THERAPIST: As I listen, it sounds as if you feel so discouraged that you just want to give up. Am I saying that right—or can you help me say it better?

In both of these examples, the therapist is giving the client an invitation for a dialogue. That is, the therapist is trying to *jointly clarify the client's meaning with this back and forth collaboration.* This approach to empathy takes the performance pressure of having to be "right" off the therapist, diminishes unwanted concerns about making mistakes or being wrong, and it furthers the working alliance. The therapist's perceptions and empathic bids certainly will be inaccurate at times. However, their sincere efforts to understand clients' experience by engaging in this process of mutual clarification demonstrates to clients that the therapist isn't just going through the motions, but sincerely wants to understand what they meant. Let's explore this interactional or collaborative approach further—it's a different way of thinking about empathy.

Client-centered therapists originally conceptualized empathy as a stable or enduring characteristic of the therapist. Instead of thinking about empathy as a personality trait, researchers have found that effective empathy comes from attempts to *collaboratively* understand the client's experience within an emerging, shared frame of reference (Barkham & Shapiro, 1986). That is, measuring across client-centered, cognitive, and dynamic therapists, *the therapeutic mode of mutual exploration—characterized by active negotiation between therapist and client—is crucial to the client's experience of being understood* (Hardy & Shapiro, 1987; Hobson, 1985). For example:

THERAPIST: What did he do when you took the risk to say that?

CLIENT: He just went on talking, as if I hadn't said anything at all.

THERAPIST: I'm sorry he wasn't able to listen better—where did that leave you inside?

CLIENT: I don't know . . . (pause) . . . I hated it.

THERAPIST: Yes, you hated it. I'll bet you did, almost like you were invisible or didn't count?

CLIENT: Yeah, something like that.

THERAPIST: Hmm, sounds like "invisible" doesn't quite capture it. What would be a better word?

CLIENT: I'm not sure . . . (pause) maybe just erased. Yeah, like my whole existence was being erased by him!

THERAPIST: "Erased." That captures it much better—you say that with a lot of feeling. Tell me more about being erased—help me understand what's that like for you.

CLIENT: Oh, I hate that feeling more than anything, and it happens over and over again. . . .

In this collaborative exploration, the therapist and client work together as partners to clarify the specific meaning this experience held for her. In this way, *accurate empathy is not a stable personality characteristic of the therapist; it is an interpersonal process characterized by mutual exploration and collaboration.*

To sum up, the therapist intervenes by validating their clients' experience, grasping the core messages, and affirming the central meaning in what they relay. In this regard, being accurately empathic is sometimes misunderstood as just being "nice" or supportive. These are benevolent, well-intended responses, yet they do not do much to help most clients change. Instead, with accurate empathy, we are looking for something more specific and far more potent—the ability to identify what is central or discern what is most important in what the client has just said. To appreciate how important accurate empathy and validation is, new therapists may recall for a moment what others have done to help them during their own crisis periods. Almost universally, helpful responses include an accurate, empathic understanding of their experience and validation of their feelings and perceptions.

Providing validation is especially important when working with minorities, gay men and lesbians, economically disadvantaged clients, and others who feel "different." These clients will bring issues of oppression, prejudice, and injustice into the therapeutic process, and their personal experiences have often been invalidated by the dominant culture. These clients, in particular, will not expect to be heard or understood by the therapist. The first step in working with all people is to listen empathically and hear what is important to them. Therapists respect the personal meaning that experiences hold for different people when they enter into and affirm the client's subjective experience. The most effective therapists, of every theoretical orientation, offer these basic human responses especially well. Pioneering clinicians as varied as Beck, Bowlby, Kohut, Rogers, Satir, and Yalom all share this respect for the client's subjective experience, and emphasize the need for therapists to respond empathically and understand the client's subjective experience.

Demonstrate Understanding

To engage a client in a working alliance, the therapist listens to the client's experience, tries to discern the most basic feeling or key concern that each successive vignette holds for this particular client, and accurately reflects back what is most significant in the client's experience. An accurately empathic reflection that captures the central meaning in the client's experience is a complex intervention; it is never just a rote parroting of what the client has said. An effective reflection or restatement is more akin to an accurate interpretation or creative reframing. It goes beyond what the client has said and communicates that the therapist understands the core message, registers the emotional meaning, or distills what is most important from the client's frame of reference. Rogers (1975) believed that communicating such understanding provides a deep acceptance that is a prerequisite for meaningful change. New therapists will find that they achieve credibility with their clients and strengthen the working alliance when they can articulate or *demonstrate* their understanding in this tangible way.

According to one unfortunate stereotype, a therapist is someone who says superficially, "I hear you," "I know what you mean," or "I understand." However, this type of global, undifferentiated response is not effective and, paradoxically, often furthers the client's sense of not being seen or heard. The therapist does not simply say, "I understand" but demonstrates that understanding by *articulating* the central meaning of what the client has said. Therapists show the client that they can be helpful when they capture and express the specific meaning in the client's experience, rather than offering well-intended but vague reassurances. And by checking with the client to see whether that understanding is accurate, the therapist is working collaboratively, which in turn strengthens the working alliance. For example

THERAPIST: Yeah, I know just what you mean—that's happened to me, too.

VERSUS

THERAPIST: Let's check this out together and see if I'm hearing what you're really saying. As soon as you realize that others are taking you seriously—listening to your ideas and thinking about what you're saying, you become anxious or unsure—worried that you're going to be dismissed or put down. That's when you start to get quiet, go along with others, and end up just hating yourself when you leave that meeting. Am I getting that right—can you help me say it better?

Let's look more closely at how clinical trainees can start to put these ideas into practice. In the following illustration, we examine a brief case study of a client who initially had the experience of not being heard by her therapist but later felt understood with a second therapist.[1]

While growing up, Marsha did not feel heard or understood by her parents. Her father was distant. He was not comfortable talking with his adolescent daughter and believed his wife should handle the children. Her mother was critical, demanding, and often intrusive. Whenever her mother felt or believed something, she expected her children to see it the same way—differing points of view were not permitted. For example, if Marsha felt something that her mother did not feel, her mother would chide her, "That's ridiculous. What's wrong with you?" As an adolescent, Marsha was painfully self-conscious and insecure. However, she didn't really understand why she felt so unhappy and often thought, "There's just something wrong with me." Marsha frequently cried alone in her room, but without really understanding why she was crying or what she was upset about.

Marsha had always thought that everything would get better when she went away to college. To her dismay, though, she found herself becoming even more depressed during her first semester away. She could not stop crying, it seemed, and she started gaining weight. More confused about herself than ever, Marsha began seeing a counselor at the Student Counseling Center. Although she did not really have words for what was wrong, Marsha tried to help her therapist understand her problems:

MARSHA: I feel empty—kind of lonely.

THERAPIST: Do you have any friends? What are your classmates like?

MARSHA: I guess I have friends. I have a roommate in the dorm.

THERAPIST: What do you do with your friends—go to the movies, shopping?

MARSHA: Yeah, I do those things. I belong to a foreign language club, too.

THERAPIST: Do you like your friends? You're new to the university; maybe you need some new friends here at school. There's lots of ways to meet new people here.

MARSHA: Well, I've had friends, but I think I just feel empty.

THERAPIST: But you've just left your family and come to college. You must miss your family and feel lonely. It's natural to feel lonely when you move away from home. Most of the other kids in the dorm feel that way, too. I know I sure did when I moved away to college.

MARSHA: Oh.

---

[1] The clinical examples in this text are based on actual cases; however, identifying information, including the gender of the client or therapist, has been altered to ensure confidentiality.

THERAPIST: This is not unusual at all. You're going to be just fine once you get through this transition.

MARSHA: I hope so. Maybe my family is different, though. In high school, I always thought that my family had more problems than my friends' did.

THERAPIST: Yeah, but, like I said, it's normal for you to be kind of emotional at this time in your life.

MARSHA: It is? But I still feel different from everybody else.

THERAPIST: Sure, it's real natural for you to feel different from everybody else. Late adolescence, moving away to college—it's a tough time in life. You're going to be just fine. Do you have a boyfriend?

MARSHA: Yes. I try to tell him what's wrong, but he just tries to fix it. And then he gets frustrated with me because nothing helps. I just don't know why I get depressed a lot (*pauses*). . . . My mother used to yell at me a lot, though.

THERAPIST: Do you have trouble eating or sleeping?

MARSHA: No trouble eating—unfortunately, I've gained 12 pounds. I wake up at night sometimes—worrying about things that don't really matter, feeling like crying.

THERAPIST: Are you eating alone? Maybe you should be eating with friends.

MARSHA: (*with resignation*) Maybe that would help.

Marsha could not say exactly why, but she did not like seeing the therapist and, after a few sessions, did not go back. However, as the semester went on, she became more depressed, continued to wonder what was wrong with her, and struggled in her classes. During advising for second semester, a concerned professor saw her distress and encouraged her to go back and try another therapist. Reluctantly, Marsha agreed to try it when the professor suggested that she could just go back for one session and, if she didn't like it, talk to the therapist about making some changes in the way they worked together or just stop.

Therapy began more slowly this time. Marsha found herself impatient with the therapist and reluctant to share much. She alternately acted blasé and then distressed, but she would never allow the therapist to remain joined with her on a specific topic for very long. The therapist was not frustrated, however, and was effective at communicating his continuing interest in her. After a few weeks, Marsha sensed that this relationship might be safe enough to risk disclosing her feelings again.

MARSHA: I feel empty.

THERAPIST: Empty. Help me understand what that feeling's like for you.

MARSHA: I don't know. I just feel empty inside.

THERAPIST: You're telling me this empty feeling is important for you. Let's work together and try to get closer to it. Is there some way you can start to tell me a little more about that emptiness—help me share it with you?

MARSHA: (*long pause; says nothing*)

THERAPIST: This is hard. Maybe you can use just adjectives, or make a gesture, that will describe the emptiness. Or maybe you've noticed when you start to feel it—who you are with or what's going on?

MARSHA: There's just an emptiness inside. The wind blows right through me. I'm always hungry—it never gets filled up.

THERAPIST: That's a good start. You just feel empty inside and nothing fills it up. Nothing takes care of this—sex, alcohol, food. The hole just stays there.

MARSHA: (*doesn't speak; nods and becomes teary; looks at therapist*)

THERAPIST: (*holding her gaze kindly*) I can see how much you are hurting right now as you share this with me. It's sad that things have happened to make you feel this empty and alone, but I'm honored that you are willing to risk sharing all of this with me.

MARSHA: (*cries harder*) There's just something wrong with me; there always has been.

THERAPIST: Tell me about it; help me understand what's always been wrong.

Marsha was heard and felt understood this time. Breaking the old relational pattern, she was not crying alone in her room anymore. In subsequent sessions, the therapist consistently listened and responded to what was most important for Marsha. Although there were ups and downs, this empathic connection forged a working alliance that ultimately allowed Marsha to resolve her lifelong dysthymia.

The concept of understanding the client sounds simple enough. All the therapist has to do is listen carefully, grasp the central feelings or core message, and find a way to articulate or reflect this understanding to the client. In actual practice, however, demonstrating our understanding in this way is not so easy to do. Most of us have been strongly socialized to "hear" in a limited, superficial way that minimizes the emotional meaning and avoids the interpersonal messages embedded in clients' remarks. This was why Marsha's first therapist could not hear what she was really saying and steered away from the sensitive feelings and concerns involved. Reflecting her resiliency, Marsha kept coming back to her core message, but the therapist neither heard the content of what she was saying nor followed the process. By reassuring her, trying to talk her out of her feelings, and orienting her toward superficial problem solving before he really knew what was wrong, he reenacted her developmental history of not being seen or heard.

Most beginning therapists are aware that they possess a "third ear"—that they already have a highly developed ability to hear the basic meaning in what people say. However, they often feel they must avoid acknowledging the true content of these underlying, and often nonverbal, messages because of their emotional content. Therapists downplay or avoid the emotional message or basic meaning for many reasons:

- Feeling concerned that they might embarrass the client
- Feeling unsure of how to respond

- Feeling reluctant to violate cultural or familial proscriptions about forthright communication
- Feeling a need to take care of others by shielding them from their own pain or discomfort
- Feeling threatened by the emotional intimacy

As a result, many therapists are adept at switching automatically to a more superficial or everyday, social level of interaction to avoid the client's vulnerability and pain—which usually emerges from empathic listening. In addition, therapists do this to move away from *embedded messages* about their relationship that most therapists do not want to hear or approach. That is, *clients are often making covert statements about the therapeutic relationship, or indirect references to what is going on between them, especially about problems or concerns they are having with the therapist right now*. For example, let's see how therapists can ineffectively dodge or avoid this personal reference and then, in contrast, see how therapists can effectively approach these embedded messages about the therapist–client relationship:

> CLIENT: Do therapists who work with clients like me ever see their own therapists for their own personal problems?
>
> INEFFECTIVE THERAPIST: Yeah, they probably do, but why don't we get back to the real concerns that you came in for.
>
> VERSUS
>
> EFFECTIVE THERAPIST: Yes, most therapists seek help with their own issues at times. I'm wondering if you may be having a question about me, or something related to our work together, that we should talk about? If so, I think it would help us out if we could talk about that.

In contrast to her first therapist, Marsha's second therapist broke the social rules and responded directly to the core affective message "I feel empty." As a result, Marsha felt understood for the first time in therapy. In a small yet significant way, this was a corrective emotional experience. The therapist joined her in her experience—which significant others had not done before. Disconfirming her expectations, he did not move away from her feelings or try to talk her out of them; in sharp contrast, he genuinely welcomed them and actively encouraged her to enter them more fully. When the therapist is able to provide such experiences consistently, trust builds as the client learns that their previous expectations do not fit this relationship—and some other relationships as well. Behavior change begins first as the client becomes more confident in the expectation to be seen and heard by the therapist and, with the therapist's help, then extends this new schema or way of behaving to other relationships in their lives. For many new therapists, stepping beyond the social norms they grew up with is an exciting yet stressful component of clinical training. *New therapists are encouraged to try responding in this more direct, empathic way; use the ability they already possess to hear what is most important to the client; and while still being diplomatic, take the risk of saying what they see and hear.* Most beginning therapists hold themselves back too much.

To summarize, therapy offers clients an opportunity to be understood more fully than they have been in other relationships. When this understanding is offered in the initial sessions, clients feel that they have been seen and are no longer invisible, alone, strange, or unimportant. At that moment, the client begins to perceive the therapist as someone who is different from many others in their lives and possibly as someone who can help. In other words, hope is engendered when the therapist understands and articulates the subjective meaning that each successive vignette holds for the client. With this in mind, let's look at guidelines to help therapists clarify the central meaning in what the client presents and be more accurately empathic.

Identify Recurrent Themes

As we have seen, the therapist's goals at the beginning of the session are to

- encourage the client's initiative,
- actively join the client in exploring her concerns,
- identify the core messages or central meaning in what the client says, and
- articulate or reflect this empathic understanding to the client.

Although this, too, may sound simple, it is not. What's so hard about it? It requires the therapist to give up a great deal of control over the direction of therapy and over the timing and content of which issues are brought up for discussion. The "letting go" of what the client discusses—not knowing what the client is going to say or do next—is challenging for most clinical trainees. Most therapists are interpersonally skilled in their everyday lives and used to being in charge or in control (albeit subtly) of many of their social interactions. However, the same social skills that facilitate interpersonal relations outside therapy (that is, subtly leading or initiating the conversation, filling awkward moments and facilitating transitions, keeping the conversation moving or staying on comfortable topics) limit a client's progress within the therapeutic relationship. In treatment, the novice therapist is placed in the new and far more demanding position of responding to the diverse and unpredictable material that the client produces, rather than subtly directing what the client will cover—which most new therapists are skillful and practiced at doing. This new therapeutic approach, which encourages the client's initiative and ownership of the treatment process, achieves our overarching treatment goal. We are enacting an interpersonal process that fosters self-efficacy by sharing more responsibility with clients for resolving their problems. However, doing so puts significant personal demands on the therapist. Specifically, *the therapist needs to have the flexibility to be able to relinquish control over what will occur next in the therapeutic relationship, tolerate the ambiguity of not knowing what clients will produce or where the current topic will lead, and make sense of the varied material that clients present.* No doubt, this is a tall order.

What assistance is available to help therapists meet the demands of working in this way? If therapists can identify *recurrent themes* in the client's vignettes or stories, it will help them make sense of the client's experience and better understand what is most important or central in the wide-ranging material the client presents. This empathic understanding, in turn, provides therapists with the only legitimate control they can have in therapeutic relationships—over their own responding—as opposed to ineffective attempts to subtly direct or control the material that clients produce. Additionally, the appropriate or legitimate inner controls that come from this understanding will allow therapists to *tolerate the ambiguity* inherent in this work.

Suppose the client tells the therapist about his reasons for coming to therapy. As the therapist follows the different recollections and descriptions the client chooses to relate, the therapist is trying to discern an *integrating focus* for the wide diversity of material that the client is presenting. The therapist can do this by identifying patterns or themes that recur throughout the different material the client presents. These integrating themes occur in three interrelated domains:

- Repetitive relational themes or interpersonal patterns
- Pathogenic beliefs, automatic thoughts, or faulty expectations
- Recurrent affective themes or central feelings

As we will see, *identifying these unifying patterns or themes helps therapists conceptualize their clients' problems and formulate a treatment focus*. To beginning therapists, the disparate material that clients present often seems unrelated and disconnected. However, as they progress in their training, new therapists will increasingly recognize that synthesizing themes are present in all three of these interpersonal, cognitive, and affective domains as they learn to listen for them.

Restating this cardinal concept, through empathic listening, therapists can learn to put the client's story into the broader context of their lives, which usually reveals integrating patterns in their thoughts, feelings, and behavior. As we will develop later, these integrating themes can be shared tentatively, as hunches or possibilities, to formulate a treatment focus:

THERAPIST: It's sounding to me as if, here again, you have to be the responsible one who takes care of everything—and wind up feeling tired and resentful in the end. I'm wondering if this sounds like a familiar pattern in your life—does something like this go on a lot?

CLIENT: It doesn't sound like "something" in my life—that *is* my life. My work, my marriage, my childhood! That's been my role my whole life—like it's my fate. But if I don't stop doing this, it's going to kill me—I'm exhausted!

To help therapists develop this essential skill, guidelines for writing process notes within an interpersonal process framework are provided in Appendix A. Keeping process notes like these after each session is one of the best ways for therapists to begin recognizing the themes that organize their clients' lives. Let's see how new therapists can begin to recognize these helpful patterns in each of the three domains.

Repetitive Relational Patterns   Although terminology varies, the bedrock of all interpersonal and contemporary psychodynamic treatment approaches is identifying the repetitive relational patterns that repeat throughout the client's symptoms and problems. Therapists listen for the interpersonal themes that keep reoccurring throughout the different narratives the client relates. For example:

- Whenever I risk trusting someone, they let me down.
- They expect so much from me, but no matter how hard I try, it's never enough.
- I always have to give up what I want in order to be close. If I try to meet my own needs or say what I want, I'm always seen as the needy or demanding one.

As beginning therapists review the process notes they write after each session, they will often observe that the varying problems that each client talks about usually come back to only two or three basic themes or patterns that keep getting expressed in different ways.

Interpersonal therapists have always focused on clients' repetitive, self-defeating relationship patterns. Since the 1980s, however, this approach has been developed more systematically as a treatment focus in short-term or time-limited treatments (Levenson, 1995; Luborsky & Crits-Christoph, 1990; Sifneos, 1987; Strupp & Binder, 1984). These treatments all focus on clients' maladaptive relational patterns, which different approaches variously label as the core conflictual relationship theme, cyclical maladaptive pattern, maladaptive transaction cycle, and others. However, in each approach, the therapist is trying to identify the central themes that keep repeating in the varying problems the client is experiencing with others—and beginning to re-create with the therapist. The therapist begins the change process by *highlighting* these relational patterns—so clients can begin to recognize these scenarios and anticipate when they are coming into play. The client's relational pattern, for example, might be to readily feel controlled by others, let down, or abandoned. Or, perhaps the client quickly feels criticized and put down, or even idealized, by others. It is compelling for clients when the therapist can identify the same interpersonal themes occurring in three different spheres: in current interactions with others; in formative relationships with family members; and in the current, here-and-now interaction with the therapist. Let's explore this basic sequence more closely—it offers therapists working within different theoretical orientations such an easy way to help clients.

First, as clients are telling their stories and relaying vignettes from current or past relationships, the therapist is trying to identify repetitive relational patterns that are causing problems in their lives, and highlight or punctuate them. As the therapist shares these observations in a tentative manner and explores them collaboratively, clients become aware of the problematic patterns and begin observing when these scenarios are occurring with others. At the same time, the therapist is actively looking for opportunities to *link* the same maladaptive relational patterns that are occurring with others to their current interaction. The therapist does this by asking about or observing how this same, problematic pattern that they have been discussing with others may be occurring between them right now. For example, suppose the client is acting in a controlling or dominating manner with the therapist; the client interrupts the therapist whenever she starts to speak; or the client acts helpless and confused and begins speaking to the therapist in a childlike tone of voice. To facilitate change, *the therapist needs to have the flexibility and willingness to work through this reenactment with the client, and help create a different and better way of relating in their relationship than the client has been doing with others.* That is, the therapist helps the client find new and more adaptive ways to respond in their relationship—so that the new behavior is not just being talked about but is being enacted right now. For example, the therapist makes a process comment and gives the client interpersonal feedback about how she is acting with the therapist right now:

> THERAPIST: Michelle, I know people don't usually talk together this directly, but I'd like to share a reaction I'm having right now—maybe there's something to it or maybe not. You've been saying that people don't take you seriously, even though you are clearly a bright and capable woman. But as I listen to you right now, you are speaking to me in a childlike tone of voice—almost like a little girl. If you are doing that with others in your life, many will just dismiss you. What do you think as I share this possibility?
>
> CLIENT: Well, it's kind of embarrassing to hear, but I've heard this before. . . . But I don't know why I talk like that, I don't want to be that way. . . .
>
> THERAPIST: Let's be partners and work together on this here in our relationship. Whenever you or I notice the "little girl voice," let's speak up and name it. And why don't you try talking to me, in here, in a stronger voice. If you can claim your own power in here with me, then we can help you start doing it out there with others. What do you think?
>
> CLIENT: Yes, I want to change this, and that little girl voice does have a lot to do with why other people just shine me on. . . .

This is in vivo or experiential relearning—the therapeutic relationship is a social learning laboratory. As this relearning occurs in the therapeutic relationship, the therapist can readily begin generalizing or transferring these new ways of behaving, and changing these same maladaptive patterns in the clients' current relationships with others. That is, as they talk about and change the way they are interacting together, the therapist and client jointly begin exploring when and how these same scenarios or patterns are being played out with others in their current relationships:

> CLIENT: I think I slip into that "little girl voice" thing whenever I feel someone has authority over me—like my boss, or my professors. I don't do it with my friends, but maybe I am doing that with my dad. . . .

As they try out and explore new ways of responding, and begin to succeed in changing these maladaptive behavior patterns with others, clients often become interested in doing a "cost-benefit analysis"—assessing what they get from this interpersonal pattern (for example, "When I talk in the 'little girl voice,' no one will get mad at me or think I'm challenging them") and how they pay for it ("People in authority don't take me seriously or respect me"). Oftentimes, clients also become interested in trying to understand where they learned to act in this way—which helps clients understand the problem and be able to maintain the new behavior after treatment ends. Throughout this text, we will focus on how therapists can use process comments to help clients recognize maladaptive relational patterns with others, and how they are recurring with the therapist. We then will find ways to change these problematic patterns—first with the therapist, and then with others in their lives. For now, however, let's go on to identify themes in the cognitive and affective domains—all three are equally important and inextricably entwined.

Pathogenic Beliefs Just as the therapist can better understand clients by identifying unifying themes in their interpersonal relations, cognitive patterns are an essential component of accurate understanding and effective intervention (Lazarus, Lazarus, & Fay, 1993; Beck, 1995). For example, clients' maladaptive relational patterns are closely linked to their problematic beliefs about themselves, inaccurate perceptions of others, and faulty expectations for what is going to occur in relationships and in the future. Ellis (1999) suggests that two of the most common dysfunctional beliefs concern being liked and loved ("I must *always* be loved and approved by the significant people in my life") and being competent ("I must *always*, in all situations, demonstrate competence"). Working in a psychodynamic tradition, Weiss and Sampson (1986) and Weiss (1993) have clarified the "pathogenic beliefs" that help create and sustain the maladaptive relational scenarios described above. For example, regarding clients who feel excessive or unrealistic guilt, the client's pathogenic belief might be as follows:

- I am being selfish whenever I say no or do what I want.
- Others are hurt by my independence.
- I must not enjoy my own success, and I need to sabotage my own goals and ambitions, or I will be disloyal.

Similarly, a pathogenic belief for clients who are prone to feeling shame or that others will ignore or reject them, might be as follows:

- I am unimportant and do not matter.
- I am inadequate, and others will see me as lacking or weak.
- Others would ignore me if they knew what I needed, or resent my emotional needs as childish and demanding.

Long ago, George Kelly (1963) introduced his theory of "personal constructs," which became the forerunner for cognitive behaviorists today who try to discern clients' "core beliefs," such as those listed here, and the "early maladaptive schemas" that we introduced in Chapter 1 (Young, 1999). In particular, Beck (1976) and Beck and Freeman (1990) have illuminated self-talk, automatic thoughts, and the dysfunctional interpretations that are central to clients' problems—and that reflect *core themes* about ourselves and our relationships. They write compellingly about the cognitive schemas people develop—that is, ways of viewing themselves, others, the world, and the future. In particular, these authors have identified a *cognitive triad* in which people come to view themselves as defective or unlovable; the world as unmanageable or overwhelming;

and the future as bleak and hopeless. We began to see in the last chapter that, as a result of formative developmental experiences, clients begin to exercise a *selective bias* in processing information. As a result of this cognitive filter or selective attention process, *clients inaccurately synthesize the same repetitive themes from diverse interpersonal experiences.* They continuously construe or interpret events to arrive at the same conclusions, such as disinterest or unfairness from others, which come to characterize their experiences. Thus, this mindset or lens for viewing others creates patterns in the client's cognition, as they keep slotting different experiences into the same narrow perspective (viewing others as wonderful or idealized, needy and dependent, disinterested and unresponsive, critical and judgmental, and so forth—even when they are not).

Similar to these cognitive schemas, we also saw that attachment theorists use the term *internal working models* to clarify how these overgeneralized beliefs and expectations are learned in repetitive interactions with caregivers, and how they may operate consciously or without awareness. As before, therapists try to

1. identify these core beliefs that provide integrating threads throughout the client's life and problems;
2. make them overt to clients or name them—especially as they are occurring in the current interaction with the therapist; and
3. question the truthfulness or utility of these beliefs as they come into play with others in current relationships.

As Beck emphasizes, automatic thoughts and dysfunctional beliefs are not so evident when clients are functioning well. It is when clients are upset or distressed (as Beck would say, in their *hot cognitions*) that pathogenic beliefs and faulty expectations will be revealed. At these crisis points, they are expressed as the client's reality and experienced as the only way that relationships have ever been and ever will be. In other words, when clients are not upset, they can imagine a wide range of interpersonal scenarios and outcomes, entertain a more realistic set of beliefs about themselves and others, and have flexible expectations about what may occur in relationships. In contrast, when distressing life circumstances have activated their conflicts, clients' beliefs and expectations often become rigid and unidimensional. Thus, it is when clients are upset or distressed that therapists can best assess their orienting schemas and how they process information, and intervene through cognitive reframing and helping clients consider this familiar, problematic scenario from a new perspective or different point of view. In closing, therapists often learn more from the question "How did you explain that to yourself?" than from asking, "How did you feel about that?"

Recurrent Affective Themes   Therapists also can identify unifying themes in clients' emotional reactions. Recognizing and responding to these *recurrent feelings* can be nothing less than a gift that therapists offer to their clients. For example, the therapist might reflect like so:

> THERAPIST: This is so hard for you. Here again, when _____ happens, you are always left feeling _____.

Often, a primary or core affect comes up again and again for the client in many different situations (Greenberg & Paivio, 2003; Mueller & Kell, 1972). As the therapist listens to the client, an overriding feeling such as sorrow, bitterness, or shame may pervade the client's mood or characterize the different experiences the client relates. For instance, the therapist might then say:

> THERAPIST: As I listen to you, it sounds as if you have felt hopelessly *burdened* all of your life.

OR

> THERAPIST: This doesn't just sound like worry or anxiety that keeps coming up; it seems much bigger than that. Am I hearing a sickening feeling of *dread*?

When the therapist can identify and accurately name this core affect—the feeling that some clients may experience as nothing less than the defining aspect of their existence—it has a profound impact. Clients feel that the therapist understands who they really are and sees what their life is like in a way that others have not been able to do. Perhaps nothing goes further toward establishing credibility and strengthening the working alliance than the therapist's ability to identify a core affect and accurately reflect the far-reaching meaning it holds for the client.

In many cases, responding to the primary affects or capturing the repetitive emotional themes that keep coming up in the client's life is one of the most significant interventions that therapists can provide. Unfortunately, new therapists often feel insecure about responding to clients' feelings. Trainees will need their supervisors to help them recognize these affective themes and respond to these primary feelings. Because this topic is so important and often receives too little attention in clinical training, we will explore these interventions in depth in Chapter 5. Completing the process notes provided in Appendix A will also help therapists identify the recurrent patterns in all three domains.

To sum up, if the therapist can identify the maladaptive interpersonal patterns, pathogenic beliefs, and primary feelings that link together the clients' experience and problems, clients will feel secure in knowing that the therapist understands them in a way that others do not. Providing this highly differentiated empathy is a gift to clients. *It is not an isolated intervention or technique but a consistent attitude that characterizes the therapist's way of being with the client throughout the course of treatment.* Therapists work together with the client to understand what a client is experiencing in this moment and to clarify the integrating themes that organize or reveal more meaning in the clients' life. When the therapist does so, the therapist helps clients find themselves in their own reflection from the therapist.

This approach to understanding clients by finding unifying themes in their behavior, cognition, and feelings applies equally well to crisis intervention and short-term therapies. Although therapists need to be more focused and limited in their goals, their principal aim in shorter- and longer-term modalities is to provide a treatment focus by identifying these enduring patterns and themes.

The primary challenge of successful time-limited therapy is that, far more quickly and accurately than in longer-term treatment, therapists are trying to

1. establish a working alliance through accurate empathy;
2. identify recurrent themes in all three domains;
3. make these patterns overt by highlighting how they may be occurring right now in the current interaction with the therapist; and
4. link the here-and-now interaction with the therapist to what clients could be doing differently with others in their lives.

Therapists in training need to learn how to enact these interpersonal processes with clients in longer-term treatment first; none of this is easy to do in the beginning. With some time and breathing room to practice, they soon will enjoy adapting these skills to effective short-term treatment.

Immediacy: Working in the Moment
Using Process Comments to Build a Working Alliance

Therapists help build a working alliance by intervening in the here-and-now and talking with clients about what may be going on between them right now. There are several closely related interventions that therapists can use to work in the moment and create *immediacy*. However, these interventions are not for the feint of heart—*they significantly increase the intensity of the therapist–client relationship*. They are powerful interventions that allow therapists to engage clients in far more meaningful ways, and they foster the working alliance by inviting a more genuine or authentic relationship.

In essence, process comments make the interaction between the therapist and client overt and put the relationship "on the table" as a topic for discussion. Process comments are not confrontations, demands, or judgments. Instead, they are simply observations about what *may* be occurring between the therapist and client at that moment. Process comments are always more effective when they are offered *tentatively*—the therapist is often wondering aloud about what might be going on between them right now, and as an invitation for further dialogue and mutual sharing of perceptions. Researchers find that process comments offered tentatively ("It sounds like . . . "; "I'm wondering if . . .") are more effective than direct challenges (Jones & Gelso, 1988). Let's illustrate how therapists can use process comments as an invitation to consider and discuss:

> THERAPIST: As you're talking about him and what went on between you two, it sounds like you might be saying something about our relationship as well. What do you think—does that same kind of frustration ever come up between us, too?

> CLIENT: Well, sometimes I do feel frustrated with you, too, like when you're so quiet and I don't really know what you're thinking.

OR

> THERAPIST: I'm wondering what it's like for you to be telling me about this sensitive issue right now. How is it for you to be sharing this with me?

CLIENT: I'm just dying inside—it's mortifying to talk about this. You're being nice—I know you have to act that way because you're a therapist—it's your job to act nice, but I know you don't respect me anymore.

Clearly, process comments bring the therapeutic relationship to life; they make it far more meaningful for both the therapist and the client. As in the dialogues here, *process comments also reveal unspoken interpersonal conflicts*— significant misunderstandings or faulty expectations that are going on in the therapeutic relationship without the therapist's awareness (for example, the first client is frustrated because he wants more responsiveness from the therapist, and the second client is convinced that the therapist doesn't respect her anymore and is just acting nice because it's his job). Unless we use process comments to reveal these unspoken problems in the therapeutic relationship, sometimes referred to as "ruptures in the working alliance," the therapist doesn't have the opportunity to address and resolve them—and that is what we are looking for in the interpersonal process approach.

Process comments are also useful when the interaction has become repetitive or has lost its direction. For example:

THERAPIST: I'm not sure where we are going with this right now. Is this the best way for us to spend our time together today, or is there another issue that might be more important? What do you think?

CLIENT: Well, I don't really have a problem I want to talk about today. Actually, something pretty exciting happened this week and I'm feeling really good, so I guess I don't know what to talk about.

THERAPIST: We don't have to talk only about problems—I want to know about what's important in your life. You've got some good news—cool. I'd love to hear about it.

Clearly, process comments bring *immediacy* to the therapeutic interchange as the therapist and client stop talking about others out there and begin talking instead about "you and me" (Robitschek & McCarthy, 1991; Binder & Strupp, 1997). Process comments also give therapists a way to identify or highlight the client's maladaptive relational patterns and a way to intervene and change these reenactments *as they are occurring with the therapist*. For example:

THERAPIST: I think that something important might be going on between us right now. Can we talk together about what just happened here? I know people don't usually talk together this way, but I think it could help us understand what's been going wrong with your wife and others as well.

CLIENT: I don't like these homework assignments we're talking about. I don't like people telling me what to do, and I guess I get kind of pissy with you, like I do with people at work.

THERAPIST: I'm so glad you are telling me this. I don't want to tell you what to do in here in therapy, and I'm learning something important about what goes wrong with your boss, and maybe your wife, too. Let's see if we can work out this control thing better here in our relationship, and I have some ideas we could start talking about that might help make this better with others, too. How does that sound?

In addition to activating relationships that have become stalled and altering maladaptive relational patterns as they begin to re-play with the therapist, process comments can be used to provide interpersonal feedback and help clients see discrepancies between what they are saying and doing. For example:

THERAPIST: What you are sharing with me right now is a very sad thing, John. Actually, it's kind of heartbreaking to hear. But you are saying it in such an off-handed manner, as if it doesn't really matter or mean much. Help me understand the two different messages I'm getting—such a painful story being told in such a half-interested, almost pleasant manner?

CLIENT: Well, no, I didn't really realize I was doing that. I do feel sad about what happened, but I sure wouldn't want anyone to think I was feeling sorry for myself—you know, being a complainer or anything like that.

OR

THERAPIST: You're telling me how furious you are about this, Janice, but your voice sounds so small or meek. Are you aware of this discrepancy between what you are saying and the way you say it?

CLIENT: Well, yeah, I guess so, but why do you bring it up?

THERAPIST: Because I think what you say matters. And if you talk to others like you are talking with me, people won't really listen to you or take you seriously.

CLIENT: (tearing) Well, that's sort of been my life story.

Across differing theoretical orientations, a variety of terms have been used to describe process comments and other related interventions that create immediacy. In particular, Kiesler discusses *therapeutic impact disclosure*. This intervention uses interpersonal feedback from the therapist to make overt how the client's maladaptive relational patterns are affecting the therapist at that moment and how they are being played out in their current interaction (Kiesler & Van Denburg, 1993). For example:

THERAPIST: Sometimes it feels like you are jumping from topic to topic, John, and I have trouble keeping up with you. I often feel like I'm getting lost or losing the point you are trying to make. Is this something that just goes on between you and me, or have others told you that they feel confused or have trouble following you sometimes?

Similarly, object relations theorists and communication theorists provide *metacommunicative feedback* as an intervention to register the unspoken emotional quality of a relationship (Cashdan, 1988). For example:

THERAPIST: Although you've never actually said anything like this, Barb, I'm afraid that if I disagree with you, you'll be angry and leave our relationship. You know, if I see something differently and disagree with you, you'll walk out the door at the end of the session and I won't see you again. What do you think about this feeling I'm having? Is this just me, or is there something to it?

CLIENT: (sheepishly) Well, maybe, my boyfriend would sure agree with that. . . .

In the counseling literature, *self-involving statements* present another type of immediacy intervention, and they may be especially useful for new therapists.

Whereas *self-disclosing statements* refer to the therapist's own past or personal experiences, self-involving statements express the counselor's current reactions to what the client has just said or done (McCarthy, 1982; Tyron & Kane, 1993). For example, the therapist might say:

THERAPIST: Right now, I'm feeling _____ as you're telling me about this.

Contrast the first, self-disclosing response with the second, self-involving comment:

THERAPIST: I have a temper too, sometimes.

VERSUS

THERAPIST: As I listen to you speak about this, I find myself feeling angry, too.

As we would expect on the basis of client response specificity, self-disclosing comments may be useful at times. Too often, however, clients' focus is shifted away from themselves and onto the therapist when therapists reveal personal information. More productively, self-involving statements keep the focus on the client and reveal information about what is happening in the relationship or how something the client has said or done is affecting the therapist. For example:

THERAPIST: I can feel my stomach tighten right now as you are reading that letter from your father.

Sharing personal reactions like this to what clients have just said or done conveys personal involvement with clients and acknowledges emotional resonance. This type of self-involving statement, and all of these process-oriented interventions that bring the therapeutic dialogue into the here-and-now, go a long way toward strengthening the working alliance. We have also seen that they give clients feedback about the impact they are having on the therapist—and others. It can be a gift when therapists use process comments to provide interpersonal feedback, and therapists can find constructive ways to help clients see themselves from others' eyes and learn about the impact they are having on others (such as regularly making others feel bored, intimidated, overwhelmed, confused, and so forth). *With these types of process comments, an effective therapist is willing to take the risk to say to clients what others often may be feeling and thinking but don't want to say*:

THERAPIST: Are you aware that you're speaking in a very loud voice right now?

Although this may be hard for new therapists to believe, it really is not a big problem if the client misunderstands their good intentions and feels put down or criticized! If this misunderstanding occurs—which it will sometimes based on the client's schemas—the therapist can just take the next step and talk through this misunderstanding. That is, the therapist can make it overt and address it with the client—which allows them to sort it through in a reparative way that does not occur in most other relationships. For example:

THERAPIST: I just said that you interrupted me when I began to speak, and that happens several times every session. You've become quiet—maybe you disagree or didn't like that I said that. Can we talk about how it was for you to hear me say that?

CLIENT: It hurt my feelings. I didn't like being told I had a big mouth like that.

THERAPIST: Thank you for being so honest with me. Let's sort this through together—I think we just had a big misunderstanding. I have never thought that you have a "big mouth," and I didn't say that to you, but I often do feel interrupted. I asked about it because if other people in your life feel interrupted like I do, it might have something to do with the loneliness and superficial friendships you've been talking about.

CLIENT: Well, I still don't like that you said that, but maybe I do interrupt people more often than I realized.

THERAPIST: OK, it still doesn't feel good that I said that, but maybe this is teaching both of us something important about what's going wrong with other people. This is a sensitive issue, but I'd like to keep talking and work it out.

CLIENT: I don't know why I interrupt people so much. I think I do it without even realizing it. I guess that's how we grew up talking in my family. . . .

In the beginning, it's hard for most new therapists to have the interpersonal confidence to intervene in these ways. It is challenging to take the risk of talking with clients about how they are interacting with you and others, and make overt and talk about what's going on right now that might be related to their problems. Most supervisors and practicum instructors agree: These interventions which bring immediacy are the most anxiety-arousing and challenging interventions for new therapists to adopt (Hill, 2004; Egan, 2002). To use them successfully, therapists need to do two things. First, they must balance the challenge of the metacommunication with being supportive and protective of client's self-esteem (Kiesler, 1988). We are going to offer process comments tentatively, as an invitation for discussion or mutual exploration, and we need to share them respectfully. Second, therapists need to check in with the client and talk about the immediacy intervention they have just made ("How was it for you when I said that?") (Safran, Muran, Samstag, & Stevens, 2002). Let's explore this matter further now and throughout this text.

Depending on the therapist's ability to communicate in a respectful manner, any of these process-oriented interventions that focus on the current, here-and-now interaction can be used ineffectively or effectively. For example, contrast the blunt, ineffective response here with the more respectful, effective response that follows:

THERAPIST: I'm not getting much out of this. Why don't we try to talk about something more important?

VERSUS

THERAPIST: I'm concerned that I'm missing you right now and not grasping the real meaning this holds for you. Can you help me get closer to what is most important for you here, or maybe there is something else that we should move on to that you would find more meaningful?

Working in the moment with process comments like these is a challenging way to intervene for most new therapists. Routinely, it is going to take a year or two before therapists feel comfortable saying what they see and broaching with clients what may be going on between them. Before going further, let's explore two situations when process comments will not work.

First, therapists should not jump in with process comments without first considering the possibility that their reaction reflects more about their own *countertransference* issues than it does about the client. As a rule of thumb, therapists usually want to wait until they have seen an interaction occur two or three times and have had time to reflect upon it or discuss it with a supervisor, before venturing this observation with the client. Differentiating the client's concerns from the therapist's own personal issues is an important topic; we will return to it many times in the chapters ahead.

Second, process comments create problems for clients when therapists promise something they cannot provide. The process comments and immediacy interventions introduced here encourage open and direct communication, which is also an invitation to have a more significant relationship. However, if the therapist then responds in impersonal, distancing, or judgmental ways that do not honor the honesty invited, clients are set back by this mixed message from the therapist. This double-binding situation will be especially problematic for clients whose caretakers gave promises of authentic emotional involvement but failed to follow through. For example:

THERAPIST: You've become so quiet—I'm wondering if there is something going on here that we should talk about?

CLIENT: Well, since you're asking about it, I'm feeling kind of discouraged about therapy. I don't know where this is going, and things aren't getting any better in my life.

THERAPIST: (*defensive tone of voice*) I don't know what you expect me to do with that. Maybe you need to work a little harder in here.

Let's redo this same conversation but, this time, the therapist follows through with his invitation for more honest communication:

THERAPIST: You've become so quiet—I'm wondering if there is something going on here that we should talk about?

CLIENT: Well, since you're asking about it, I'm feeling kind of discouraged about therapy. I don't know where this is going, and things aren't getting any better in my life.

THERAPIST: (*in a nondefensive, welcoming tone of voice*) Thanks for the straight talk. I like the part of you that is willing to speak up like that—it's a strength of yours. I've been feeling that we're not moving forward, either. It seems to me that we start out on issues that seem important at first, but they don't go very far and just sort of fall away. I've been hoping that we would clarify more of a focus or direction that would keep going further, but we haven't been able to do that yet. What do you think we could change or do differently in here to try to make this work better?

CLIENT: What you just said sounds like the story of my life. I'm always starting things—schools, relationships, playing the drums—but I don't stay with any of them. Nothing stays alive for me—things never get finished. . . .

To sum up, process comments often are the most powerful interventions that therapists can make. When used effectively, they greatly facilitate the working alliance, reveal important new issues that are central to the client's problems, and provide the most effective way to alter maladaptive relational patterns that are being reenacted with the therapist. However, as we have emphasized, these types of process-oriented interventions that create immediacy are new and very different for most beginning therapists. Trainees do not need to try out process comments until they feel ready to try communicating in this forthright way, and are ready to meet the client in the authentic manner that process comments invite. For many, it will take a year or two of practice with a supportive supervisor before these process interventions become automatic or second nature. To learn more about how to work in the here-and-now with these immediacy interventions, see the references in the section Suggestions for Further Reading at the end of this chapter.

Using Process Comments to Repair Ruptures

Carl Rogers legitimized concern about the quality of the therapist–client relationship. His contribution to the field of counseling cannot be overstated: He "changed the game" (Kahn, 1997). Rogers's core conditions of genuineness, positive regard, and accurate empathy in particular, is at the heart of a strong working alliance. However, these core conditions are not easily achieved with some clients—for example, those who are hostile, demanding, or critical; and these core conditions are challenging to maintain consistently throughout treatment with most clients. Researchers have been reporting for decades that even highly experienced therapists do not do a very good job of remaining empathic, or even neutral, with "negative" clients who are difficult to be with because they are angry, challenging, or controlling toward the therapist (Binder & Strupp, 1997). As we will explore in this section, process comments provide therapists an effective way to address the problems that occur with these challenging clients and, more importantly, to resolve the misunderstandings that occur with almost every client at times—and that can lead to treatment failure. In the working alliance literature, this is the all-important work of "rupture and repair" (Safran & Muran, 1995).

To facilitate the working alliance, the therapist's role is to *sustain* an empathic, respectful, and interested attitude toward the client (Binder & Strupp, 1997). This alliance-fostering therapeutic stance is not provided consistently, however, as recurrent "ruptures" in the working alliance inevitably occur during the course of treatment (Safran et al., 2002). Alliance ruptures occur for three different reasons: covert or overt hostility (referred to as *client negativity*), reenactments or interpersonal scenarios that ensnare or embroil the therapist, and the simple human misunderstandings that occur in every meaningful relationship.

First, let's consider client negativity. Irritability is a defining diagnostic feature for most clients who are anxious and depressed. Similarly, demandingness and hostility are characteristic features for most clients with histrionic, borderline, and narcissistic disorders, and so forth. At times, these clients are going to be dominating, demanding, critical, and so on, with the therapist—just as they are with others in their lives. For example:

CLIENT: No, that's not right, either. Do I have to explain this to you again?

Second, regarding maladaptive relational patterns, many clients are going to "pull" therapists into hostile, critical, competitive, controlling, rescuing, or other types of conflictual exchanges that reenact problematic scenarios in their lives. For example:

CLIENT: I don't think you understand me as well as my first therapist—he was so sensitive and perceptive. He really made me feel cared about. So, uhh, you're a graduate student. How many clients have they let you see now?

Disturbingly, researchers consistently find that most therapists do not deal with this type of common and expectable client hostility or negativity in a therapeutic manner. Instead, they usually respond "in kind" with their own anger, judgmentalism, and criticism toward the client (Kiesler & Watkins, 1989; Safran & Muran, 1995; Najavits & Strupp, 1994). For example, in a series of studies known as Vanderbilt I and II, clients' maladaptive relational patterns routinely elicited counterhostility and control from seasoned therapists, and when these "complementary" responses occurred, it led to poor therapeutic outcomes. That is, *clients with angry, distrustful, and rigid interpersonal styles tended to evoke countertherapeutic hostility and control—even in a carefully selected sample of highly trained and experienced therapists* (Henry, Schacht, & Strupp, 1990; Strupp & Hadley, 1979).

Third, even with clients who are "nice," simple misunderstandings and problems with treatment and/or the therapist occur at some point in most therapeutic relationships. Unfortunately, *clients often do not voice their concerns or negative sentiments, and therapists often avoid or do not ask about them* (Hill, Thompson, Cogar, & Denman, 1993; Rennie, 1992). It seems that many therapists are unskilled or insufficiently trained in restoring disrupted relationships and resolving personal impasses between them (Watson & Greenberg, 2000). Unfortunately, unresolved misunderstandings between client and therapist undermine the working alliance and lead to poor treatment outcomes (Elliott et al., 1994; Johnson, Taylor, D'elia, Tzanetos, Rhodes, & Geller, 1995).

What's the solution to client negativity and misunderstandings? Process comments provide an effective way for therapists to repair or restore ruptures in the working alliance. That is, rather than avoiding the interpersonal conflict as if it hadn't occurred or responding in kind with countercriticism or anger, the most effective intervention is to address the problem by talking directly with the client about it as it is occurring (Rhodes, Thompson, & Elliott, 1994). That is, by using process comments and the metacommunicative feedback presented earlier, therapists can neutrally observe or wonder aloud about problems or misunderstandings that may be occurring between them. Let's consider some examples:

THERAPIST: I'm wondering if you might have experienced my last comment as critical or judgmental. Can we talk about that possibility?

OR

THERAPIST: Right now, it feels to me as though we're almost arguing with each other. I'm wondering if both of us are feeling misunderstood by the other. Let's work this out. What do you see going on between us?

OR

THERAPIST: I think I was feeling impatient there and pressing you to make a decision before you were ready. What was going on for you right then?

OR

THERAPIST: When you spoke over me just then, in that loud voice, I felt kind of squashed. Maybe something wasn't feeling very good to you, either. Let's have an honest conversation about what's going on for both of us here and sort this through.

Impasses result and treatment cannot progress unless therapist and clients can talk openly about problems in the relationship and what went wrong between them (Petersen et al., 1998). Working out problems that arise between them provides an invaluable social laboratory where clients can learn how to resolve conflict with others in their lives. However, most trainees do not receive enough professional training or preparation for responding to conflict with their clients. Furthermore, on a personal level, many beginning therapists find it far easier to be empathic or supportive with clients than to approach or address interpersonal conflict and try to work through problems directly. In their families of origin, most therapists did not do this. Many families held strong but unspoken rules against discussing conflict openly and honestly—and for some, it was simply taboo. Working within every theoretical orientation, treatment success depends on the therapist's ability to restore the working alliance when it has been disrupted, and process comments give therapists the tools they need to do this effectively.

New Therapists Struggle with Performance Anxieties

Let's let therapists be human, too. They cannot always hear or fully assimilate what they have just been told, or be accurately empathic. Understandably, all therapists will miss the feeling and meaning in the client's experience at times. This section further examines performance anxieties and new therapists' worries about "mistakes," which in turn makes therapists preoccupied and less present or empathic with their clients.

Therapists will not be as effective with clients when they are burdened by their own excessive performance demands. Commonly, beginning therapists are trying too hard (to be helpful, to win approval from a supervisor, to prove their own adequacy to themselves, to be liked by the client, and so forth). When therapists are trying too hard in one of these ways, it is almost impossible to decenter, enter the client's subjective worldview, and be emotionally present.

Too often, novice therapists are thinking about where the interview should be going next, wondering how best to phrase what they are going to say next, preparing advice or suggestions, or worrying about what their supervisors would expect them to do at this point in the interview. Such self-critical monitoring can immobilize student therapists, block their own creativity, and keep them from enjoying this rewarding work. Moreover, it prevents therapists from stopping their own inner chatter, listening receptively to their clients, and understanding as fully as they could. Thus, beginning therapists are encouraged to slow down their own self-monitoring process and focus more on the moment— on their current interaction with the client, and on grasping the meaning that this particular experience seems to hold for this client right now.

Identifying recurrent themes, as described earlier, can itself become a disabling performance demand. If therapists try too hard to find the essential meaning in the narratives that the client relays, they may lose the broader context of the client's life that clarifies the patterns therapists are trying to find. It is better for therapists to be at ease, be patient, and help clients unfold their story as they experience it. Undistracted by performance demands, therapists will be better able to hear the repetitive themes and central meaning in clients' experiences.

When they feel anxious about their helping abilities, beginning therapists may develop a need to *do* something to make the client change. Novice therapists often believe they are supposed to make the client think, feel, or act differently within the initial therapy session. As a result, beginning therapists often change the topic abruptly (at least from the client's point of view) and pull clients away from their own concerns to what the therapist thinks is most important. Although therapists certainly will want to help clients make bridges to new topics they haven't considered, these attempts will usually fail when they are driven by the therapist's own internal pressure to fix the problem. Paradoxically, when the therapist acts on these unrealistic expectations, clients are more likely to drop out of treatment because they feel that the therapist doesn't understand them or just isn't helpful.

To sum up, therapists with excessive performance anxieties or who are too preoccupied about making mistakes are less effective because they are responding more to their own internal needs (to be helpful, competent, or liked) than to the client's need to be understood. The most appropriate way for therapists to manage their realistic concerns about their performance is to discuss them openly with a respected supervisor. Therapists do not need to labor under the burden of always knowing the right thing to do or what is going on with the client. What is asked, instead, is only to keep trying to maintain a collaborative partnership with the client in order to learn together what is occurring.

Care and Understanding
as Preconditions of Change

The therapist does need to help the client change, of course, but change is most likely to occur if the client first experiences the therapist as someone who understands and cares. In the initial sessions, the therapist's primary goal is to establish a working alliance. The therapist does this by establishing an emotional connection with the client through empathy and respect. Facilitating a collaborative, working alliance is more important than obtaining specific information about the client or effecting any change. In this regard, one of the best ways to evaluate the success of an initial session is for therapists to ask themselves, "Do I feel like I made contact with the client and have a feeling for who this person is?" Therapists are building an affective bridge to the client and creating the interpersonal context necessary for change, which will not begin until clients feel that the therapist is available, concerned, and sincerely trying to understand. If therapists provide this type of responsiveness throughout the initial session, they are already being helpful to the client. In attachment terms, anxiety, depression, and other symptoms may even abate as the therapist's attuned responsiveness provides a secure holding environment that contains the client's distress (Bowlby, 1988).

Finally, we must address the most important component of what it means to understand the client. Therapists want to actively extend themselves to the client and communicate their interest in the client's well-being. That is, therapists want to articulate their understanding of the client's experience in a way that also communicates their compassion.

This ability to articulate the client's experience in an accurate and caring way is illustrated elegantly in a classic article, "Ghosts in the Nursery," by Selma Fraiberg and her colleagues (Frailberg, Adelson, & Shapiro, 1975). The therapist in this case study is working with a depressed young mother whom social services has judged to be at high risk for physically abusing her infant daughter. Early in treatment, the therapist is disconcerted as she observes the mother holding her crying baby in her arms for 5 minutes without trying to soothe it. The mother does not murmur comforting things in the baby's ear or rock it; she just looks away absently from the crying baby. The therapist asks herself, "Why can't this mother hear her baby's cries?"

As the young mother's own abusive history began to come out in treatment, the therapist realized that no one had ever heard or responded to the mother's own profound cries as a child. The therapist hypothesized that the mother "had closed the door on the weeping child within herself as surely as she had closed the door upon her own crying baby" (p. 392). This conceptual understanding led the therapist to a clinical hypothesis and basic treatment plan: When this mother's own cries are heard, she will hear her child's cries.

The therapist set about trying to hear and articulate compassionately the mother's own childhood experience. When the mother was 5 years old, her own mother had died; when she was 11, her custodial aunt "went away."

Responding to these profound losses and the mother's resultant feeling that "nobody wanted me," the therapist listened and put into words the feelings of the mother as a child:

How hard this must have been. . . . This must have hurt
deeply. . . . Of course you needed your mother. There was
no one to turn to. . . . Yes. Sometimes grown-ups don't
understand what all this means to a child. You must have
needed to cry. . . . There was no one to hear you. (p. 396)

At different well-timed points in treatment, the therapist accurately captured the mother's experience in a way that gave her permission to feel her feelings. As a result, the mother's grief and anguish for herself as a cast-off and abused child began to emerge. The mother sobbed; the therapist understood and comforted. In just a few more sessions, something remarkable happened. For the first time, when the baby cried, the mother gathered the baby in her arms, held him close, and crooned in his ear. The therapist's hypothesis had been correct: When the mother's own cries were heard, she could hear her baby's cries. The risk for abuse ended as this beginning attachment flourished.

This poignant case study illustrates how the therapist can use the therapeutic relationship to resolve the client's conflict. A corrective emotional experience occurred when the therapist responded to the mother's pain in a caring manner. This case study illustrates how powerful it is when the therapist articulates an understanding of the client's experience in a way that also communicates the therapist's compassion for the client.

Using role models such as this, beginning therapists are encouraged to explore their own ways of communicating that they are moved by the client's pain and are concerned about the client's well-being. Clients' inability to care about themselves is central to many of their problems, and most clients cannot care about themselves until they feel someone's caring for them (Gilligan, 1982). Therapists provide this care when they recognize what is important to the client, express their genuine concern about the client's distress, and communicate that the client is someone of worth and will be treated with dignity in this relationship. These are the therapist's goals in the initial stage of therapy.

Closing

The interpersonal process approach is integrative and draws on concepts and techniques from different theoretical traditions. In each chapter, we will highlight the theoretical framework that has been primary in that chapter. The ideas in this chapter have been grounded primarily in Rogers's client-centered approach and, especially, accurate empathy.

The interpersonal process approach tries to resolve problems in a way that leaves clients with a greater sense of their own self-efficacy. This independence-fostering approach to therapy can be achieved only through a collaborative, working alliance. The client needs to be an active participant throughout each phase of treatment—not a good patient who waits to be cured or told what to do by the doctor. This process dimension of how the therapist and client work together is more important than the content of what they discuss or the theoretical orientation of the therapist.

In this chapter, we also have explored the profound therapeutic impact of listening with presence and working collaboratively to understand what is most important to the client. Offering the deep understanding described here is a gift to clients—one of the most important interventions that therapists of every theoretical orientation can offer. Perhaps because they are so simple, these basic human responses are too easily overlooked. They are the foundation of every helping relationship, however, and the basis for establishing a working alliance.

Suggestions for Further Reading

1. To help readers learn more about how to establish and maintain a strong working with their clients, items from the *Working Alliance Inventory* have been included in Chapter 2 of the Student Workbook that accompanies this text.

2. For helpful illustrations of how process comments facilitate group interaction, see the classic text by Irvin Yalom (1995), *Theory and Practice of Group Psychotherapy*. Practical guidelines for identifying themes, using self-involving statements, and creating immediacy are provided in Chapter 11 of Egan (2002), *The Skilled Helper*. Readers can also learn more about immediacy interventions in Chapter 16 of Hill (2004), *Helping Skills: Facilitating Exploration, Insight, and Action*.

3. *Presence* is an important but elusive therapeutic dimension. For a compelling illustration of a therapist's presence with an adolescent sexual abuse survivor, see the extended case study of "Sheila: An African American Adolescent Coming to Terms with Sexual Abuse and Depression," in Chapter 3 of *Casebook in Child and Adolescent Treatment: Cultural and Familial Contexts* (McClure & Teyber, 2003).

4. Readers are encouraged to read the chapter "Therapist Variables," in *Handbook of Psychological Change: Psychotherapy Processes and Practices for the 21st Century* (Teyber & McClure, 2000). This chapter summarizes the research literature on the working alliance and characteristics of effective therapists.

5. Illuminating ideas about the effects of multigenerational family relations on individual personality are found in two classic articles on triangular coalitions: Haley's (1976) "Toward a Theory of Pathological Systems" and especially Bowen's (1966) "The Use of Family Theory in Clinical Practice." Both of these pioneering articles will help therapists apply family systems concepts to the practice of individual psychotherapy.

CHAPTER THREE

Honoring the Client's Resistance

Joan was nervous about seeing her first client, but the initial session seemed to have gone well. The client talked at length about his concerns and, to her relief, she found it was easy to talk with him. The client shared some difficult feelings and Joan felt that she understood what was going on for him. She was thinking they had begun a good therapeutic alliance; the client seemed so friendly and appreciative when he left. One week later, however, Joan received a brief telephone message from the client saying only that he was "unable to continue therapy at this time." Confused and dismayed, Joan sat alone in her office wondering what had gone wrong.

Conceptual Overview

Just when the therapist feels that something important is getting started, some clients put their feet on the brakes. The client cancels the second appointment, shows up 25 minutes late, or asks to reschedule for Sunday at 7:00 A.M. This *resistance* is puzzling and frustrating for the novice therapist: "Why didn't that client return? We had a great first session!" Although most clients will not be resistant in this particular way, other forms of resistance will occur regularly throughout therapy—even for motivated, responsible clients who are trying hard to change. Thus working with resistance is a normal part of treatment, an expectable component of forming a therapeutic alliance. The purpose of this chapter is to help beginning therapists recognize and learn how to respond effectively to client resistance.

All clients have both positive and negative feelings about entering therapy, although the positive feelings are usually more apparent at first. Clients seek therapy in order to gain relief from their suffering. However, we must look further into the complexity of their feelings. Often, as clients seek help and genuinely try to change, they simultaneously resist or work against the very change they are trying to attain. How can we understand this paradoxical, shadow side of the change process? For many clients, *shame* is associated with having an emotional problem they cannot solve on their own. In some cultures, revealing problems to people outside one's family meets with disapproval or is seen as *disloyal*. Other clients feel *guilt* about asking for help or meeting their own needs. Still other clients, because of their cognitive schemas, have *anxiety* evoked by the expectation that others will respond in hurtful ways. Thus, if clients have a problem or need to ask for help, then shame, guilt, anxiety, and other unwanted emotions are often evoked, as well as the deeply held belief that the therapist too will ultimately respond in the same problematic ways that others have in the past. In addition to difficulties in entering treatment, we will see that *ambivalent* feelings are often triggered later in treatment when clients feel better, improve in treatment, and make successful changes in their lives. Thus therapists are trying to identify and work with the specific issues that make it difficult for this particular client to enter treatment in the beginning, and to sustain change later in treatment. Following the principle of client response specificity, specific concerns must be clarified for each individual client, but the following are common themes:

- If I let myself depend on the therapist, he might leave me (or take advantage of me, or try to control me) as others have done when I needed them.
- I cannot ask for help because I must be independent and in control all the time.
- I don't deserve to be helped by anyone, I don't really matter very much.
- I cannot need anything from others—if I'm not perfect I will bring shame to my family.
- Asking for help is admitting that there really is a problem, and that proves there really is something wrong with me.
- I am afraid of what I will see or what a perceptive therapist will learn about me if I stop and look inside myself.
- If I cannot handle this by myself, it means that I really am weak, just as they always said.
- If I don't see a therapist, my spouse will leave me.
- I'm a therapist. I can't have problems—I'm supposed to have the answers.

As they enter treatment, many clients are struggling with difficult feelings at the prospect of losing control over their emotions, being controlled or judged, and so forth. Although there will be many different reasons, *clients' resistance and defenses are driven most frequently by shame*—as seen in many of the examples above. To begin to have some understanding and more compassion for this resistance, think about your own life during a time of real crisis. Ask yourself, "What has it been like for me to have a problem or emotional need and to ask for help?" For many, sadly, it's not easy.

This topic leads many places but, in particular, it takes us to the heart of the attachment story. What does it really mean to be securely attached? *The securely attached child is secure in the expectation that when they have a problem their caregiver will try to help them solve it.* In other words, they are not alone with their problems—they have a trustworthy partner or dependable ally who wants to know what is wrong and who wants to try to help them with whatever problem they are having. With resistance, in contrast, we are usually working with insecure attachment histories. Based on their real-life experience, many clients have learned to hold the expectation that the therapist and others will not be interested in or capable of responding to them when they have an emotional need. At different points in treatment, and especially at the beginning, this dysfunctional belief will often come into play with the therapist as well.

It is easy to respond to the approach side of clients' feelings—the pain that motivates them to seek help and enter treatment. If you just scratch the surface, it is usually plain to see. One part of the client wants you to respond to it but, often, another part doesn't. This shadow side—the ambivalent feelings and the unwanted or feared interpersonal consequences that have resulted in the past for having a problem and needing help—act as a countervailing force. If unaddressed, this resistance will keep some clients from being able to successfully engage in treatment, as in our opening vignette.

Reluctance to Address Resistance

Imagine your client has missed, come very late for, or twice rescheduled the second appointment. Perhaps this has no significant psychological meaning at all. Cars do break down; traffic jams occur; children get sick; employees get called for work at the last minute. However, if this behavior reflects the client's ambivalence about some aspect of being in therapy, and not just reality-based constraints, the therapist needs to find an effective way to address this or the client is far more likely to terminate prematurely. Initially, therapists usually do not know whether the client's behavior is reality based, psychologically motivated, or both. As we will see, however, by finding nonthreatening ways to explore this behavior with the client, therapists can greatly increase the chances of the client's continued participation. In these explorations, therapists' curious, nonjudgmental manner is as important as what they say. Otherwise, the client may misunderstand their observations as blame or criticism, which never help and only serve to exacerbate client resistance.

More specifically, the model proposed here suggests that therapists first enter supportively into clients' frame of reference, *affirm or validate the reality-based constraints that they perceive,* and do whatever possible to flexibly accommodate or help resolve the problem. Only after taking seriously clients' concerns, as they see them, can therapists begin to inquire about other conflicted feelings or meanings that entering treatment may hold. The therapist can begin this joint exploration by wondering aloud, in a tentative manner, whether entering treatment may be evoking other more psychological concerns as well. If therapists approach resistance without following this two-step sequence, most clients will feel misunderstood or blamed, impeding their ability to engage in treatment.

The Therapist's Reluctance

Resistance is not a welcome concept to most beginning therapists. To a greater or lesser degree, however, every client will be ambivalent, defensive, or resistant. This push-pull may occur at the beginning of treatment and it can continue to wax and wane throughout treatment. The clients most likely to drop out of treatment after the initial telephone contact or first session are people of color, men, and clients who have been vague or evasive about their problems. Resistance also will be a significant issue in managed-care settings or Employee Assistance Programs (EAPs) where clients are assigned therapists and cannot choose their own. Resistance will be a far bigger issue when courts or others mandate clients to attend treatment. Although mandated clients often can be helped, they usually are not good training cases for beginning therapists—their reluctance to comply with the demand to attend is challenging to work with. With every client seeking treatment, however, therapists need to address signs of potential client resistance. Let us examine three reasons why many beginning—and experienced—therapists often find this difficult.

First, many beginning therapists—especially if they have never been in therapy themselves—are unaware of the multiple meanings and conflicted feelings associated with the decision to enter treatment. These therapists are often surprised to find that, in the initial sessions, clients actually resist the help they are overtly seeking. Other therapists don't want to work with resistance because they fear that it has to be addressed in a blaming or critical way. Or, they believe that resistance has to be addressed in a manner that puts therapists in a superior position and denies the validity of the client's own experience—which they never want to do. This occurs, for example, when a therapist says to a working, single mother of three, "Hmm, I notice that you are 5 minutes late for our appointment today." Ineffectively, this therapist is ignoring the reality-based constraints of her life context. Resistance does not have to be associated with this type of hierarchical therapist–client relationship. Therapists can work with resistance and still be sensitive to the social context or reality-based constraints of the client's life. For example, poor clients who use public transportation will not be able to arrive on time consistently. Clients from other cultural contexts may not think that being 5 minutes late is of any significance. For some, it may even be impolite to come on time; arriving late allows the host extra time to prepare for the visit. Therapists can educate clients about the bounds and procedures of therapy and, as we will see, respond to resistance in collaborative ways that strengthen the client's commitment to therapy, enhance the therapeutic alliance, and empower rather than invalidate the client.

A second reason why therapists are reluctant to address their clients' resistance is more personal. Most new therapists have strong needs for their clients to like them. Beginning therapists want very much for their new clients to find them helpful and to keep coming to therapy. If the client does not show up or comes late, beginning therapists, in particular, often feel painfully that they have failed. When this occurs, therapists may become increasingly concerned about their ability to help, place even more performance demands on themselves, and become overly invested in pleasing or performing adequately for the next client.

Third, therapists often do not inquire about signs of potential resistance in order to ward off unwanted criticism. Not wanting to hear what they may be doing wrong, it becomes especially hard for new therapists to invite the (seemingly) bad news:

THERAPIST: What's it been like for you to talk with me today?

OR

THERAPIST: Is there anything about our work together that doesn't feel quite right to you? If so, I would welcome hearing about it. You know, so we could make any changes we needed to make this work better for you.

Like their clients, most therapists are not eager to approach issues that arouse their anxiety or to discuss interpersonal conflict. Thus, new therapists may be hesitant to invite clients to express negative reactions they may be having toward the therapist's response to them, the conflicts evoked for them by having to ask for help, or their ambivalent feelings about needing to be in treatment. Sure, it is difficult for most people to invite critical feedback or approach interpersonal conflict, but it is so very helpful when the therapist can be *non-defensive* enough to do so.

This becomes far more important if the client alludes to some conflict or difficulty about being in treatment:

CLIENT: In my culture, we don't really talk about our problems to others outside the family.

Similarly, it becomes essential for the therapist to address potential problems or misunderstandings in their relationship when the client implies that there may be some problem or concern with the therapist:

CLIENT: So how old are you anyway? Have you been practicing long?

Even though we may want to gloss over or avoid these subtle ("embedded") messages that something is awry between us, treatment won't progress until we address the issue and sort it through with the client. For many new therapists, learning to address interpersonal conflict in this straightforward, non-defensive way is the single biggest challenge in their clinical training. The paradox is, *if therapists allow their own anxiety to keep them from asking clients about such potential signs of resistance, their clients will be far more likely to act on these concerns and drop out of treatment prematurely.* That is, new therapists want to take a breath, gather themselves to respond nondefensively, and take the risk to meet this concern head-on. For example, they do this effectively by simply inviting the client to discuss this concern more fully:

THERAPIST: I'm wondering what it's like for you to come and talk to someone like me, you know, someone who is not a member of your family?

OR

THERAPIST: I'm glad you asked—I'm a 1st-year student in a master's counseling program, and it makes sense to me that you've been wondering if I have enough experience to help you. How has it been for you to talk to me so far? What's been helpful and what hasn't?

Thus, an effective approach is to make a process comment that suggests the possibility that something about entering treatment may be difficult for the client and invites the therapist and client to explore this possibility together. Before too long, new therapists will find that they can do this in an affirming way that does not make the client feel accused or blamed, but actually enhances the working alliance.

THERAPIST: You seem discouraged today, and you've had some trouble getting here. I'm wondering if something isn't going right here. Any ideas?

CLIENT: I don't know what you mean.

THERAPIST: I'm wondering if something about our relationship—or being in treatment—isn't working very well right now. We seem to be missing each other in some way. Any thoughts about what could be going wrong between us, or what I could do to make this work better for you?

CLIENT: Well, maybe you could talk a little more or give me some more feedback. You're pretty quiet. I don't really know what you're thinking most of the time.

THERAPIST: I'm glad you're telling me this—thanks for your honesty. Sure, I'll be happy to share more of my thoughts with you—that'll be easy to change. I'm wondering what's it like for you when you don't know what I'm thinking?

CLIENT: Well, maybe when your quiet it's like your judging me or something.

In this way, taking the risk to explore the client's resistance will often reveal key concerns (that is, "judging") that both impede the client's participation in treatment and are central to his problems.

The Client's Reluctance

Unfortunately, the client usually shares the therapist's reluctance to address resistance. Clients are often unaware of their resistance and externalize it to others or outside events. For example, the client may say, "Yeah, I'm late for our sessions, but the traffic is always so bad." Although there may be truth to this statement, the consistency or frequency of the behavior suggests the need for further exploration. In order to acknowledge both the possible reality-based constraints and the probable psychologically based resistance, the therapist can ask whether leaving earlier for the appointment is possible or whether an alternative session time needs to be negotiated. In tandem, the therapist also asks whether they can discuss any thoughts or concerns the client might be having about treatment:

THERAPIST: What were you thinking about our session last week as you walked out to your car and drove home?

OR

THERAPIST: I'm wondering if you had any thoughts about me, or concerns about our work together, during the week?

OR

THERAPIST: I'm wondering what people in your family, or friends at church, might think about you coming to counseling and talking about personal problems?

There is no criticism or blame here. And, even if the client is not ready to talk about whatever psychologically based resistance may be operating, the therapist has invited the client to consider the possibility of other psychological factors and laid the groundwork for further discussions. This is part of educating the client about the treatment process and making it normal and safe to talk about their relationship. For example, the therapist can go on to say:

THERAPIST: I know other people don't usually talk this directly, but one of the best ways I can help you is if we can talk together about the feelings and concerns you have about our relationship and the way we're working together.

In contrast, when clients see themselves resisting, they often feel frustrated and confused by their own contradictory behavior. For example, clients often exclaim with dismay:

- Why do I go to all the trouble and expense of coming here to see you, when I can't think of anything to say as soon as I walk in the door?
- Why would I forget our next appointment after we had such a great session last week? It doesn't make sense!
- Why do I keep asking you for advice, and then say, "Yes, but . . ." to whatever you suggest? What's wrong with me?

If it becomes clear to clients that they are resisting in some way and sabotaging their own efforts, they typically evaluate themselves harshly—as being bad in some way—and assume that the therapist shares this critical attitude. Thus, when the therapist begins to inquire about resistance, most clients want to avoid the topic because they are afraid the therapist is going to *blame* them— for not really trying, being unmotivated, and so on. Because it is not therapeutic for the client to feel blamed or criticized, the therapist tries to make this self-critical attitude overt and to help clients reframe their critical attitude toward their own resistance. How do we do this? First, *both the therapist and client want to honor the client's resistance as an outdated coping strategy that originally served a self-preservative and adaptive function.* That is, the therapist's intention is to help the client appreciate that this coping strategy was the best possible response she had available at an earlier time in her development. Second, the therapist helps the client to realize that she no longer needs this coping strategy in many current relationships, such as the present relationship with the therapist. Instead, she can work with the therapist to learn more flexible ways of responding, first with the therapist and then with others in her life. Reframing the client's resistance with this understanding and acceptance is a big concept; let's examine it more closely.

Clients' resistance, and even their symptoms, once served as a survival mechanism that was necessary, adaptive, and often creative. For example, anxiety signals danger—perhaps not a current threat but a past or developmental problem that really did exist. Depressive symptoms may reflect a client's attempt to cope with painfully unfulfilled longing for protection or love from a caregiver, or a way to cope with the exploitation or derision that she experienced and came to expect from others. In other words, formative experiences with caretakers and current interactions with significant others have given clients very good reasons for not wanting to ask for help, share a vulnerable feeling, or risk trusting someone. If the therapist and client explore patterns in how significant others are responding now and have responded in the past, the client's resistance will make sense and become understandable—it's not "irrational." The feelings that underlie the client's resistance always make sense developmentally, although they may well be no longer necessary or adaptive in many current relationships. In this process approach, we are working with resistance as a relationship issue rather than a fixed personality trait or individual behavior (Miller & Rollnick, 2002).

Suppose that a client is having trouble entering therapy. The therapist is encouraged to *use immediacy questions and work in the moment to highlight maladaptive relational patterns*:

- How have others responded to you in the past when you have asked for help or needed someone? Is it similar or different with me?
- If you and I begin to work together on your problems, what could go wrong between us?
- What might I do to hurt you or make things worse if you seek help from me?
- How could our religious and ethnic differences get in the way and be a problem for you and me—for our work together—here in therapy?

Routinely, questions like these reveal some of the most meaningful issues in clients' lives. In particular they highlight what the client is afraid of yet expects— that is, they make clear what's really wrong.

Such process-oriented questions also help the therapist and the client identify the aversive consequences that followed when the client asked for help in the past. Routinely, clients will have very specific answers to these questions and describe in detail the interpersonal scenarios that typically transpired in the past. When they asked for help in their family, for example, they were ignored and felt powerless, were made fun of and felt ashamed, or were told that they were selfish and too demanding. They may have been invalidated and told that they didn't really need or want what they asked for. They may have learned that anything they received brought unwanted obligations or that anything they needed or asked for seemed to burden their caregiver, which made them feel guilty. In this way, therapists and clients together can clarify together the very good reasons why clients originally learned to respond in these ways.

Thus, exploring the client's resistance often reveals an *outdated coping strategy*—not a lack of motivation. To see how the therapist can begin to change this outdated or maladaptive coping strategy, I briefly introduce a four-step sequence that therapists can follow.

First, the therapist is trying to help clients *identify* their outdated coping strategies with others. For example, the client always takes care of everyone else, at the expense of her own needs, and feels burned out at work and resentful at home. As part of this sequence, the therapist also helps clients understand why they originally needed to cope or protect themselves in these ways. For example, the client's parent was alcoholic or depressed, and she grew up taking care of her parent rather than being taken care of. Another aspect of identifying this outdated coping strategy is to clarify how clients may be continuing to do this now in their current interaction with the therapist. For example, the client doesn't share difficult feelings because she doesn't want to "bother" the therapist.

Second, the therapist *validates* the protection this interpersonal coping strategy once provided. The outdated coping strategy (always taking care of others, not meeting her own need) is now causing problems in the therapeutic relationship and with others. However, once it was a strong and adaptive response to a reality-based problem (for example, "taking care of" preserved her attachment to her unresponsive caregiver). The original coping strategies that clients identify (such as withdrawing, provoking, avoiding, pleasing, intellectualizing, and so on) minimized the hurt and allowed clients to cope with the unwanted responses they received. Affirming the validity of this coping strategy is often an empowering liberation to clients as this new understanding leads them to feel more self-acceptance rather than self-blame.

Third, the therapist holds a steady intention to track the process dimension and ensure that the therapeutic relationship does not repeat this maladaptive pattern. The therapist can provide a corrective emotional experience—rather than a reenactment—by checking this out with the client and asking:

THERAPIST: At work you take care of everybody and everything, at home you do the same—and understandably feel resentful and exhausted. Does it ever feel like you are taking care of me in some way, too?

Fourth, the therapists' ultimate goal is to help clients transfer this new experience or relearning with the therapist and apply it with others in their life:

CLIENT: No, you're different than my boss and husband and daughter—I don't have to take care of you. It's nice, I can be responded to.

THERAPIST: Good, I'm glad this is one place where you can let that depleting role go and allow yourself to have your own needs met. Who else in your life can you break this pattern with?

CLIENT: Well, I could probably do more of this with my sister, and some of my coworkers are very responsible.

THERAPIST: Let's try to build on those relationships and find others where things could be more balanced. But it sounds like home—with your family—is the most pressing place for you right now. I'm wondering if we could work together and try to change your part in this pattern with your husband and daughter.

CLIENT: What do you mean?

THERAPIST: As I listen to you, it often sounds like you do everything for everybody before they have a chance to pitch in—you're so capable and you respond so quickly. What would happen if you asked your daughter to clean up her own room or your husband to help with the dishes?

CLIENT: Oh, I've done that—it's hopeless.

THERAPIST: Well, maybe it is, but let's walk through the sequence together and see how it usually goes. Maybe we can spot some places to try something new— like waiting a moment and giving them time to respond before you jump in and take care of it.

By following this four-step sequence, over and over again, therapists provide the client with the experience of no longer needing to keep using the outdated coping strategy. More accurately, they learn to *discriminate* when they still need to respond in this old way (for instance, at work with their rigid and demanding supervisor) and with whom they don't (their supportive friend or coworker).

There is a great deal of complex but important information here—we'll return to this pivotal topic and work further with this four-step sequence of change. To sum up for now, we have seen that clients are far more likely to become stalled in therapy, or to drop out prematurely, if their resistance is not addressed in a supportive and collaborative manner. One of the most important ways to engage clients successfully in a therapeutic alliance, and to identify key conflict areas in their lives, is to invite clients to discuss any potential problems they may have about any aspect of being in treatment or negative reactions toward anything the therapist does. In this regard, researchers find that satisfied clients had therapists who asked them how they were feeling about what was going on in the therapeutic relationship (Rhodes et al., 1994). *The guiding principle is that if clients talk about their ambivalence, dissatisfaction or potential misunderstandings, they will be less likely to act on them and drop out.* Before going on to illustrate specific ways of implementing this principle, we need to learn how to identify clients' resistance and distinguish it from reality-based constraints that must be affirmed.

Identifying Resistance

Especially as they begin their training, most new therapists will have clients that drop out of treatment within the first few sessions. Usually, this doesn't feel very good but we can do a lot to prevent these unwanted premature terminations. Too often, however, the therapist might blame the client and conclude that he wasn't really motivated or wasn't ready to look at his problems. Or the therapist may blame herself and think that she unwittingly made some irrevocable mistake in the previous session. However, researchers who have interviewed clients after treatment find that *clients often report having negative feelings toward the therapist which they hide from the therapist* (Hill et al., 1992, 1993). Too often it doesn't feel safe for clients to bring it up on their own and, too often, therapists do not approach interpersonal conflict and ask about potential problems or misunderstandings between them. As a result, many clients prematurely terminate because they are acting on, rather than talking about, their conflicted feelings about entering therapy, or misunderstandings or difficulties with the therapist that they have not addressed and talked through. This unfortunate but common outcome is not necessary—there is so much therapists can easily do to help clients successfully engage in treatment. Specifically, the therapist

- Helps clients identify when resistance may be occurring,
- Approaches this issue in a noncritical manner that allows clients to express their concerns more fully, and
- Helps them resolve these concerns by responding flexibly—as described next.

But how does the therapist know when a client is resisting—how do we identify when it is occurring? Therapists may never really know what any behavior means for a particular client, but resistance may be operating when clients *consistently* have difficulty participating in treatment. For example, after reality-based constraints have been considered (such as scheduling conflicts), psychologically based resistance (for example, getting help is inconsistent with the familial role of caretaker) is probably occurring when any of the following occur repeatedly or in combination:

- The client misses appointments or comes late.
- The client needs to reschedule appointments frequently.
- The client has very limited hours available for therapy.
- The client cannot make a firm commitment to attend the next session.

It is important to stress that the same behavior often means very different things to different clients. However, when clients have trouble attending sessions, some form of resistance is often at work. In such cases, the therapist can begin to generate "working hypotheses" about the possible meaning this resistance holds for the client.

Formulating Working Hypotheses

Therapists are far more effective when they can formulate case conceptualizations to guide intervention strategies, and they often need to write treatment plans for third-party or HMO insurance payments. Looking for greater specificity, some HMOs will require operationalized and measurable treatment goals. In this regard, helpful texts are available to help therapists clarify where they are going in treatment and how they plan to get there (Bloom, Fischer, & Orme, 2003; Fonagy, 2002). The first step in clarifying treatment goals is to formulate tentative working hypotheses about the client's maladaptive relational patterns, faulty expectations and pathogenic beliefs, and core conflicted feelings. One systematic way to do this is to begin formulating answers to three questions during the initial contact. (Further guidelines are provided in Appendixes A and B.)

1. What does the client elicit from others? Even in the initial telephone contact, the client's interpersonal style—and what it tends to *elicit* or *pull* from others—is an important source of information. For example, if the client sounds helpless and confused on the telephone, might he adopt a victim stance and invite others' rescuing behavior? If the client presents in an angry and demanding way, might others often withdraw or tend to become embroiled in power struggles?

If so, perhaps this client is testing whether the therapist can tolerate the angry challenges without being run over and still see the underlying need or pain. As therapy progresses and the therapist learns more about each client, many of these initial hypotheses will prove inaccurate and will need to be discarded. As therapists learn to trust themselves and attend to the feelings and reactions evoked in them by the client, however, they will find that many hypotheses do indeed fit the client and can be further developed and refined. Thus, to begin conceptualizing the client and developing initial treatment plans, therapists can follow these four steps:

1. Formulate working hypotheses about his interpersonal style and what it tends to elicit in others (for example, his loud and aggressive demeanor tends to elicit a fight-or-flight response).
2. Evaluate these tentative initial hypotheses as the therapist learns more about the client and they interact further (as the therapist listens to his narratives, it sounds as if coworkers, his family, and others also seem to react to his dominating style with anger or intimidation).
3. Revise these hypotheses over time to more accurately reflect each client's personality and problems (he takes over and tries to intimidate when he is uncertain or feels insecure).
4. Formulate intervention plans (provide interpersonal feedback, and help the client become more aware of the problematic impact his abrasive style has on others).

For example:

THERAPIST: You're speaking to me in a harsh, loud voice right now. How do the people you love respond when you talk to them like this? What does your wife say and do, or can you describe the look on your son's face when you sound this angry?

Although these four steps may feel challenging right now, they can be learned with time and practice. To help, we will return to this important concept of "eliciting behavior" in the chapters ahead.

2. What is the threat? On the basis of information that the therapist has started to gather about the client's current and past relationships, the therapist can begin speculating about the different ways that entering treatment could be aversive for the client. Therapists can formulate working hypotheses such as the following about the issues or concerns that are likely to make it difficult for this particular client to enter treatment.

- Is it guilt inducing for this highly responsible parent, who grew up taking care of her own anxious and insecure mother, to ask for something for herself?
- Does this Christian client believe that her inability to resolve her problems through prayer means that she has failed in her faith?
- Is it incongruent for this older Latino male to seek help from a younger female therapist?
- Is it shameful evidence of his own inadequacy if this blue-collar worker has an emotional problem that he cannot solve by himself?
- Is it disloyal for this Asian client to discuss personal problems with someone outside the family?

Therapists may also keep in mind that resistance and defense are attempts to manage unwanted feelings of shame, guilt, and anxiety. Although clients may be unaware of it, they often struggle with the worry that, if they continue in treatment, the therapist will hurt them in a way similar to how others have in the past. This schema won't come up in everyday interactions, but it will be activated when the client is in distress (the "hot cognitions" activated during crisis). We will see how therapists can use process comments to help clients clarify such concerns and learn how to assess whether the therapist is indeed responding in these familiar but unwanted ways (reenacting) or is offering a new and safer relationship than they have come to expect (resolving).

Let's step back for a moment and consider one of the bigger meanings conveyed in the question "What is the threat or danger if you talked with me about that—what would go wrong between us?" Therapists don't want to "play it safe" with clients and keep things on the surface. Instead, therapists want to take the risk of going beyond just empathic understanding, exploration and clarification. They want to use this opportunity to be someone different in the client's life and meet the challenge by asking about the deeper meaning. In other words, *therapists need to be willing to risk more intensity in the therapeutic relationship*. Exploring the client's resistance, and asking these here-and-now, immediacy questions (such as "What's the threat here? What's going to go wrong between us if you disagree with me?") is not for the feint of heart. It will put the client's core concerns on the table and bring real intensity to the therapeutic relationship. In this way, new therapists are encouraged to take the risk of responding in this stronger, more forthright manner with clients in every phase of treatment. If the therapist has held a consistently empathic and respectful attitude toward the client, the client will welcome such honest communication. Like all of the ideas suggested here, try it out, evaluate how your client responds, and decide for yourself.

3. How will the client express resistance? Finally, therapists should formulate a third set of hypotheses that anticipate how clients are most likely to enact these concerns with the therapist. For example, suppose a therapist has observed that her depressed client feels guilty and believes he is selfish whenever he does something for himself or someone does something for him. The therapist then hypothesizes that this client may withdraw from treatment as soon as he starts to feel better, or emotionally disengage from the therapist when he realizes that she is genuinely interested in his problems and cares about his well-being.

To illustrate further how the therapist generates working hypotheses on the basis of these three questions, let's consider another example. Suppose that a young male client whose problem is substance abuse telephones the therapist to schedule an initial appointment. During this telephone call, the client tells the therapist that his alcohol and drug usage is more problematic than it has ever been and that his mother believes that he needs to be in therapy. On the basis of the helplessness and confusion expressed in his telephone call, and his disclosures that his mother wants him in therapy and how problematic his drug usage now is, the therapist begins formulating several tentative hypotheses. First, the therapist hypothesizes (the therapist doesn't "know"; she is just considering possibilities) that an important part of the client's interpersonal style is to let others know that he is hurting, to elicit help from them, but then to avoid taking responsibility for his own behavior.

Second, the therapist hypothesizes that this client may become resistant to treatment when the therapist addresses the issue of how he meets his emotional needs. Or, the client may become resistant when he is given reality-based feedback about how he eschews responsibility for his own decisions (for example, to enter therapy or to continue drinking). Third, the therapist hypothesizes that this client is impulsive and may act out his resistance by bolting out of therapy as soon as these anxiety-arousing issues are addressed (just as, perhaps, he has used alcohol and drugs to avoid dealing with certain feelings, interpersonal problems, or responsibilities he feels inadequate to meet).

The therapist knows that these initial hypotheses may not be accurate and is ready to revise or discard them as more is learned about the client. However, the therapist will be able to respond more effectively if she can try to anticipate the concerns that might cause each client to drop out oft treatment and the way each client is likely to express these concerns. Prepared with working hypotheses such as these, therapists will be able to "get it" more quickly and be prepared to recognize in the moment what is happening between them—not 2 hours later when they are reviewing a tape of the session. Empowered to understand more readily, therapists will be able to respond more effectively and diminish acting out by helping clients anticipate their own resistance before they act on it.

As therapy proceeds, the therapist will learn much more about the client and many of these initial hypotheses will need to be discarded. Some will be accurate, however, and can be elaborated to better understand this particular client. In this way, the therapist identifies the patterns and enduring issues that keep arising for this client. The therapist will then be better prepared to center treatment on these repeated themes, which provide the *treatment focus* necessary for the ongoing course of treatment. As we will explore further, this is part of the ongoing process of formulating and refining case conceptualizations and treatment plans.

Responding to Resistance

The tired cliché is true: life is hard. We need defenses in our lives; they often are adaptive and necessary to cope. We certainly needed these defensive coping strategies when we first learned them as children, and, in a reality-based way, we often need them now in some situations and relationships as adults. The therapist's goal is to help clients learn to *discriminate* when defensive coping strategies are needed and when they are outdated and no longer apply—most clients do not do this well. Throughout this text, we will see how the therapist is trying to help clients assess when their coping strategies are still valid and in which situations they are no longer necessary or effective. For example, becoming quiet, looking almost invisible, and compliantly going along with a hostile, dominating supervisor at work may be adaptive, but it is no longer needed in the current relationship with the therapist, and it is causing significant problems in his relationship with his girlfriend.

In this section, we see how therapists can respond to common types of resistance. Sample therapist–client dialogues illustrate effective and ineffective responses at three critical points: during a telephone conversation in which the client has difficulty scheduling the initial appointment; at the end of the first session; and during subsequent sessions with a client who has difficulty keeping appointments. As before, student therapists should find their own words to express the principles embodied in these dialogues.

Addressing Resistance during the Initial Telephone Contact

In trying to schedule the initial appointment, the therapist wishes to obtain a firm commitment from the client. The client, however, may be ambivalent about entering therapy and, overtly or implicitly, express this during the initial telephone contact. The therapist's intention is to attend to potential signs of resistance and to talk openly with the client about this issue if it seems to arise.

An Uncertain Commitment   In the following example, the client is only somewhat uncertain about attending the first session.

> THERAPIST: It's been good talking with you, and I'll see you on Tuesday at 4 P.M.
>
> CLIENT: OK, I guess I'll see you then.
>
> THERAPIST: You sound a bit uncertain; maybe there's nothing to it, but I'm wondering if we should talk a little more about how it feels to come in?
>
> CLIENT: Well, maybe a bit uncertain, but I do want to come.
>
> THERAPIST: Good, I'm looking forward to meeting with you. We can talk then about that little bit of uncertainty if you'd like, and any questions you may have about me or treatment. See you next Tuesday at 4 P.M.

An effective response to the client's ambiguity ("I guess . . .") is modeled in this dialogue. The therapist hears the indecision in the client's commitment and addresses it directly. An ineffective response would be for the therapist to let it pass by on the assumption that it probably doesn't mean much. Perhaps it doesn't, and chances are that the client will still arrive at the appointed hour. However, the chances of the client not arriving are far greater if the therapist does not ask about the signs of uncertain commitment and invite a dialogue. Furthermore, although this ambiguous comment may seem subtle or small to the beginning therapist, it is a statement about the counseling relationship—about "you and me." Treatment benefits when the therapist tracks clients' comments about the therapeutic relationship and offers an open-ended bid to explore them further.

In addition, the therapist can mentally note this indecision as a possible sign of resistance—that something about beginning may not feel quite right. This is an instance in which the therapist can begin to generate working hypotheses—derived from the first of our three orienting questions—about the possible meaning of this "uncertainty." As the therapist meets this client and listens to the narratives of his life, she can begin to consider tentative working hypotheses about other material related to "uncertainty"—or dismiss it if she doesn't hear this theme in other arenas of the client's life.

A More Ambivalent Client   Now imagine the same telephone conversation, but with a more difficult client:

> CLIENT: OK, I guess I'll see you then.
>
> THERAPIST: You sound a bit uncertain; can we talk a little more about coming in?
>
> CLIENT: Well, in my family, we don't really talk to others about family problems.
>
> THERAPIST: I respect your concern about being loyal to your family. Maybe one of the things we can talk about is how you can ask for help from others and get your own needs met without feeling that you are being disloyal.
>
> CLIENT: I'd like that. My family is really important to me, and some people don't understand that.
>
> THERAPIST: Yes, it makes sense to me that your family is very important. When we meet, we can work together to try and find ways to sort out your own needs along with respecting your feelings about your family. I'll look forward to beginning together on Tuesday at 4 P.M.

An ineffective response would be to accept the client's uncertain commitment ("OK, I guess I'll see you then") and avoid the possibility that an important issue about what it means for this client to have a problem, ask for help, or enter treatment is being played out. In this dialogue, the therapist has tried to understand and respond to the client's concern instead. By affirming the client's ambivalent feelings and being respectful of his conflict, the therapist has done much to help this client take the first step and enter treatment.

Why would the therapist take such a stance? Wouldn't it be more supportive to let clients leave the appointment a little bit tentative, if that is what they need to do? No. Little therapeutic change will occur until clients take responsibility for the decision to enter therapy and work on their problems. Therapists need to be flexible. As we will see shortly, for example, clients can make a very limited commitment—for one session or just a few sessions—to get a better sense of whether treatment, or this therapist, is right for them. Without this commitment, however, the therapist has little to work with. In fact, it is better for the client to remain out of therapy than to enter without commitment and have an unsuccessful therapeutic experience. As Yalom (1981) emphasizes, the therapist must be concerned about preventing failed hope in certain clients. If clients have one or more unsuccessful therapy experiences, it may discourage them from trying to seek help with their problems again.

Clients Who Try to Make the Therapist Responsible   Let's return to the telephone conversation between the therapist and a prospective client. Some clients will try to make the therapist take responsibility for their decisions.

> CLIENT: I'm not sure if I should start therapy or not. What do you think I should do?

Although it is important to reach out to clients and welcome them into treatment, the therapist does not want to resolve clients' ambivalence by assuming responsibility for their decision or by trying to provide the necessary motivation for therapy. With certain clients, a helpful response might sound something like this:

> THERAPIST: From what you've just been telling me, it really does sound like this has been a very difficult time for you, and I would like to try and help. Would you like to come in for one appointment and see how it feels to talk with me? As we talk about these things, I think you'll get a better sense of treatment, and of me, and be able to decide for yourself whether you'd like to continue.

> OR

> THERAPIST: I can't offer you any guarantees, of course, but, yes, I do believe that therapy may be of help. I'd like to try and work with you. Let's agree to just one session. After we've talked together a little more, I think you'll be able to see for yourself if I can be helpful.

As we would expect on the basis of client response specificity, such direct invitations may be useful at times. When clients have felt unwanted while growing up or believe that others are uninterested in them, it may be more important for therapists to express genuine interest in working with them. However, even with clients such as these, therapists still intend to respond in a way that ensures clients take responsibility for their own decision to enter treatment; therapists do not want to cajole, coerce, or win clients into therapy. At the same time, therapists do not want to be aloof or indifferent. The therapist needs to meet clients where they are—working supportively with their ambivalence or conflicts about entering, but without taking responsibility for the clients. In many cases, an effective response to the client's question may be as follows:

> THERAPIST: It seems as if one part of you wants to be in therapy, but another part of you is hesitant. Tell me about both sides of your feelings.

> OR

THERAPIST: It sounds like a part of you wants to try this and a part of you doesn't. Why don't we agree to meet for just one or two sessions? Let's see if we can work together to sort through both sides of this, so you will be able to decide for yourself what you'd like to do.

Responding to the shadow side (the ambivalence or resistance), the therapist is inviting clients to express their concerns about entering therapy. That is, the therapist is trying to give clients the interpersonal safety to name their concerns—to give them permission to talk freely about their reticence. In both responses, the therapist is working with the client to reach a decision but leaves the client responsible for her own decision. Here again, we are working with the process dimension. Although this process distinction is subtle, it will make all the difference in the eventual outcome of treatment, regardless of the theoretical orientation of the therapist.

The client in this dialogue is providing the therapist with potentially important information. As before, the therapist should begin generating working hypotheses about what this behavior may mean. For example, clients like this who are "uncertain" or who try to get others to make decisions for them may not have been supported in the past when they tried to take initiative or assume responsibility for their own actions. These clients may now try to avoid the aversive consequences they have experienced in the past by getting the therapist and others to take responsibility for them. If so, this could make it difficult for clients to initiate new activities (such as applying for college) or to follow through on their own interests (such as selecting the major or career they are most interested in). Being held back in this way may leave clients feeling resentful and frustrated.

Having generated such hypotheses, therapists can be alert for evidence of these themes in other areas of the client's life. If these themes about "uncertainty" or avoiding personal responsibility prove to be enduring issues that keep coming up as relational themes in the client's life, the therapist can use them as a focus to provide direction for treatment.

THERAPIST: You know, Marie, I'm hearing the same theme again with your boss that you were talking about with your boyfriend—you become quiet and "go along," rather than speak up for yourself, when you disagree. What do you think about this possibility?

The therapist does not want to become wed to these early hypotheses, however, and needs to be ready to discard them if the theme does not keep coming up or hold real feeling for the client. As therapists become more experienced, their initial hypotheses will become more fruitful. This developing skill will become an especially important aid to counselors as they begin working in short-term and crisis intervention modalities that require quicker, more accurate assessment during the initial phase of treatment.

As emphasized in Chapter 1, the therapist's intention is to keep from reenacting the client's maladaptive relational patterns in the therapeutic relationship. However, this can easily occur in a metaphorical or encapsulated way in the initial negotiations between therapist and client. This is another reason why the therapist should not take over and tell the client seeking advice what to do. For example, the following type of response is not recommended:

THERAPIST: On the basis of what you've told me, I'm confident that this can help you, and I think that you should begin therapy now.

Assuming such a directive stance often reenacts problematic relational scenarios for clients and confirms a faulty belief. For example, many clients believe that they cannot act on what they want; others have learned that they need to act dependently and allow others to take responsibility for their decisions. Although such clients also believe that they need to ask the therapist to tell them what to do, paradoxically, they resist the control they have just elicited. Clients with these compliance issues then have to resist the therapist and reject the good help the therapist may be offering—or they are further complying and losing even more of themselves by being helped and getting better in treatment! As this scenario continues, clients come to feel badly about themselves—confused as to why they are rejecting the help they have just asked for; and guilty for rejecting the therapist who was trying to help. In contrast, therapeutic responses that give the client responsibility for deciding whether to follow through and enter treatment prevent this reenactment from occurring, and will provide some clients with a new or corrective response that begins to resolve an old conflict.

Exploring Resistance at the End of the First Session

At the end of the first session, no matter how well it seems to have gone, the therapist is encouraged to ask clients how the session felt to them and whether they have any concerns about the treatment process or the therapist. Unless the therapist is willing to ask about potential problems, these concerns are probably going to remain unspoken. Recall that researchers who follow up with clients after treatment often find that clients do indeed have some negative feelings or interpersonal conflicts with the therapist, but they usually do not bring them up or address them on their own (Hill, 1992). Failing to inquire about, and then deal with or talk through the interpersonal conflict with the client, is an opportunity lost. Many clients will feel some reticence about entering treatment. And, at times, almost all clients will feel the therapist misunderstood what they just said, either because of their own distortions based on their cognitive schemas or because the therapist simply misunderstood or did not grasp something that was meaningful to them. Similarly, most clients will be uncomfortable at some points in treatment with how the therapist is responding to them (feeling, for example, that the therapist is being too quiet, too directive, and so forth). All of these clients will be far less likely to drop out prematurely and will be better able to change in treatment, if the therapist has

- given them overtly spoken permission to talk about these expectable concerns as they arise,
- approached interpersonal conflicts and listened to them nondefensively, and

- taken their concerns seriously and tried to sort through misunderstandings and problems.

Addressing Interpersonal Conflict   Five or 10 minutes before the end of the first session (to allow clients enough time to talk about their reactions to the therapist and the session), the therapist can check in with the client:

THERAPIST: How was it to be here today?

Most clients will answer, "Fine." Thinking that the therapist may need some reassurance, clients may go on to say how helpful the session has been. If so, the therapist might respond as follows:

THERAPIST: Good, I'm glad you've found our first session helpful. You've told me a lot about yourself today, and I feel we have gotten off to a good start, too. I would like for us to be able to talk about our relationship more directly than others usually do—what does and doesn't feel good about how we work together. Was there anything about coming today, or anything about our work together, that did not feel quite right? If so, it would help if we could talk about that.

CLIENT: Oh, no. I was eager to begin, and you have been very nice and understanding.

THERAPIST: Good. In the future, though, if something does come up—and it probably will—I would welcome it if we could talk together about it. How would that be for you?

In this vignette, the therapist is trying to establish an important set of expectations for what is going to occur in treatment. The therapist is telling the client that, in contrast to many others, the therapist wants to hear about the client's dissatisfaction and address problems between them straightforwardly. Many clients will be surprised by this invitation but welcome it because honest communication or talking directly about interpersonal conflict was not allowed in their families of origin. With these invitations, the therapist is laying the groundwork for a different kind of relationship that is more direct and authentic. Therapists want to establish these important expectations with their actions, and not just their words, early in treatment. As emphasized before, there will be misunderstandings or "ruptures" in the therapeutic alliance—it's a normative part of the treatment process, not a mistake (Safran & Muran, 2003). Effective therapists repair them by making process comments, asking about potential problems, and talking them through.

THERAPIST: You seem more distant or preoccupied today, any ideas about what may be going on between us?

CLIENT: When you said that, I think I felt judged by you.

THERAPIST: I'm glad you can tell me that—I respect your willingness to work on the problems that come up between us. Feeling judged wouldn't feel very good to me, either. Let's sort this through together—help me understand what I said that made you feel judged.

Therapeutic relationships will remain superficial—they cannot develop very far—unless therapists are willing to broach these expectable ruptures and take the risk to talk them through. More help with this core issue of "rupture and repair" follows—it is essential for successful treatment.

A More Assertive Client   Now imagine that the therapist raises the same type of query, but this time the client is critical of the therapist.

> THERAPIST: How has it been for you to talk with me today?
>
> CLIENT: Well, OK for the most part, but I did feel you were trying to hurry me up a few times.

Clinical training holds many challenges but, for many, none is bigger than learning to approach the real-life, interpersonal conflict that is occurring right now between the therapist and the client. We do not want to deny or avoid it by moving on to another topic, truncate it with a punitive response, or express hurt and communicate that the client's criticism is in any way unwanted. The best way to approach the conflict is to remain as *nondefensive* as you can and, as hard as it may be to do at times, encourage the client to express her concern more fully.

> THERAPIST: Thanks for being honest and bringing that up. Let's look at it together and see if we can understand what happened for each of us. Tell me more about your feeling of being pushed or hurried up.

Of course, it is often difficult for new therapists to do this—just as it is for many supervisors in the workplace, teachers in the classroom, and highly trained and experienced therapists. Such direct, here-and-now conversations about conflict between "you and me" may arouse the therapist's own anxiety, especially when the therapist has strong needs for the client's approval. If they are to approach such anxiety arousing ruptures and restore the therapeutic alliance, however, therapists first need to become aware of their own characteristic responses to interpersonal conflict. For example, some therapists will want to deny or avoid the client's complaint and move on as if it wasn't said. Others will readily agree with the client and quickly apologize in order to abate the criticism. Still other therapists will automatically begin to justify themselves and offer lengthy explanations. Some therapists will counter with their own hostility and criticize the client (Binder & Strupp, 1997). Therapists begin to learn better ways of responding by becoming more aware of how they typically react to criticism or unwanted confrontations.

One of the best ways to do this is to recall how your own family of origin dealt with interpersonal conflict. Regardless of age, your initial reaction to conflict usually follows closely how your parents dealt with conflict in their marriage and how each parent dealt with conflict with you as a child. Rather than automatically following their initial response propensity, therapists can learn to approach interpersonal conflict with more openness. With a little help from a supportive supervisor, new therapists can learn to do this effectively and, in a year or two, feel comfortable and actually welcome the conflict as a therapeutic opportunity.

After inviting clients to express fully their concerns, the therapist is not trying to issue a judgment and disagree with the client's perception. Perhaps the client is right and the therapist was "rushing" the client. The therapist's intention is to honestly examine her own behavior and see whether the client's perception is accurate. If so, the therapist does not need to be afraid of acknowledging this to the client.

> THERAPIST: You know, I think you might be right. I was aware of the time going by, and I wanted to touch on a few more issues before we had to stop. I probably was hurrying you there, and I can see why that bothered you. I'm glad you made me aware of that; I'll try not to do it again. You know, I like your honesty—it's a real strength of yours.

Throughout their first years of training, many beginning therapists suffer greatly over their fears of making mistakes, hurting clients, and failing to help. Sadly, these worries keep many from enjoying their clinical training, and others from taking personal risks and trying out new ways of thinking about and responding to clients. Many caring and responsible students do not enjoy their clinical training because of these worries. It is liberating as new therapists learn that they do not need to be afraid of making mistakes; they need only to be able to acknowledge them. In this example, the therapist is responding to the client in a reality-based and egalitarian way. The therapist validates the client's perception, which tells the client that the therapist is willing to have a genuine dialogue and a real relationship. It also tells the client that the therapist will respond to the client's concerns with respect. Too often, therapists and clients enact a relationship in which the client fulfills the role of the weak or needy one and the therapist is the healthy one who does not make mistakes or have problems. The response suggested here discourages that illusion. The therapist's willingness to risk having an authentic relationship is a gift to many clients. The client finds that, at least sometimes, problems with others can be talked through and resolved. The relearning from such corrective experiences with the therapist often has a powerful effect on clients and accelerates change with others in their lives. For example, clients routinely come back to the next session with reports of acting stronger with others like they have just done with the therapist, such as having their own voice and speaking up for themselves with someone who previously had been intimidating.

In contrast, therapists can handle the confrontation in a way that keeps the relationship superficial or puts them in a superior position.

> THERAPIST: Oh, you thought I was hurrying you up. How was that for you?

OR

> THERAPIST: That's interesting. Do you ever feel this way in other relationships, too?

These responses are a profound misuse of the client-centered reflection or analytic neutrality. With this type of *deflection*, the message to the client is, "It's always your problem. I will not look at my own contribution here. This will not be an honest relationship." Such a response limits the relationship to a superficial encounter and sets up a covert power battle between client and therapist. Little therapeutic gain will be realized as long as the interpersonal process continues in this problematic mode.

But what if the therapist does not agree with the client's complaint? We don't want to say anything we don't really mean. Thus, the therapist can still accept the validity of the client's perception, yet without agreeing to the comment.

THERAPIST: I'm sorry you saw me as being impatient with you; that wouldn't feel very good to me, either. I wasn't aware of being in a hurry or trying to push you, but I'll be alerted to that in the future. If you ever feel that happens again, stop me right then and we'll look at it together.

Therapists do not want to disavow the client's perceptions, and it's not helpful for therapists to agree to something they don't believe is true. In this example, the therapist is telling the client:

- I will take your concerns seriously even if I do not see it the same way.
- We can have differences between us and still work together.
- Your feelings about our relationship are important to me.

This type of response doesn't just tell clients—it behaviorally shows them, that problems between them can be resolved. In turn, this lends hope that the therapist and client may be able to resolve the client's problems with others as well. By addressing conflicts between the therapist and the client directly, the therapist provides the client with an important model of having interpersonal conflicts resolved in a new and more constructive way. Different theoretical orientations offer varying names for this type of experiential relearning: exposure trials, modeling, in vivo relearning, corrective emotional experience, and others. Especially for clients who have learned to avoid interpersonal conflict, such experiential relearning is empowering indeed. With most, it leads quickly to related behavioral changes with others in their lives.

CLIENT: Guess what? I spoke up at my big meeting at work! I didn't even think about it—I just jumped in and said what I thought. It was great. My boss seemed really surprised, and he has treated me a little differently since.

Finally, it is also possible that the client is systematically misperceiving the therapist's behavior in line with their cognitive schemas. For example, this client may readily experience the therapist (and most other authority figures) as demanding more than the client would like to produce, and then being dissatisfied with whatever the client does produce. If so, this client may have grown up with a parent who repeatedly demanded that the client do everything on the parent's timetable or ignored the child's own accomplishments. Even if the therapist has gained further evidence to support such a hypothesis, this type of historical or transference interpretation should not be offered before the therapist and client have resolved the dispute in their real-life relationship. Therapists stand to lose a lot if clients see them as sidestepping a reality-based confrontation. However, the therapist can use this information to generate working hypotheses about their cognitive schemas and faulty expectations that can be utilized later.

Clients Who Test the Therapist's Adequacy   Because it is so challenging for most new therapists to respond to interpersonal conflicts with their clients and work with ruptures in the therapeutic alliance, let's look at another example.

In the following situation, the client challenges the therapist's competence. This cuts deeply into an anxiety-arousing area for most new therapists. However, the therapist is able to remain nondefensive enough to approach the issue directly and invite the client to express her concern more fully.

> THERAPIST: I'm wondering if there is something about our relationship, or being in therapy, that just isn't feeling good for you right now?
>
> CLIENT: Well, you did say you were a student here. Isn't that what you said when we first spoke on the phone?
>
> THERAPIST: Yes, I'm a 2nd-year graduate student, working on my master's degree.
>
> CLIENT: Well, I don't really feel like a guinea pig or anything, but you really are just practicing on me, aren't you? I don't want to sound unkind or anything, but do you think that you've had enough experience to help me?
>
> THERAPIST: Are you worried that if you go to all of the trouble of coming here and talking about difficult problems with me that I just won't know enough to be able to help you?
>
> CLIENT: Well, sort of. After all, you really are just a beginner, and I must be 15 years older than you. What do you think? Can you help me?
>
> THERAPIST: I cannot offer you any guarantees, of course, but I will do my best. As we continue to meet together a few times, you'll get a better sense of me—and what it's like to work with me—and be able to decide for yourself whether this is the right match for you. But for now, let's see if there are any questions I can answer for you about my training status, and how our age difference might get in the way.

In this vignette, the therapist was able to approach the client's concern directly. She responded effectively because she remained nondefensive, even though this situation was anxiety arousing and difficult for her to do. The therapist did not act on her initial impulse, which was to try to reassure the client of her ability to help. Instead of responding to her own personal need to seem competent to the client, the therapist was able to respond to the client's concern—by inviting the client to express her reservations directly and fully. The therapist tolerated her own personal discomfort well enough to be able to discuss the client's concerns and, in so doing, behaviorally demonstrated her competence. This is always more effective than offering verbal reassurances, which would only sound hollow to the client and still leave the burden of proof on the therapist.

Clients' Concerns about Entering Therapy   In the previous two examples, the therapist inquired about resistance, and the client expressed concerns that personally challenged the therapist. This is tough going for a new therapist. In the following example, the therapist asks about the client's potential resistance, and the client expresses concerns about entering treatment. Such concerns are far more common than challenges to the therapist and, of course, much easier for the therapist to address.

CLIENT: It's hard for me to ask for help. I guess I'm not used to talking about myself and my problems.

OR

CLIENT: I feel a little awkward talking to a stranger. In our family, we always kept problems to ourselves.

The therapist's aim is to let clients know that she has heard their concerns—that she is trying her best to grasp what they are saying and "get" the core message or central meaning. The therapist's intention is also to demonstrate that she takes the client's concerns seriously and is willing to be flexible and try to do something about them if possible. Again, the best way to behaviorally demonstrate this empathic responsiveness is to invite clients to discuss their concerns more fully.

THERAPIST: That really does sound like an important concern for us. Let's talk some more about it, so that we can both understand it better. What does it mean for you to talk to me about your problems and risk asking me for help?

OR

THERAPIST: I'm glad you can tell me about this awkward feeling. I think you're right: It is hard to talk to someone you don't really know about personal things. So let's try to help with that by being sure to go at your pace in here. You decide when and how much you wish to share with me. I'd be very comfortable with that way of working. In fact, I'd even prefer it.

Listening and responding to the client's concerns with such affirming responsiveness is also important when ethnic or cultural differences exist between the therapist and client (Pinderhughes, 1989; McClure & Teyber, 2003). Cultural differences can be a sensitive issue; neither client nor therapist may feel comfortable addressing them. It sometimes seems as if there is an unwritten societal rule against recognizing differences. As before, however, the best way to work with these important differences—which may be central to the client's identity and sense of self—is to acknowledge them in an open-ended way. Therapists want to give clients overtly spoken permission to be able to express any concerns they may have. For example, suppose that the client and the therapist clearly are of different ethnicity, age, socioeconomic status, or religion.

THERAPIST: I wanted to check in with you about what it's like to meet with me. Was there anything about coming today, or talking with me, that could get in the way of our work together?

CLIENT: Well, I don't mean to be disrespectful, but I don't know if you can understand my situation because you're white.

THERAPIST: I'm glad we're talking about this because our cultural backgrounds really are different. At times, there probably will be important things about you that I won't understand as well as I want to. When that happens, help me out and tell me when I'm not getting it. We'll talk it through together until we've got it right. Was there something today that I missed or didn't understand that we should revisit?

The therapist responds affirmingly by accepting the client's concern, validating his point of view, and being willing to explore it further. *Different* is an important word as therapists work with people of color, gay men and lesbians, and other clients who feel on the outside. For example, some minorities may feel different in a predominantly white school or workplace—especially if few attempts are made to reach out and include them. In some settings, people who are different may be regarded pejoratively as deficient. The therapist's willingness to acknowledge differences and invite clients to express any concerns or misgivings they may have will go a long way toward establishing a therapeutic alliance and keeping clients who may feel different from dropping out prematurely.

In all three of the examples here, the therapist is able to approach the client's concerns. As the therapist and client discuss the issue, some of the client's concerns will abate with a more complete understanding of the treatment process. Other concerns will be resolved as the client's faulty beliefs and expectations about the therapist are corrected. However, some of the concerns that clients express cannot be assuaged so easily. The therapist can offer to be sensitive to these concerns and express a willingness to work with the client on them over the course of therapy. One of the best ways to do this is by enlisting clients' help in better understanding their background and experience. As we have just seen, the therapist can invite clients to tell the therapist when they are feeling misunderstood, that treatment is not progressing and they don't feel like they are being helped, or there is a problem of some sort in their relationship. The therapist can then ask clients to share their concerns more fully and begin a mutual dialogue to understand and change the problem.

As we have seen, when obvious ethnic, religious, or other cultural differences exist, this approach of enlisting the client's help is even more apt. Ethnic, class, gender, and religious differences are complex, and it is counterproductive for therapists to labor under the misconception that it is their responsibility to understand everything about a client from a different culture. However, therapists do have a responsibility to educate themselves about clients from different social contexts, for example, by reading and consulting with informed colleagues. In particular, the therapist can invite the client to educate the therapist whenever the client feels the therapist is misinformed or does not understand. By welcoming this clarification instead of feeling threatened by it, the therapist signals acceptance of their very real differences, openness to a genuine dialogue, and offers an invitation for a real relationship. In so doing, the therapist will often earn credibility with the client and strengthen the working alliance.

The therapist also needs to explore more fully with clients how they believe their cultural or social context influences their current problems and shapes their subjective experience.

THERAPIST: Would you help me understand this problem in terms of the expectations from your culture?

CLIENT: I can't discipline my children the way I want without being disrespectful to my mother. I'm a Latina and my mother lives with us—she thinks I'm still her daughter, you know, and that it's her role to help out by taking care of the children. I love her very much and would never want to hurt her feelings, but I think she's spoiling them.

Even when obvious cultural differences are not evident, therapists will be more effective whenever they try to understand the clients' subjective worldview more fully (Speight, Myers, Cox, & Highlen, 1991, p. 29).

To sum up, the therapist's intention is to encourage clients to express any concerns about treatment or the therapist that arise, and then sincerely try to accommodate those concerns. This will go a long way toward resolving them and allowing many more clients to enter treatment and successfully receive help.

Resistance during Subsequent Sessions

Understanding Clients' Inability to Change: The Three R's   As treatment progresses, some motivated clients will have phases when they find it hard to come to sessions or simply can't make progress in treatment. Too often, therapists give up on clients at this point—usually doubting whether the client is "motivated" or questioning whether this client "is ready to make the hard decisions necessary to really change." Instead, more effective therapists welcome these expectable problems and use them as opportunities to take treatment further. Rather than blame or judge the client, effective therapists invite the client to join them in a collaborative exploration to understand the resistance or block. As we would expect from client response specificity, the reasons for clients' resistance to entering treatment, and then to change later in treatment, are complex and varied. However, when therapy has stalled or clients have stopped making progress on a problem, therapists should consider three likely issues: ruptures, reenactments, and resistance.

First, we have already seen that *ruptures* occur when there has been a misunderstanding or interpersonal conflict between the therapist and client that has disrupted the working alliance. The therapist repairs or restores these expectable ruptures by talking directly with the client about the interpersonal conflict or problem between them. For example,

> THERAPIST: You sound irritated with me, James. Maybe there's a problem between us? If so, I'd sure like to try to talk it through and work it out together.

Again, one of the biggest problems in the counseling field, cutting across practitioners from varying theoretical orientations, is therapists' reluctance to address the misunderstandings and misperceptions that so commonly occur and disrupt the therapeutic alliance.

Second, *reenactments* also hold back clients from being able to make changes later in treatment. As emphasized earlier, therapists should consider the possibility that in some actual or metaphorical way, the same problem that the client is having with others has been triggered or is being played out between the therapist and the client. That is, the client's treatment block is occurring because the interpersonal process between the therapist and client is reenacting, rather than resolving, some aspect of the client's problem. Based on her cognitive schemas, for example, the client believes the therapist has been critical of her or become disappointed in her—just as her spouse has been. As before, the therapist can intervene by working in the moment with the current interaction between the therapist and client:

> THERAPIST: What do you think I am thinking about you right now, Joan, as you're telling me about this?

Or, suppose that another therapist chose to self-disclose a problem of his own—without realizing how pervasively this particular client had been parentified while growing up. Now, this client feels she is in her old familial role again and has to "take care of" the therapist, too. When such reenactments occur—as they predictably do for all therapists—the therapist can use process comments to help change the therapeutic interaction once it becomes clear that they have become stuck or progress has slowed. By being able to name and talk about what is occurring between them right now, the therapist and client can restore their working alliance and resume productive change.

Early in their training, most beginning therapists will not possess the objectivity necessary to identify these predictable but subtle reenactments and will need assistance from their supervisors. All therapists can also benefit from personal exploration in their own therapy, so that they can become more aware of their own response styles and better anticipate how these styles may interact with their clients' interpersonal patterns. When such reenactments are occurring between the therapist and the client, clients do not have the interpersonal safety they need to explore their problems and treatment becomes stalled on a more superficial or intellectualized level. Once the ship has been righted and the therapeutic relationship is enacting a solution to clients' maladaptive relational patterns, rather than a repetition of them, clients regain the interpersonal safety they need to resume progress and change.

Third, *resistance* will be reevoked throughout treatment. As different issues are explored in treatment, clients will not be fully aware of all of the varying conflicts and difficult feelings that have been activated by a certain topic. Especially if their feelings and perceptions have been consistently invalidated, people eventually lose awareness or become "confused" about many aspects of their experience. Resistant clients are not lying or deceiving the therapist, and they have not lost their motivation—these are different treatment issues. Resistance occurs when clients are simply unaware of the multiple and often contradictory feelings that have been activated by seeking help, exploring difficult topics, or successfully making changes and getting better. For example:

- Getting further help is relieving but may also arouse detested feelings of humiliation.
- Feeling cared about by the therapist's genuine concern is comforting, but it can also evoke sadness about the many times this need went unfulfilled.
- Being listened to and understood in a consistent, reliable manner is empowering, but it also may trigger guilt over being given to rather than taking care of others.
- Becoming more successful at work is exciting, but it evokes fears of being envied or expectations of being undermined.

Unless addressed and clarified, these commonly occurring reactions will lead to resistance. They will cause some clients to leave treatment prematurely, and others to have to sabotage the success or undo the changes they have achieved. Commonly, when these concerns are activated later in treatment but go unaddressed, therapy will become stalled. Treatment becomes repetitious or intellectualized as clients are unable to make progress with their problems or sustain positive changes they have made. At the beginning of their clinical training, however, many therapists are not familiar with the ambivalent, push-pull, or paradoxical nature of these conflicts. Let's explore this further by recalling Marsha from Chapter 2.

It was profoundly reassuring for Marsha when her second therapist approached her feelings directly and validated her experience. At the same time, however, this very positive experience also aroused other contradictory feelings that she "hated." Marsha now felt heard and understood but, at times, also felt

- sad, as years of unacknowledged loneliness welled up;
- angry, at not being heard so many times;
- guilty, for feeling angry at her parents and being disloyal to them; and
- anxious, for breaking familial rules and talking to someone about their problems and what was really wrong in her family.

Marsha did not want to experience any of these feelings and, without being very aware of it, it was hard for her to stay with treatment when these feelings were strong. She did not understand them and felt threatened by them, yet she could not make them go away. Fortunately, her therapist was comfortable with the push-pull nature of conflict and consistently helped her understand the different, contradictory aspects of her experience that emerged. His accepting response helped her to integrate these ambivalent feelings and gradually resolve them. In these ways, the second therapist was providing Marsha with *containment*. In attachment terms, this is a "holding environment" that provided her with the psychological safety she needed to address—explore, experience, share—these sensitive issues for the first time. As these changes occurred in the therapeutic relationship, Marsha began telling the therapist that she found herself feeling "more confident" than she had before and that she was getting better grades and making new friends at school.

At different points in treatment, clients will often be unaware of, and threatened by, the conflicted feelings evoked in the change process. As a result, they may feel misunderstood or even blamed when the therapist inquires about potential signs of resistance. Sometimes clients do feel that the therapist's questions imply that they are doing something wrong or are not trying hard enough. Concerned that the therapist is frustrated or disappointed with them, clients may try to justify their good intentions:

> CLIENT: Oh, no, you don't understand. I really do want to see you and get here on time. It's just that. . . .

As already emphasized, therapists do not want clients to feel blamed or judged. Rather, *the therapist is inviting the client to join in a collaborative exploration to understand the threat or danger that some aspect of treatment has aroused.* Once clients understand that the therapist's intention is to explore the threat or danger, and not blame or judge, they readily join in a productive exploration.

Thus, therapists can best approach potential resistance in a spirit of mutual exploration. Although new therapists are often concerned about being too blunt or being misperceived as critical, the problem actually goes the other way. *Because so many therapists are too worried about "hurting" their clients' feelings or evoking their disapproval, they do not respond as clearly and forthrightly as they could.* Consequently, many beginning therapists dilute what they say and do with clients to mute their impact. If therapists are worried that their explorations may seem intrusive or unwanted to the client, they can shift gears and find another way to respond. Better yet, they can check this out by just asking the client directly.

THERAPIST: I like talking about these issues that come up between us, I think it helps our work together. But I wanted to check in and make sure that you're feeling comfortable with this. How is it for you to talk about things between us so forthrightly?

CLIENT: It was different at first, I wasn't sure what you were up to. But now I like it—I feel like I'm finally talking to somebody about what's really going on. My wife likes it, too; she says I talk about things in a more straightforward way than I used to.

The therapist's intentions are to inquire about potential signs of client resistance in a respectful manner that invites a collaborative dialogue. As before, a *tentative* approach is needed. Therapists can merely wonder aloud with clients about the possible meanings certain behavior may hold; there is no insistence on exploring a client's resistance or any other topic. Therapists can approach resistance without arousing defensiveness through a series of gradual steps that are progressively more direct; a three-step sequence is often effective.

In the first step, the therapist offers a *permission-giving and educative response* to encourage clients to talk about the positive and negative reactions that come up for them about treatment or toward the therapist. In particular, the therapist explains that problems or misunderstandings with the therapist are inevitable, and the client can help by bringing them up. For many, this is all that is needed. When this invitation is insufficient, the therapist can take a second step and encourage the client to explore the potential threat these feelings hold; this focuses on the *defense*. For example:

THERAPIST: I'm wondering what the threat or danger might be for you if you didn't change the topic but stayed with this?

CLIENT: I don't know. Maybe I'd start crying or something.

THERAPIST: Something about crying with me isn't safe or OK. Help me understand that. If you were sad and started crying, what might I do that would be unwanted or make things worse for you?

CLIENT: Well, I'm not really sure why I'm thinking this, but I'm afraid you might put me down or something . . . .

In the third step, if the client continues to show signs of resistance but cannot talk about it, the therapist can draw on previous working hypotheses and try to interpret the *content*. For example:

THERAPIST: It seems like it's been harder for you to get here and really talk to me since we began talking about your wife's illness. She's got cancer, Bob, and that's been really hard to face. What do you think about what I'm suggesting?

CLIENT: I just can't stand to think that those kids could lose their mother. . . .

THERAPIST: Yes, of course, it's been just too heartbreaking for you.

Let's explore each of these three steps more closely.

Step 1: Permission-Giving and Educative Response

THERAPIST: Can we talk about how you feel about coming to see me? I noticed that you were 20 minutes late today and had to reschedule your appointment last week. Maybe it doesn't mean anything, but I was wondering if something doesn't feel right for you. Any thoughts come to mind?

CLIENT: Oh, no, I really do want to be here. And you've been very helpful already. It's just that my boss called.

THERAPIST: OK, but if in the future you ever find yourself having any concerns about the way we are working together, I would like you to tell me about them—especially any problems or misunderstandings that might come up between us. How would it be for you to tell me about something I did that you didn't like, or something about being in treatment that was uncomfortable for you?

CLIENT: Well, that might be kind of hard for me to do.

THERAPIST: What would be the hardest thing about that for you?

CLIENT: Well, I wouldn't want to hurt your feelings or anything. I think I try real hard to be nice to everybody all the time. Maybe I do that too much, huh?

At this point the therapist and client are off and running together. The therapist can help clients work through the issues that would prevent them from expressing dissatisfaction or disagreement. In this case, for example, they have brought out the clients' concern that the therapist (and others) will not like him unless he is always "nice." Not only will this help keep clients engaged in treatment; it opens up important new problems that need to be addressed—in this example, the client's outdated coping style of compulsively pleasing others.

In this first example, the therapist offered a permission-giving and educative response in the hope that the client would then feel free to talk about any ambivalent feelings about treatment or interpersonal conflicts with the therapist that may emerge. If the client continues to show signs of resistance without being able to talk about it (missing, canceling, rescheduling, coming late, discussing only superficial issues), the therapist moves to the second step by exploring the interpersonal threat.

Step 2: Explore the Danger/Identify the Threat

THERAPIST: It's important for our work together that you and I are able to talk about any difficulties that come up between us—they're bound to occur in any relationship. Would you be able to tell me if something about our relationship or work together was troubling you?

CLIENT: I think so.

THERAPIST: Well, maybe so, but based on what you've told me about some of your other relationships, I guess I have to say that I'm not convinced of that yet. I'm wondering if it might be pretty difficult for you to tell me if I did something that you didn't like, or if there was something about being in treatment that wasn't feeling very good for you. Sometimes people believe good clients or nice people never have frustrated or disappointed feelings, or at least they don't ever express them.

CLIENT: Well, yes, that's probably true.

THERAPIST: Let's try to figure this out together. It seems that you are having some trouble being in treatment but don't feel comfortable talking with me about that yet. I'm wondering what the threat or danger might be for you if you were less worried about being nice and took the risk of telling me what doesn't feel right. What's gone wrong with others in the past, or what might go wrong between us?

CLIENT: You probably wouldn't want to see me anymore—I'd be too much trouble, and you'd rather I just went away and didn't bother you anymore.

In this example, the therapist is trying to identify the threat that keeps the client from expressing conflicted feelings. *The therapist is not trying to find out why the client is not showing up for therapy but why the client is having trouble talking about it.* Here, the client reveals her pathogenic belief that if she disagrees with the therapist or speaks up, the therapist (and others) will leave or reject her. The potential for change is present at this moment as the therapist clarifies that he does not wish to end their relationship as soon as they have a disagreement. As this faulty expectation in their relationship is highlighted and disconfirmed, it routinely brings up how it was in fact reality based in other relationships. That is, this client begins recalling and describing how this rejection she expects is what she actually experienced growing up in her family and in her first marriage ("go along or go"). By highlighting how this outdated coping strategy has come into play in the therapeutic relationship, and working together to change it in their real-life relationship, the therapist can then take the next step in the change process. That is, the therapist can help this client begin exploring current relationships with others in her life where she is also anxiously pleasing and afraid to say what she wants, and begin to change this problematic way of relating with them as well. For example:

THERAPIST: Is this "pleasing" a problem with others in your life, too?

CLIENT: I think it's hard for me to say no and discipline my kids because I don't want them to be unhappy with me—so, yeah, I guess I am pleasing them, too.

THERAPIST: Now that we're starting to understand this, let's see if we can help you change it. You know, take a stronger stand with them, tolerate their disapproval, and follow through and enforce the rules you set. How does it usually go when you try to discipline them?

In most situations, exploring the threat or danger in this way will be most effective. If focusing on the threat does not work, however, it's time to move to the third step. As another alternative, the therapist can try to interpret the content of the client's resistance. For example, if the therapist thinks the client's resistance is about to pull the client out of treatment, and the client cannot talk about the difficulty, the therapist can draw on working hypotheses and interpret the resistance.

Step 3: Tentatively Interpret the Reason for the Defense

THERAPIST: It's been hard for you to get here; you've missed the last two sessions. Maybe you just forgot, but I'm wondering if there is something about being in therapy, or something that we have been talking about, that just doesn't feel safe to you. Thinking about some of the things we've been talking about, I'm wondering if you're concerned that I'm going to "judge" you, too, as others in your life have been doing?

CLIENT: Yeah, that's probably true. You wouldn't respect me either if you knew more about me.

THERAPIST: "I wouldn't respect you." I'm glad you're taking the risk to tell me this. Has it ever felt like I was judging you or didn't treat you respectfully?

CLIENT: Well, no, not yet.

THERAPIST: I'm glad that hasn't happened, and I don't think it's going to. But I hear you—you're afraid I'm going to lose respect for you if you tell me more about yourself. I can sure see how that would make it harder for you to be in therapy and talk about what's going on.

CLIENT: Yeah, I think that respect thing has gotten in the way for me my whole life. . . .

Why does the therapist take this more direct interpretive stance only as a last resort? The interpretation, accurate or not, is the therapist's issue and puts the ball in the therapist's court. Whenever possible, it is better to try to follow the client's lead or focus on the therapeutic process rather than pull the client along in the therapist's direction. In other words, it is usually more effective to explore why the client is resisting, or make a process comment and address the current interaction, than to make interpretations or tell the client what to do. Why? Although the interpretation was productive in this case, the goal is to maintain a working partnership and keep the relationship as collaborative as possible.

Shame Fuels Resistance

Resistance, defense, and ambivalence will occur throughout treatment—it's just part of the work. Perhaps the most common source of resistance is shame, which often begins to appear as clients enter more fully into their problems. Only in recent years have therapists begun to fully appreciate the pervasive role of shame in clients' symptoms and problems. Shame is more apt to generate resistance and hold back change than any other issue. In an unfortunate interaction, it is also the feeling that therapists are most likely to avoid. It can be so very difficult to bear sitting with clients as they suffer such an excruciating feeling—therapists often feel inadequate to help, and it brings up their own shame as well. As therapists read and learn more about this cardinal issue, they will develop the ability to hear the *shame motifs* that pervade so many of their clients' narratives. They will also become better able to recognize the many different faces of shame that clients will present (perfectionism, blaming, temper outbursts, contempt, low self-esteem and depression, eating disorders and other addictions, preoccupation with appearance, social withdrawal, and many others). Soon, therapists in training will be able to detect shame-related themes in their clients' concerns about entering therapy, as well as in the issues and concerns that block their progress later in treatment.

Shame is different than guilt—and a far more significant problem to deal with. With guilt, clients feel only that they have done something wrong or bad to hurt another. With shame, on the other hand, it's not that I have done something bad; it's that I *am* bad. A more primary and pervasive feeling than guilt, shame is a total or all-encompassing feeling about one's self. With shame, clients often report feeling worthless, inadequate, or defective in the core of their being—essentially flawed in who they really are. To feel shame is to feel that the true self with all of its defects is exposed, naked, and vulnerable to the judgment and criticism of others (Spiegel & Alpert, 2000). Clients may be without words to communicate their experience when they are suffering a profound shame reaction; sometimes all they can say is something like "I'm just hideous," "I don't matter," "There's something wrong with me," or simply, "I hate myself." A full-blown shame reaction is agonizing to suffer and very difficult for therapists to bear witness and share. During these annihilating moments, some clients may express a wish to die.

It is universal to feel embarrassed in awkward situations, just as it is common for many to feel more deeply ashamed in certain moments. However, in contrast to these situation-specific feelings, clients who have been repeatedly shamed in their important relationships develop a *shame-based sense of self* and are *shame-prone*. Shame-prone individuals, for example, will not be able to accept constructive criticism or good-natured teasing because it threatens to expose their deeper, long-standing feelings of shame. Humiliating feelings of being profoundly diminished or exposed are readily triggered for shame-prone clients. These intense reactions are triggered in many different situations in their daily lives that would not evoke such strong reactions in others. In response to the intense feelings of shame that are continuously evoked, shame-prone clients employ a wide variety of coping strategies to keep their shame concealed from others and, perhaps more importantly, from themselves. Routinely, such clients protect themselves from having their shame-worthiness revealed by

- acting arrogantly and self-righteously toward others;
- intimidating, controlling, or inducing shame in others;

- adopting perfectionistic standards around their own work, cleanliness, or religious practices;
- developing eating disorders or other addictions.

Therapists will observe other faulty coping strategies that clients have developed to cope with these dreaded feelings. Perhaps most disturbing to witness is the *shame-rage cycle* that characterizes most battering and physical abuse episodes (Dutton, 1995; Simon, 2002). The individual explodes in rage in the seconds after feeling diminished or demeaned by someone, often in response to a completely innocuous, benign comment such as his wife asking him to take out the trash. The rageful response is a desperate but futile attempt to disprove their shame-worthiness ("I'm not weak"; "I'm not a piece of garbage"; "Nobody can boss me around"). This is an artificial or defensive attempt to "restore" a sense of personal power—often through contemptuously diminishing others or aggression ("You can't tell me what to do. I'll show you who's really the trash here and who's really the boss! I'm not weak—I'm powerful!").

Clients who have grown up suffering *contempt* from caregivers (for example, a parent who, in a disgusted tone of voice, said, "What's wrong with you? You make me sick"), and who have developed a shame-based sense of self as a result, may not remember the original developmental experiences that were so crushing of their self-esteem or self-worth. For example, the client whose drunken father made him lick the kitchen floor, while siblings looked on helplessly, can usually remember what happened and recount this humiliating experience to the therapist. However, he usually does so without being able to feel the shameful affect or experience the unwanted emotions that accompany this abuse. Similarly, this client often cannot recognize or identify the current experience (feeling "bossed around" or told what to do) that has just triggered his temper outburst. In the following dialogue, the therapist is helping this highly shame-prone client make the connection between the triggering event (feeling put down or "dissed" by someone) and his presenting problem (losing relationships because he keeps losing his temper).

THERAPIST: As I listen to the sequence here, it sounds like you lost your temper just after your wife asked you to take out the trash. I wonder if that's telling us that you lose your temper when it seems like others have just put you down or dominated you in some way?

CLIENT: I hadn't thought of that, maybe. . . . You know, that reminds me, I almost got in another fight this week, when this worm flipped me off on the freeway. I wanted to kill him—my buddy said my face was so red he could see the veins sticking out of my neck as I was screaming at him.

THERAPIST: That fits with my suggestion that when someone flips you off, or even just tells you what to do, you feel shamed—and then the rocket goes off. It's like it almost becomes a matter of life and death for you—as if you're saying they just can't do this to you again.

CLIENT: Yeah, that's right. I hadn't thought of the word *shame*, but I think I do go kind of crazy when someone puts me down.

THERAPIST: Yes, and maybe you feel shamed sometimes even when they really aren't putting you down. It sounds to me like your wife really was just asking you to take out the trash. Now that you think about it, do you think she was trying to boss you around or put you down?

CLIENT: No, not really—she can be bossy but she's a good person. I've got to change this—I'm ruining my life.

THERAPIST: Let's keep talking about "shame"—that's the feeling that seems to trigger or set off your temper. When I say "shame," what comes to mind?

CLIENT: Well, my stepfather used to beat us. He's dead, but I still hate him. . . .

For other types of clients who are also shame-prone, intense anxiety can be evoked in a host of different situations that hold the ever-present threat of revealing their basic unworthiness, inadequacy, or inherently flawed and defective self. As the dreaded feeling of shame is evoked yet again in this new situation, therapists will commonly witness *shame-anxiety*. For example, something as innocuous as asking someone for a date, applying for a job, needing to ask for help with a flat tire, getting lost and needing to ask for directions, or making a simple mistake at work evokes the anxiety that their shame will be exposed. That is, strong anxiety is evoked by the threat of having one's badness or flawed self revealed. In many shame-prone clients, this occurs so frequently that they live with a pervasive, anxious vigilance that manifests itself as the presenting symptom of anxiety attacks or a generalized anxiety disorder. In other words, the anxiety signals that the client's defenses against her shame are being threatened in this situation, and raising the specter that her shame-worthiness will be revealed. To manage this anxiety and protect against the potential exposure of her shame, the client may cope by withdrawing or avoiding, being quiet and "good," trying harder to be perfect, never needing anybody or anything, and so forth, which becomes a characteristic or habitual coping response.

As we will explore later, therapists can help clients recognize and change these problematic coping strategies for defending against their shame (for example, hitting and yelling, withdrawing and avoiding, controlling others, inhibiting oneself, complying with others and going along when they want to say no, becoming intoxicated or getting high, disassociating, and so forth). We will learn how therapists can help clients allow themselves to actually experience the self-hatred or contempt they feel toward themselves, and change by beginning to have some empathy or compassion for themselves for the plight they once were in. The biggest problem with these interventions, however, is that therapists often have as much difficulty witnessing and responding to the raw experience of shame as clients have in bearing it. As we will see, some therapists are themselves shame-prone and many have their own shame dynamics activated by the client's shame—and therapists often try to cope by reassuring the client, minimizing the problem or avoiding the topic altogether, offering intellectual explanations, and providing other well-intended but ineffective responses. Guidelines for responding to shame and helping these clients "contain" these and other painful feelings will be provided in Chapter 5.

For the time being, therapists need to begin listening for the presence of shame in their client's narratives and to begin considering the possibility that *resistance often serves to protect clients from their shame and having seemingly unacceptable parts of themselves revealed.* This is a simple concept to say but, because of our own shame dynamics and cultural prohibitions against approaching such sensitive issues, takes most new therapists several years to learn. Supportive supervisors are necessary to help new therapists begin to work with shame dynamics.

> Susie's boyfriend did not show up for their date. Feeling desperate and empty, she found herself in the kitchen spooning peanut butter out of the jar and into her mouth until the jar was empty. The next day in therapy, telling her therapist about what had happened, she became overwhelmed with shame—feeling like a "disgusting pig." Susie's therapist wanted to look away; her mind was racing trying to think how she could make her client feel better and stop this sickening pain. By asking about the binging episode, Susie's therapist felt she was causing her client an incredible self-loathing that only seemed to be accelerating. Quickly, the therapist's shame about her own weight and appearance began to flood her. The therapist wanted to look away, get away, and do anything to change the conversation and end this excruciating moment—it was too painful and too close. Fortunately, however, she was able to recall her supervisor's words of support and take a slow, deep breath instead. Rather than flee from her own shame and stop the client from feeling hers, the therapist was able to provide a profound corrective emotional experience by responding, "I'm honored you are able to share such a vulnerable part of you with me. It is a privilege to be with you right now." Susie cried hard and, 3 months later, had not experienced another binge-eating episode.

Learning to work with clients' shame is not easy, but therapists will find their clients remain in treatment and make real progress with their problems as they become more effective in working with shame. More specifically, clients resolve their shame as therapists respond to it empathically, as Susie's therapist demonstrates so effectively. It's exciting to see the far-reaching changes that result as the client's shame-based sense of self improves ("low self-esteem" is often a euphemism for shame). For example, clients who were too preoccupied with their appearance or weight are more self-accepting; others who didn't care enough about how they looked take pride as they lose weight, begin fixing their hair, or dress better. Anxiety over having their shame-based sense of self revealed, and depression over feeling so bad about oneself, diminish markedly as the client finds that the therapist can see this wounded part of them and still respond compassionately.

These behavioral changes are not an end point in treatment, however. Continuing with the push-pull nature of change, many clients will experience conflict over succeeding in treatment and making these positive behavioral changes. This paradox or self-defeating behavior is perplexing to most new therapists. In other words, resistance predictably returns, even as clients make healthy new changes. For example, some clients will experience binding feelings of guilt or disloyalty over making progress and having success in treatment. Therapists will observe that some clients will retreat from progress in treatment or undo successful changes they have just made. Often, clients cannot sustain positive changes they have just achieved because the healthy new behavior is inconsistent with their cognitive schemas. Following attachment theory and internal working models, acting in this successful new manner (that is, stopping binge-eating episodes, leaving an unrewarding relationship, graduating from college) threatens their attachment ties to internalized caregivers who did not support their independence and success. Becoming stronger or improving in therapy makes some clients feel cut off from parental approval and affection, disloyal to caregivers, or guilty about hurting, leaving, or surpassing the parent. As they make significant behavioral changes, or report feeling better about themselves, the therapist can expect some clients to report feeling either (1) alone, empty, or disconnected following meaningful change or (2) guilty, selfish, or bad. When these reactions occur, clients need to be reassured of the therapist's continued presence and support for this stronger self. If not, they will have to manage their guilt or restore their insecure attachment ties by sabotaging their own success. For example, the therapist challenges the old schema, disconfirms the problematic expectation, and provides a corrective emotional experience by saying:

> THERAPIST: You've just had a big success at work—and I think that has something to do with why you feel so anxious and alone right now. I am very happy for your new promotion and feel close to you right now as you are telling me about it.

These commonly occurring sources of resistance and impediments to change will be explored further in the chapters ahead.

Closing

Listening to the client with presence and respect is the most important intervention the therapist can make in the beginning of treatment. Such listening and empathic understanding are the basic tools that the therapist uses to establish a working alliance. The next step in the treatment process is to address ambivalence or resistance to therapy in order to maintain the relationship that has just begun. Resistance will be more of an issue with some clients than with others, but it will occur to some extent for every client. As we have seen, if the client can talk about these conflicted feelings, and the therapist responds affirmingly, they will be far less likely to pull the client out of treatment prematurely. Resistance will wax and wane throughout therapy, often reemerging as clients enter more deeply into difficult issues and as they try out new behavior and make successful changes. To sustain progress in treatment, therapists need to continue responding effectively to the many types of resistance that occur. If therapists formulate working hypotheses about potential client resistance, it will help them to anticipate when it is likely to occur and to respond more effectively when it does. In sum, the therapist's goals are

- to encourage the client's lead and ownership of the treatment process,
- to provide accurate empathy to foster a working alliance,
- to remain alert for potential signs of resistance and address it through process comments, and
- to begin formulating treatment plans by generating working hypotheses.

Resistance, ambivalence, and defense provide a window to observe the fascinating workings of internal conflict. People do not possess a unified self; we are so complex and multifaceted that it often feels as if one part of ourselves is working against another part. By resolving an internal conflict like the ones discussed here, a person becomes a little more integrated or whole. Integrating disparate parts of the self is usually an important part of behavior change and the avenue to greater self-efficacy. Working successfully with resistance shows clients that they possess the internal resources, and have the helping person they need, to make the changes in their lives they choose.

Suggestions for Further Reading

1. A case study illustrating how therapists can respond to a client's shame is provided in Chapter 3 of the Student Workbook: "Shame Dynamics: Case Study of Hank."

2. Shame probably plays a bigger role in resistance and preventing change than any other single issue, but it is the most sensitive affect for therapists to approach. Readers are encouraged to read Karen's (1992) informative article "Shame." A highly informative article about shame intervention in couples therapy is "The Systemic Treatment of Shame in Couples" by Balcom, Lee, and Tager (1995). In particular, this article illuminates how couples "blame" the other to externalize or defend against their own shame.

3. In the interpersonal-process approach, we are working with resistance as a rela-

tionship issue rather than a fixed personality trait or isolated behavior. Readers are encouraged to explore other texts that similarly reframe the client's "resistance" as an outdated interpersonal coping strategy (see Miller & Rollnick, 2002).

4. A process-oriented approach to multicultural issues in counseling appears in Chapter 1 of the casebook in *Child and Adolescent Treatment: Cultural and Familial Contexts* (McClure & Teyber, 2003). An excellent discussion of ethnographic and cross-cultural issues in counseling can be found in Pinderhughes's (1989) *Understanding Race, Ethnicity, and Power*.

CHAPTER FOUR

An Internal Focus
for Change

Dana was presenting a new client she was seeing to her first-year practicum class. She showed a videotaped segment from her previous session in which her client was complaining about her past two husbands, her mother, her previous therapist, and others. Lost in trying to grasp what all of this meant, Dana half-pleaded with her practicum instructor, "What should I do?" With her usual good sense of humor, her instructor jokingly replied, "OK, you guys are always asking me what to do, and this time I'm actually going to tell you. Next week, Dana, tell your client that she has two choices. Either everyone she has ever known needs to be in therapy—so she should refer every person in her life here to the clinic for treatment. Or, she can just meet with you herself and begin to explore the decisions and choices she makes and her contributions to these problems." The class laughed and got the point.

Conceptual Overview

The first stage of therapy is complete when therapist and client have established a working alliance and begun to work together on the client's problems. This collaborative partnership is a necessary prerequisite to the second stage of therapy: the client's journey inward. In order to change, clients need to become less preoccupied with the problematic behavior of others and begin to explore their own role in problems. As real and compelling as these problems with others usually are, clients need to stop focusing exclusively on historical events, past relationships, and the problematic behavior of others in their lives. Instead, clients change when they begin to clarify their own thoughts, feelings, and reactions to the troubles they are having. Why? Clients will usually fail in their attempts to change others in their lives, whereas they can often resolve problems by changing themselves—the way they respond and their own participation in problems. In this process, clients examine their habitual response patterns in problematic situations, evaluate their usefulness in their current lives, and begin trying out new and more adaptive ways of responding. This internal focus on changing their own responses, rather than trying to change the other person, leads clients toward greater self-efficacy. In particular, presenting symptoms of anxiety and depression often diminish as clients are empowered to act more purposefully and better live in the ways they choose.

Thus the therapist's task is to help clients make the transition from seeing the source and resolution of problems in others to adopting an internal focus for change. This is a two-fold process. First, the therapist helps clients begin to look within. That is, without invalidating clients or using language that makes them feel "blamed" in any way, the therapist focuses clients away from complaining about or trying to change others, and toward understanding and changing their own problematic reactions. Second, the therapist's intention is to help clients assume more responsibility for change—more ownership of the treatment process—by becoming active agents in their own therapeutic work. This occurs as clients become more aware of how their cognitive schemas are not always accurate and their usual coping styles are not effective in many current relationships. With greater awareness of their own responses, and how they may be participating in or contributing to the problem, clients increasingly recognize that they do not have to keep responding in their old maladaptive ways. With this new perspective, clients often join readily with the therapist in exploring new behavioral options and working actively to change the way they are responding to problematic others. Throughout this chapter, we will see that *when clients change their part in problematic scenarios, the interaction sequence changes—even if the other person doesn't change and keeps giving the same hurtful or disappointing response.* This new option of changing their participation or role in problematic interactions, coupled with the therapist's support for the clients' own self-direction and initiative within the therapeutic relationship, will empower clients to change.

Self-efficacy develops out of this collaborative process between therapist and client. The therapist is not fixing or curing the client—the client is sharing ownership of the treatment process. Why is this interpersonal process so important? The therapist's goal is to empower clients, not just give answers or tell them how to live their lives. Our goal is to help clients know their own mind—to have their own voice—and become better able to make their own decisions. When given answers or prescribed solutions, clients do not make them their own—they cannot apply the advice in the next situation or learn how to solve future problems without having someone else tell them what to do. By helping clients look inside at what they are feeling, expecting, and doing, we are teaching them skills that they can carry over to their everyday lives and use with themselves when treatment is over. In this way, enduring change results when clients participate actively in treatment and feel ownership of the change process.

For example, clients "own" insights, and can successfully translate new understandings into new behavior, when they have shared responsibility for constructing them. That is, the "aha" experience that can lead to new behavior comes about when the client has been an active participant in achieving insight, not when the therapist has just explained a connection or interpreted what something means. Treatment gains are maintained after therapy has stopped when the client helped to make them happen. However, it may be difficult for some clients to shift their focus away from others and begin looking at their own contribution or role in problems. In order to take this rewarding journey inward, clients need a relationship with a therapist who is both affirming, on the one hand, and, on the other, willing to take the risk to challenge them to look within. In this chapter, we will see how therapists can do both.

Shifting to an Internal Focus
A Prerequisite for Change

Early in treatment, many clients see the source of their problems in others. Clients often want to spend more time describing others' problematic behavior than discussing their own experience of and response to the problems. For example, many clients begin the first few therapy sessions by announcing that the problem is really with another person:

- My husband won't pay any attention to me.
- My wife is always on my back.
- My children are impossible; they won't do anything I say.
- My boss is a demanding tyrant. If he doesn't back off, I'm going to have a heart attack.
- My mother won't stop criticizing me; I can't do anything right according to her.
- I'm 27 years old, and my father treats me like a child.
- My boyfriend has a drinking problem.

It is essential for the therapist to affirm these complaints as genuinely troubling concerns and communicate overtly that the therapist wants to help the client improve this situation. The therapist's intention is to address the problem that is most pressing or salient for the client right now—to find and begin with the client's "point of urgency." Even though these concerns and complaints may include overreactions or strong transference distortions, from the client's point of view, they are the truth. Thus, as emphasized in Chapter 2, the therapist joins the client there, validating and even helping clients articulate more clearly how the problematic behavior of others is indeed troublesome.

In some cases, however, such affirmation would not be genuine for the therapist. If not, therapists can at least affirm the subjective reality of the client's perceptions. For example:

THERAPIST: So, from your point of view, it seems that they are being unfair again.

If the therapist does not provide this affirmation first, many clients will feel that the therapist doesn't understand the reality of the other person's problematic behavior, doesn't believe them, or doesn't care about what's really troubling them. This unwanted outcome is counterproductive and will often reenact clients' developmental history—they are not listened to or believed again. Thus, to keep this potential invalidation from occurring, the therapist's first aim is to hear and validate the client's concerns. Throughout treatment, there is a consistent intention here: we are repeatedly trying "to meet the client where she is at." Only after making this genuine empathic connection, which often requires the therapist to decenter and see issues from the client's subjective viewpoint, can the therapist take the next step.

The second step is to begin focusing clients back on their own thoughts, feelings, and reactions to the problematic behavior of others, rather than joining the client in focusing exclusively on the external problem. Often, it will be easy for therapists to pair these first two steps together.

> THERAPIST: I'm sorry your father does that to you. I think it would be upsetting, too. What do you say and do when he . . . ?

OR

> THERAPIST: It sounds so frustrating to have your boss respond that way over and over again, no matter what you do. How do you feel, what's going on inside, when he . . . ?

The same principle applies when working with people of different race or ethnicity, women, gay men and lesbians, and others who accurately explain the social inequities that realistically contribute to their problems. Once again, the therapist wants to join such clients in the reality of their social context, as they experience it, before focusing inward on their reactions to, and ways of coping with, these inequities.

Why is therapy more productive when therapists focus clients inward? In many cases, clients who feel anxious, distrustful, angry, or helpless because of the behavior of another person will try to enlist the therapist in blaming, criticizing, or trying to change the other person. However, *therapy will not progress very far if the therapist merely joins clients in focusing on the other person's behavior, no matter how problematic this behavior may be.* Why? As the existential therapists inform us, clients' attempts to change the other person will usually fail; clients are much more likely to resolve the problem by changing their own way of responding (Wheelis, 1974). Yalom (2003) emphasizes that once clients recognize their own role in creating their life predicament, they realize they have the power to change it. Working in a supportive and nonblaming way, therapists are trying to help clients consider the question, How do I contribute to my own distress?

Thus, the therapist's task is to expand clients' focus beyond the other person and include themselves as well. Note that the therapist should be *flexible* in adopting an internal focus; the therapist is trying to clarify both the reality of what the other is doing *and* the client's ineffective response to it. Most clients will benefit greatly from the therapist's help in understanding the behavior or potential intentions of problematic others. For example, if a client has grown up being befuddled by a parent with a borderline or narcissistic personality disorder, it may be deeply validating for the client to read the descriptions of these disorders in the *Diagnostic and Statistical Manual*, fourth edition (DSM-IV). This type of external validation may help this client realize that all of this was not his fault and did not happen solely to him:

CLIENT: My father really was unpredictable and self-centered—it wasn't just me!

In tandem, however, therapists are working to increase clients' awareness of their own thoughts, feelings, and reactions in problematic situations. Recall the externalizing comments cited at the beginning of this section. Each of the following questions can be used to focus those clients back on themselves and begin to consider their role in the problem as well:

- What do you find yourself thinking when your husband is ignoring you?
- How do you feel when your wife is nagging you?
- What do you do when your children disobey you?
- What thoughts were you having as your mother was criticizing you?
- How do you react to your father when he diminishes you like that?
- What is most troublesome for you about your boyfriend's drinking?

Simple inquiries of this type serve two important functions. First, they tell clients that the therapist is listening to their concerns and is taking them seriously. The therapist has not changed the topic or brought up something that was discussed 10 minutes ago but is responding directly to clients' concerns as clients see them. We will lose clients readily if we are not helping them with the problems that they see as most important.

Second, while inviting clients to say more about their concerns, the therapist is also shifting clients' focus away from others and encouraging them to look more closely at themselves. Inviting clients to become more aware of their own reactions is a critical step toward understanding and resolving their problems. As we will see, the simple response of focusing clients on their own experience is a powerful intervention that will elicit strong feelings and reveal important links and connections.

In many cases, clients will welcome the therapist's offer to talk more about themselves. When clients begin to focus inward and learn more about themselves, they begin to change. How? Clients' active exploration of their own expectations and responses in problematic situations will lead to *increased awareness of the narrow range of responses they repeatedly employ*, and, with the therapist's help, they can begin to identify and try out a wider range of options that are actually available to them. This increased self-awareness of their old response patterns, coupled with learning new or more flexible ways of responding, leads to more problem resolution skills and increased feelings of self-efficacy. However, not all clients will respond so positively to the therapist's initial invitation to look more closely at themselves. Some clients may avoid or actively reject an internal focus and continue to talk about the problem out there in others. In that case, the therapist continues to inquire about the personal meaning that this particular situation holds for the client.

THERAPIST: Where would you like to begin this morning?

CLIENT: My wife is impossible to live with. She complains constantly; nothing ever pleases her.

THERAPIST: It sounds like you've had a difficult week with her. What's been the hardest thing for you?

CLIENT: Do you know how hard it is to live with an angry, demanding wife who keeps trying to tell you what to do all the time?

THERAPIST: You're very angry at her. And I can see why; it would be hard to have someone after you like that. Tell me, how do you respond to her when she is doing this?

CLIENT: I don't know; I just hate it. I guess I yell back sometimes or just try to get away from them. It's not just her, you know; her whole family is like that.

THERAPIST: They really are very critical of you, and I can imagine how hard that would be to live with. But it seems like criticism, in particular, really gets under your skin. What's it like for you to be criticized so much?

CLIENT: I hate it. I just hate it. They make me feel like I can't do anything right—that I'm doing it wrong and failing all the time.

THERAPIST: "Doing it wrong and failing"—ouch! Sounds like her family's constant criticism taps into your own painful feelings about yourself of not measuring up or not being enough. And having those shameful feelings of inadequacy aroused all the time could be infuriating.

CLIENT: Yeah, I hate them for making me feel this way. If I could just make them stop, everything would be OK.

THERAPIST: This is very important for you, and we need to work together on it. I would like to understand better what you do when they criticize, so that I can help you learn some more assertive, limit-setting responses. But your own feeling of not measuring up is also a part of this problem that we need to work on, too, so that you can stop charging at the red flag they are waving. If we can change the internal part of you that overreacts, as if you believe what they are saying is true, it would be much easier for you to handle this than it has been in the past.

CLIENT: What do you mean, "overreacts"?

THERAPIST: Sounds like their criticism makes you feel you can never do anything right. Is that accurate?

CLIENT: Yes, but that's not true. They just expect too much, and I can't do it all. I can never do it good enough for them.

THERAPIST: I believe you—no matter what you do, they will never be pleased. But I don't think that's the whole problem—their expectations still upset you so much. It's the way you end up feeling inside, about yourself, that becomes the bigger problem.

CLIENT: Yeah, maybe so.

THERAPIST: So being able to work on your feelings about yourself would be helpful?

CLIENT: I see what you mean, but I'm not sure what to do.

THERAPIST: Tell me more about your feeling of not measuring up.

CLIENT: Well, I guess I've always sort of felt like I'm not really good enough.

In this dialogue, the therapist validated the client's experience while at the same time encouraging the client to look at his own reactions and role in the problem. Although the client kept trying to focus on the problematic behavior of his wife and her family, the therapist's repeated but patient invitations to have the client look at himself as well soon slowed his defensive, externalizing stance. As a result, the client moved closer to his own problematic feelings and beliefs about himself that contributed to the marital conflict. By gaining a better understanding of his own feelings of inadequacy, the client will become less shame-prone and reactive to others' criticisms. This capability, in turn, will be an important part of learning to respond more assertively and set limits more effectively, rather than merely yelling back or withdrawing—which only exacerbated the conflict with his wife.

Many clients have not been encouraged to focus on themselves before. As in this example, some clients may resist this internal focus initially, perhaps because it asks them to come face to face with difficult feelings that they have not been able to understand or resolve alone (for example, "I feel inadequate."). For many others, focusing on the other fends off the *blame* they have learned to expect from others—perhaps having heard too much criticism in the past. However, the paradox is that as long as clients avoid the internal or personal aspects of their relational conflicts, by externalizing their problems onto others, they will feel powerless and frustrated—without any control in the situation.

Many clients who enter treatment are overly invested in changing others as a means of managing their own problems or insecurities. In attempting to shape or direct others, the client is seeking an external solution to a problem that can best be changed internally. The reality is that, beyond clearly expressing our preferences and personal limits ("I would like . . . "; "I will not . . . "), we cannot readily influence how others think, feel, and act. In contrast, however, we can change interpersonal scenarios that keep playing out by changing our own responses to problematic others. For example, clients routinely fail in their attempts to make their spouse stop drinking, smoking, or overeating. Similarly, some clients try for decades to win the approval or recognition from others that they never received from a parent. Others try unsuccessfully for years to have their grown offspring choose a different mate, religion, or career. As a result of these failed attempts to change others, many clients enter therapy with feelings of helplessness, hopelessness, and depression.

To gain a greater sense of self-efficacy (and feel less anxious or depressed), a therapist can offer clients the more productive alternative of decreasing their attempts to change others and changing the way they respond instead. This internal focus is often a necessary prerequisite before clients can adopt new, more effective responses to old problems. Only when clients begin to focus on understanding and changing themselves will they begin to feel in charge of their lives and capable of change. Clients experience these feelings as empowerment. Thus, *a consistent therapeutic intervention is to pair an empathic or affirming response with an invitation to focus clients back onto themselves* (such as "I'm sorry that happened. How would you like to be able to respond the next time he does that?"). The following types of questions will help clients explore their own responses:

- What is the main feeling you are left with when . . . ?
- What were the thoughts you were having when . . . ?
- What was the most difficult thing for you when . . . ?
- How would you like to be able to respond when . . . ?

Focusing Clients Inward

Focusing clients on their own behavior often reveals how they are contributing to, or participating in, their own problems. Often, clients will not have realized their own role in these conflicts. As we will see, clients who can focus on themselves and see their own participation in a conflict are usually motivated to change their own part in it. This, in turn, often allows the other person to respond differently as well. To illustrate, let's return to our example of the husband who complained about his critical wife and in-laws.

Although the client did begin to explore his own feelings of inadequacy and the ineffective ways he responded to his wife, he kept complaining about his "obnoxious" wife and "superior" in-laws. As before, the therapist was empathic, validated his experience, and affirmed his anger. At the same time, however, the therapist did not join the client in focusing exclusively on his wife or blaming her as the sole source of his problems. Instead, the therapist continued to focus the client away from his preoccupation with his wife's behavior and toward his own reactions to her. For example:

> THERAPIST: Rather than insulting her, what could you say instead the next time she criticizes you like that? What would be a realistic way for you to speak up for yourself and say what you don't like?

What would have happened if the therapist had not taken this approach? If the therapist had responded to the client's *eliciting pull* to blame his wife, the client would have remained an angry but helpless victim. In contrast, if the therapist had emphasized just the client's contribution to his marital conflict, without first validating his experience, it would have repeated the problematic relational pattern for him of always being blamed. Exploring the client's feelings of inadequacy revealed that these shameful feelings stemmed from rarely being supported by his parents and consistently being blamed for whatever went wrong—in his recollections, "everything was my fault." Thus, by validating the client's experience first and, second, focusing him inward, the therapist helped the client adopt a new and more assertive response to the old problem of excessive criticism from his wife. For example, the client rehearsed with the therapist and later tried saying to his wife:

CLIENT: You're saying, in so many words, that I did it wrong, but I disagree. I guess we just see this differently.

AND

CLIENT: I feel put down when you talk like that. Please stop, or I'm leaving the room.

By coupling the therapist's validation with a consistent invitation to adopt an internal focus and recognize his own contribution to the problem, the client readily became more aware of how his responses were contributing to the marital conflict. For example, the client learned that he was quiet and unresponsive to his wife when he arrived home from work. The therapist helped him see that his silence increased his wife's attempts to gain his attention, and, in response, he became even less communicative. This interaction cycle escalated as his wife became more demanding of his attention and he withdrew even further—by taking a nap, for example. By focusing on his own emotional reactions at that moment, the client realized that his wife's increasing requests for attention made him feel "overwhelmed" by the prospect of trying to meet all of her needs. He had held the faulty belief that if he did not always respond to her, he was failing as a husband and again being inadequate.

While encouraging the client to become more aware of his own internal and interpersonal responses to his wife, the therapist was also suggesting different ways he could respond that would change his part in their marital conflict. Specifically, the husband began to express more directly to his wife when he would like to talk or interact together and when he would prefer to be alone. For example, he clarified with his wife that he wanted 30 minutes alone to unwind when he returned home from work, after which he would like to sit down together and share their day.

More important, as the husband gained a better understanding of his tendency to feel overwhelmed by others' demands, he became less reactive to his wife's requests. The biggest change occurred as he increasingly realized that he felt shamefully ruled by her because he could neither say no to her requests nor express his own wishes. On the one hand, if his wife wanted something from him, he believed that he was failing if he didn't comply and do it her way. But, on the other hand, he felt ashamed of being "told what to do" by her and responded angrily. And because he couldn't set limits or ask for what he wanted, he resented the "unfairness" in their relationship. The therapist was also able to help the client see that he expressed this resentment to his wife indirectly— in his own critical and withholding ways. Over the next few months, this new awareness allowed him to respond more often to his wife's requests without feeling like he was complying with her or being dominated. No longer feeling that he must comply with them, his resentment toward her abated. This was not a simple "aha" experience that changed his life, of course, but his situation gradually improved as he continued to set limits with her and speak up for himself as the therapist actively encouraged him to do. As the client set these limits, he began to feel better about himself and, in turn, was less angry and better able to respond to his wife in friendlier and more respectful ways.

Thus, as clients begin to focus on their own behavior, they often start to feel less "stuck" or powerless. As therapists help them see that more behavior alternatives are available to them, clients begin to feel more hopeful and in charge of their lives. Thus, the therapist's intention is to try and orient clients away from helplessly complaining about or trying to shape and control others' behavior. To emphasize, however, therapists will be more successful if they do this in tandem with affirming clients' legitimate concerns about the problematic behavior of others.

Reluctance to Adopt an Internal Focus

Therapists' Reluctance   Most new therapists feel comfortable in the role of trying to assuage their clients' fears or concerns—a benevolent social skill at which they have long been adept. In sharp contrast, however, *focusing the client inward often means that the therapist is drawing out and highlighting—rather than reassuring or minimizing—the client's fears and concerns*:

THERAPIST: Don't worry, your doctor will figure out what's wrong—it'll be alright.

VERSUS

THERAPIST: Your doctor can't make a diagnosis yet and tell you what's wrong— you just don't know what to expect or prepare for. It makes sense that you're scared right now.

For many different reasons, therapists often have difficulty approaching and clarifying client's concerns rather than trying to minimize or assuage them—as occurs in most social interactions. For some, such directness goes against unspoken familial rules or cultural norms. In treatment, however, therapists want to do more than follow these limiting social rules and norms they may have grown up with. Instead, the therapist's intention is to capture the core meaning or central feeling in what the client just said. *With every turn of the conversation, the therapist's intention is to register/reflect/highlight the most important meaning or key concern—what's really being said here—rather than keeping things conversational or on the surface.* By focusing inward, therapists are inviting clients to share in a dialogue that reaches for the deeper meaning— again and again, as a consistent aim throughout treatment. When therapists can repeatedly capture the most important feeling or key issue in what the client just said, clients leave the session with the feeling "We're really getting down to it" or "I don't know what to do yet, but we're sure talking about what's important." This is not so much about leading clients toward more threatening material or pressing for difficult feelings; rather it is more about not backing away from the significance of the material that clients volunteer or already choose to present on their own. Perhaps the biggest impediment to taking an internal focus is the therapist's reluctance to engage the client consistently in this substantive manner. As we will see, this approach brings meaning and intensity to the therapeutic interaction but often leaves new therapists unsure of "what to do" as clients get closer to their real concerns.

Other reasons occur as well. Some therapists may equate being forthright with being intrusive or exposing because they have seen directness utilized only in these hurtful ways. Some therapists are reluctant to hear the emotional message in what their clients are saying or to help clients explore their participation in interpersonal conflicts because they don't want to make others feel bad. These therapists try to make clients feel better by reassuring them about their insecurities and by emphasizing only their strengths and successes. However, stepping back from the accurate reflection of clients' concerns or distress in these ways is usually ineffective. Although well intended, it keeps things on the surface and prevents clients from being able to look at what's really wrong. When this happens, clients lose the opportunity to sort out what's really wrong, explore what they want to do about it, and develop the self-efficacy that comes from being able to address and resolve problems.

Still other therapists may avoid an internal focus in order to please clients and maintain their approval. These therapists fear that clients will become angry as the therapist invites them to explore the difficult feelings or choices they have been avoiding. Therapists are also concerned at times that they may not be able to remain present and responsive to the client when strong emotions are expressed. For all of these reasons, therapists often join with clients in looking away from the internal aspect or the client's participation in the problem. Therapists collude with the client in externalizing problems when they *repeatedly*

- give advice and tell the client what to do or how to respond to others,
- interpret or explain what something mean,
- reassure clients that their problems will go away or are not something to be concerned about, or
- disclose what the therapist has done to cope with a similar person or problem.

As we would expect from client response specificity, each of these responses will be effective at times. However, if they come to characterize the ongoing course of treatment, it will hold clients back from learning about themselves and, ultimately, from being able to resolve their own problems. Thus, throughout the course of treatment, *therapists are repeatedly trying to help clients become more aware of their own internal reactions and interpersonal responses in problem situations.* Cognitive behaviorists call these "self-monitoring techniques." As we will see, however, this can be challenging for therapist and client alike. As clients focus inward, unwanted feelings, faulty beliefs about themselves, and expectations of hurtful responses from the therapist and others will need to be dealt with.

Clients' Reluctance   Why do clients tend to avoid an internal focus? As we have seen, a shift away from their focus on others can make some clients feel that the therapist does not really understand them. Clients can become distressed because it seems as if the therapist is not grasping how difficult the other person really is behaving or is just not sympathetic to their concerns. Other clients fear that if they give up their attempts to change others, they will have to either accept the blame for the problem and be the "bad" one, or remain forever resigned to "defeat." Still other clients, not yet feeling that a working alliance is solidly in place, do not have the security they need to approach the difficult feelings or choices that looking inside entails. Clients also will be unable to enter certain issues if the therapist and client are reenacting some aspect of the client's conflict in their interpersonal process. When some type of reenactment is occurring, the therapist and client will usually need to change or resolve it before the client is able to progress further.

Cultural factors may also impede an internal focus. For example, Native American, Asian, and other clients from communally centered cultures may initially react negatively to looking within because it may sound self-centered. If clients perceive a difference in values and goals, they may question the therapist's credibility. Therapists can help by educating clients about the treatment process and, more important, by responding in a manner that is congruent with the clients' worldview.

> THERAPIST: Let's look at their role in this—and your role, too—and see if we can make more sense of what's going on for everyone here.

When clients are not ready to look at their own behavior, or shared participation in problems, this presents another opportunity to work with resistance. Here, instead of pressing the client to focus on his own behavior, the therapist can work with him to explore his reasons for not wanting to look within. How can the therapist best respond when clients resist an internal focus? One thing the therapist can do is to follow clients along for a while, while offering clients repeated invitations to say a little more about themselves. In most cases, clients will gradually accept the therapist's invitation and soon come to welcome the opportunity. However, if the client remains unable to respond, another option is to give the client permission to disclose or proceed at her own pace. For example:

> THERAPIST: It doesn't seem comfortable for you to share very much of yourself yet, but I do think it's important that you choose how much you share.

If the client's externalizing stance does not begin to change, however, the therapist can make a process comment that simply describes their interaction:

> THERAPIST: I've noticed that you talk very easily about your husband and your daughter, but you don't say very much about yourself. Are you aware that that happens?

OR

> THERAPIST: I think we've been missing each other the last two sessions. I keep asking what you were thinking about or trying to do in a particular situation, and you keep responding by telling me more about the other person. What do you see going on between us?

These process comments will help clients become aware of their external-izing style with the therapist, and learn that they probably respond in the same way to other people as well. For example, clients who cannot share themselves in a personal way with others will often be perceived as boring or aloof. In that case, part of the clients' presenting problem—for example, clients' sense of loneliness or lack of meaningful relationships with others—may also be en-acted in their relationship with the therapist. Although it is often discouraging for clients to realize that they are recapitulating their conflict in the therapy ses-sion, this reenactment provides the opportunity to begin resolving the problem by changing the distancing pattern within the therapeutic relationship. By pro-viding this kind of interpersonal feedback, the therapist helps clients become aware of how they interact with others and how others experience them. Then, the therapist offers a new and more effective way of relating in their relation-ship. Using *self-involving statements*, for example, the therapist can describe their current interaction and invite a more meaningful relationship:

THERAPIST: I feel that I am being held away from you when you talk about oth-ers rather than yourself. I feel like I am missing you, and I don't want that. I would like to learn more about you, or your reservations in talking about yourself. Can we work together on this?

When the therapist uses questions or process comments to focus clients in-ward, *clients often reveal a variety of new and important concerns that they have not talked about before.* For example:

- Well, I guess there really isn't very much about me that people would want to know.
- I'm not used to telling people what I'm thinking or feeling.
- Every time I try to get close to someone, I get hurt in the end.
- You wouldn't like me very much if you knew what I was really like.

In this way, the therapist's process comment has brought out important new concerns that can now be addressed in treatment. The therapeutic rela-tionship has intensified, and significant new information that is central to the client's presenting problems has been revealed. *This process of helping clients focus inward and explore their resistance to looking within will provide some of the most important material to be addressed in therapy.* These important new concerns are unlikely to be brought out in treatment unless the therapist uncovers them by working in the moment with different types of process com-ments and focusing clients inward. Thus, one of the most important reasons for adopting an internal focus is to bring out key issues that are contributing to the client's problems and make them accessible for treatment.

Placing the Locus of Change with Clients

The first component of the internal focus was to help clients look within and become more aware of their own responses. The second component is to help clients adopt an internal locus for change and help them to begin to act from within. In this way, we now examine how clients can gain a greater sense of self-efficacy in their lives by becoming more active agents in their own change process (Bandura, 1997). In this section, we first will see how the therapist can use the therapeutic relationship to foster the client's own initiative. Then, we will explore interventions that encourage the client's participation in and sense of *ownership* of the change process. Throughout, the therapist's goal is to foster clients' initiative by responding to their interests and concerns.

Fostering the Client's Initiative

In the course of treatment, a therapist has the opportunity to convey to clients the broader message that they are able to take charge of their lives, make their own decisions, and live their lives more as they choose. Effective therapists of every theoretical orientation nurture clients' own sense of personal dignity and self-efficacy. More important than verbally encouraging clients' initiative, therapists can give clients the experience of *acting* more effectively during the therapy session. Once clients experience greater agency in their relationship with the therapist, it is relatively easy to help clients generalize this greater sense of effectance to other relationships beyond the therapeutic setting (Bandura, 1997). Let's see how therapists can help clients become more active, initiating partners in the change process.

The first way to help clients feel more responsible for and capable of change is to *encourage them to shape their own agenda in treatment and talk about the issues they feel are most important or salient right now.*

THERAPIST: Where would you like to begin today?

CLIENT: I'm not sure—what do you think would be best?

THERAPIST: I'd like to join you in exploring whatever you think is most important to work on right now. What would that be?

CLIENT: I'm not sure.

THERAPIST: Let's sit a moment, let you take a breath and settle in to yourself—and then see what seems most alive for you right now.

CLIENT: I think I want to talk about my wife. We had the best weekend together we've had in a long time.

THERAPIST: I'm happy for you. What was different this weekend—were you doing anything differently?

CLIENT: I think I talked to her more like you talk to me. I asked about what she thought and what she wanted more, and I think she liked it. . . .

Once clients begin pursuing their own concerns in this way, the therapist can be an active participant who helps clients to explore their own concerns more fully, to understand their problems better, and to generate potential solutions and behavioral alternatives.

Why is it essential to encourage clients' own initiative and meet them at their "point of urgency," even in the time-limited modalities that increasingly dominate the therapeutic milieu? In one way or another, most clients have been unable to act on their own interests or pursue their own goals. In past relationships, these clients have not had significant others support their own interests or, simply put, help them do what they want. As a result, these clients will feel encouraged, but often anxious as well, if the therapist supports their own self-direction *and cares about what seems to matter most to them*. Regardless of treatment length or the therapist's theoretical orientation, therapy becomes a more intense and productive experience when the therapist can successfully engage clients in pursuing their own interests and exploring the issues that matter most to them. Clearly, this is where clients take ownership of the change process and invest more fully in the working alliance.

When the therapist can help clients explore and understand the material they choose to focus on, rather than merely direct them to the therapist's own agenda, change usually begins to occur and presenting problems improve. When clients provide the momentum and direction for therapy, they often experience a corrective relationship where, for the first time, they can have their own opinions, act more effectively, and be more in charge of their own lives. Once a client has begun to initiate, the therapist actively participates by providing the types of responses that this particular client utilizes most productively. Commonly, this will include interventions from a wide variety of theoretical orientations— contributing information, explanations, or interpretations; constructing behavioral alternatives; suggesting possibilities about what problematic others may be doing or intending; providing interpersonal feedback about how others may be perceiving or reacting to the client; and so forth. Whatever interventions or techniques the therapist uses, however, they will be far more effective when given in response to the clients' own interests or concerns, rather than the agenda of the therapist, employer, or spouse. This process dimension is one of the most important characteristics of the therapeutic relationship and we will explore it further in the chapters ahead.

Avoiding a Hierarchical Relationship   We have already noted that clients may resist the therapist's attempts to make them active participants in treatment. Why? Frequently, clients will be conflicted about expressing their own wishes, acting on their own initiative, and achieving or enjoying success. Many clients have not been supported in having their own voice or becoming stronger, and feel guilty or anxious if they speak on their own behalf, act more boldly or demonstrate more self-direction, or achieve success. As a result, they continually elicit advice and direction from the therapist and others. Therapy will not be productive if this helper–helpee mode comes to characterize the therapeutic process. This mode shifts responsibility away from the client and onto the therapist and creates the hierarchical relationship discussed in Chapter 2. Clients will not be able to feel greater self-efficacy and adopt a stronger stance in their lives as long as this dependency is fostered—their *compliance* is being reenacted in the therapeutic relationship rather than being challenged or questioned.

Too often, therapists unwittingly comply with the clients' subtle or overt requests to tell them what to talk about in therapy and what to do in their lives.

It is flattering to think that we know what is best and can tell others what to do; it appeals to the narcissism in every therapist. As we would expect on the basis of client response specificity, there will be circumstances in which directives, advice, opinions, and so forth, are very helpful for a client. For example, they will often be helpful for clients who grew up with caretakers who were not able or were not interested in helping them solve their problems—perhaps because they were involved with drugs, depressed, or otherwise self-absorbed. However, when directive or prescriptive responses *characterize* the ongoing process of therapy, clients will usually remain dependent on helpers to manage their lives. In addition, researchers find that many clients tend to be resistant and uncooperative when therapists are too directive, so therapists want to assess carefully how this particular client is responding to the therapist's advice or suggestions (Bischoff & Tracey, 1995; Mahalik, 1994). Because this problematic process occurs subtly (but frequently), we need to examine further this dependency-fostering approach to therapy.

Supporting Clients' Own Autonomy and Initiative   Effective therapy of any treatment length should foster the client's self-efficacy—we want the client to develop a greater sense of agency through the treatment process (Cervone, 2000; Galassi & Bruch, 1992). *Therapists cannot just talk with clients about choice, responsibility, and personal power, however; they want to cocreate a relationship in which clients are behaving in stronger ways with the therapist.* When clients can first do this with the therapist, the therapist can readily help them transfer this stronger way of acting to the rest of their lives. Thus, the therapist first gives clients permission to follow their own interests and actively encourages them to introduce the material that seems most relevant to them. As detailed in Chapter 2, the therapist then tries to capture the emotional message or identify the core meaning in their narrative and encourages clients to elaborate it further. In response to clients' increasingly specific exploration, the therapist actively helps clients clarify the repetitive themes that bring the central problems into focus—and work together to understand this theme or issue. They explore how aspects of this theme or problem may be evident in the way they interact and, together, begin to generate more effective ways to respond and change this pattern with others in their lives. When this mutual interchange continues throughout several sessions, most clients become committed to the treatment process and will report that meaningful changes with others are beginning to occur. Facilitating this growth process may be the primary challenge and satisfaction of being a therapist. Let's see how this sequence might actually sound with a client:

> CLIENT: I need more direction from you. What should I do here? You're the expert.
>
> THERAPIST: I have a couple of ideas to share with you, but I think it would work best if I heard yours first.
>
> CLIENT: (*impatiently*) If I knew what to do, I'd just do it and wouldn't ask you or be coming here to see you!

THERAPIST: All right, let's swap ideas. You tell me what you think is going on; then I'll tell you what I see occurring; and let's see what we can put together.

CLIENT: Like I just said, I don't know—I don't really have any ideas.

THERAPIST: Let's wait for just a minute and see if anything comes to you. If not, I'll be happy to go first and start us off.

CLIENT: (*20-second pause*) Well, maybe I'm afraid of being alone or something like that.

THERAPIST: That's interesting and fits with what I'm thinking about. Tell me more about what it means to be "alone"; it sounds important.

CLIENT: I think I've always been worried about that.

THERAPIST: From other things you've said, it makes me wonder if your parents cut off from you emotionally whenever you disappointed them by doing what you wanted rather than what they expected. I wonder if that created the feeling of being all alone, even though others were physically present—which would have been very confusing to a child.

CLIENT: Yeah, I think that's right, and the worst thing about it was that it seemed like it was all my fault. They were going away because I let them down. What do you think?

THERAPIST: I think you have very good ideas, but something often seems to hold you back from expressing them in the strong way you have just been doing with me. I have noticed that happening in here with me sometimes, and I hear it in your relationships with others as well. I wonder if there is some connection between holding yourself back in this way and being afraid of being left. What comes to mind as I suggest this?

This collaborative interaction, in which the therapist and client each build on what the other has just produced, is an *independence-fostering* approach to treatment. Clients have the experience of sharing responsibility for the course of treatment, as the therapist actively encourages them to define and address their own concerns. However, as therapists support clients in achieving this more active stance, they are simultaneously contributing their own ideas and suggestions in a way that creates a working partnership. *The therapist in the dialogue above is highly engaging and active—but not very directive.* She is following the client's lead, but she is not being nondirective, either. The interpersonal process we are striving to achieve is this collaborative effort—which has not been developed very well in the counseling literature.

Taking this point further, a common misconception is that longer-term, dynamic, or relationship-based therapies are dependency fostering, whereas short-term, problem-solving, or strictly behavioral approaches are not. Actually, whether the therapy fosters dependence or independence is determined by the therapeutic process and not by the length of treatment or the theoretical orientation of the clinician. In short- or longer-term therapy, the client's dependency is inappropriately fostered when the therapist *repeatedly* directs the course of therapy, gives advice, and prescribes solutions for the client. Let's examine this collaborative alliance further—it offers an effective middle ground between the less productive polarities of directive and nondirective control.

Shared Control in the Therapist–Client Relationship   As we have begun to see, treatment will be most successful when the therapist and client share control over the agenda and direction of therapy. In most cases, it is overly controlling for the therapist to play the predominant role in structuring and directing the course of treatment, and it is ineffective to nondirectively abandon clients to their own confusion and conflicts. In a more productive relationship, therapists will encourage the client to take the lead but will actively contribute their own understanding and guidance about the material that the client has produced. It will be therapeutic for many clients to experience a relationship in which both participants share responsibility rather than a relationship in which controls are held by one party or the two parties compete. This collaboration in the therapeutic alliance is the key to greater self-efficacy and provides clients with the support to act in new and stronger ways with others in their lives. As discussed previously, the therapist is encouraging the client to take the lead and initiate whatever she would find most useful to discuss.

> THERAPIST: I would like to begin each session by having you bring up what you want to talk about. I would like to join you in working on what seems most important to you. I will be sharing my own ideas and suggestions as well, but I think this is the best way to begin a good partnership.

Some therapists will be frustrated by this approach and direct clients toward taking action and finding solutions, especially if they are feeling pressed to fix the problem quickly because of the limited number of sessions available. Although a highly prescriptive stance will be helpful at times, such as when clients are in crisis, it imposes several important limitations. First, as we have seen, a directive approach prevents the therapist and client from discovering key issues that were not evident in the client's presenting problem. Second, the goal of therapy is not merely to fix the client's presenting problem, but to do so in a way that leaves clients with a greater sense of their own personal resources and ability to live their own lives meaningfully. Working collaboratively, the client gains an increasing sense of self-efficacy by participating in treatment progress, which directive approaches do not provide. That is, the client shares ownership of the success and change that is achieved—so different than feeling that the smarter or stronger therapist fixed them.

Responding to these arguments, the solution-oriented or directive therapist might counter, "What if you follow the client's lead and it takes you nowhere?" Certainly, a nondirective approach requires an unrealistic amount of time from the therapist and often results in a disorganized therapy that is lacking in focus. Moreover, a purely nondirective approach is likely to reenact problematic relational patterns for clients who grew up in a permissive home, just as a purely directive approach may reenact such patterns for clients who grew up in an authoritarian home. As we have seen, however, there is an effective middle ground of shared therapist–client control, although it has not been delineated well in the counseling literature. For example, the therapist below does not wait nondirectively for the client to bring up more relevant material or directively lead the client to a new topic. Instead, the therapist makes a process comment about their current interaction and invites the client to join in reshaping their interaction:

> THERAPIST: This doesn't seem to be taking us anywhere right now. Is there a better way to use our time—or another way to talk about this—that could get us closer to what's really wrong or matters most to you?

Therapists need treatment plans and intervention strategies with short- and longer-term goals. However, these are most effective in producing enduring or sustainable change when they develop out of a collaborative interaction; the client shares responsibility for successes and disappointments and feels ownership of the treatment process. To implement this interpersonal process, *the therapist needs to be able to tolerate ambiguity and refrain from subtly controlling the interaction.* Instead, the therapist is trying to create opportunities for clients to voice their own concerns and to act on their own initiative. The therapist's task, then, is to be able to decenter; enter into the client's subjective worldview; and empathically highlight the core meaning, central feeling, or relational pattern that this issue seems to hold for the client.

This approach requires an active therapist who is neither directive nor nondirective *but has the flexibility to tolerate ambiguity and share control* (that is, not know what the client is going to say or do next but be open to whatever he presents). In this approach, therapists respond actively by

- communicating their understanding and affirming the client's experience;
- joining the client in exploring further the key meaning in what the client has just produced;
- using self-involving comments or their own experience of the client to provide interpersonal feedback about how the client affects the therapist— and may be contributing to problems with others;
- affirming new behavior that steps out of old patterns and demonstrates change with the therapist or with others; and
- providing a focus for treatment by highlighting the repetitive interaction sequences, painful feelings, and faulty beliefs that repeat throughout the client's narratives.

Furthermore, the therapist is using process comments to address treatment impediments—such as resistance or an externalizing focus—and to prevent unwanted reenactments between the therapist and client (for example, the client is silent about her faulty belief that the therapist is impatient, bored, or burdened by her—as others often have been). Therapists do this by working in the here-and-now and asking about what may be going on between them in their relationship or current interaction. For example:

> THERAPIST: What's it like to be talking with me about this?

> CLIENT: Well, you're probably feeling impatient with me—like everybody else does—because I'm still stuck on this.

In this way, the therapist intervenes by taking the client's problems out of abstract discussions about others "then and there." Instead, *the therapist is looking for opportunities to create immediacy by linking the client's problems with others to what is occurring between them in the way they are interacting together right now.* For example:

> THERAPIST: No, I'm not feeling "impatient" with you at all. I can see this is a complicated problem for you, and I appreciate how hard you are trying to do something about it.

> CLIENT: Really, you're not frustrated with me. You know, that makes me almost want to cry—it feels like I've been a disappointment to just about everybody in my life. . . .

Working in the "here and now" with clients is a big new step for most therapists. It brings therapists and clients out of the mode of everyday social interaction, and it quickly takes us to the heart of what's wrong. Significant feelings come up and key issues that the therapist didn't know even mattered to the client now take center stage. And, just as important, the relationship becomes more intense and begins to matter more to both the therapist and the client.

As we will see in Chapter 6, an authoritative parenting style that encompasses both parental warmth and firm limits reflects a more effective middle ground between the better known and more widely adopted authoritarian and permissive parenting styles. In parallel, therapists will find that shared control offers most clients a more productive alternative than either directive or nondirective approaches. New therapists need role models of how therapists and clients can work together in this collaborative way, so let's examine interventions that foster a strong working alliance and illustrate this middle ground of shared control.

Interventions That Place Clients at the Fulcrum of Change

As we have seen, the therapist's goal in the interpersonal process approach is to encourage the client's lead while still participating actively in shaping the course of treatment. The key is for the therapist to intervene without taking the impetus away from the client. The following examples illustrate effective and ineffective ways to do this.

Ineffective Interventions   Suppose that the client is filling the therapy session with seemingly irrelevant storytelling. The therapist cannot find a common theme to any of the client's narratives or understand the personal meaning that these vignettes hold for the client. It seems as if nothing significant is occurring. At this point, it is easy for the therapist to stop the client and direct her toward a specific topic that the therapist thinks would be more fruitful. This refocusing will certainly be effective at times, but therapists will succeed more often if they revitalize the therapeutic interaction without shifting the impetus away from the client and onto the therapist. In the following dialogue, the onus for therapy comes to rest with the therapist, and the client loses an internal focus for change:

THERAPIST: I'm wondering where this is taking us. Have I lost the focus here?

CLIENT: I'm not sure where I'm going with this, either. What do you think I should talk about?

THERAPIST: You've had trouble asserting yourself in the past, and I think we need to look more closely at that. Last week, you said you wanted to ask your boss for three weeks of vacation instead of two. How are you going to handle that confrontation?

CLIENT: I'm not sure. What do you think I should say?

THERAPIST: To begin with, you need to arrange a face-to-face meeting with him. It is important that only the two of you are present, so that you can have his full attention and there is less threat for either of you to lose face. Then use the "I" statements we have practiced to directly state what you want.

CLIENT: Sounds good, but what would you say to him first?

Assessing the client's responses in this dialogue, it is clear that this intervention is not being productive, even though the therapist has moved the client to a more salient topic and provided useful information about effective confrontations. The fulcrum of therapeutic movement has tipped from the client to the therapist, and a hierarchical teacher–student process has been established. Most clients will not be able to utilize the therapist's useful information until they pick up the momentum and begin to actively participate again. Furthermore, because the client remains in a passive role as the therapist continues to inform, the client does not gain the increasing sense of self-efficacy that comes from participating in success or progress in treatment. A more productive intervention might be to ask the client whether he feels another topic might be more relevant, and to wait until the client becomes actively involved again before offering additional information or further shaping the direction of treatment. However, if the client does not become actively engaged again, therapists do not want to just sit and wait nondirectively. Instead, they can intervene by simply observing this with a process comment. For example,

THERAPIST: I'm having difficulty following you right now. Wherever you go, it just seems to trail off and get lost for me. What do you see happening here between us?

Effective Interventions   Let's look at how the therapist can refocus the client toward more productive material—but in a way that keeps the momentum for treatment with the client. Again, process comments that make the current interaction between therapist and client an overt topic for discussion are often effective in these circumstances.

THERAPIST: I don't have the feeling that what you're talking about is really very important to you. Am I missing the point here, or is this not really so important? Help me out.

CLIENT: Yeah, you're right, I'm not sure where I'm going with this, either. What would you like me to talk about?

THERAPIST: I think that we should try to identify what would be most important to you and talk about that. What might that be right now?

CLIENT: I'm not really sure.

THERAPIST: Let's just sit together quietly for a moment, then, and see what comes to you.

CLIENT: (*pauses*) Do you think it would be a sin if I was bisexual, or if I had sexual feelings toward men sometimes?

New therapists often feel uncomfortable with silences and may find themselves filling them. If therapists can refrain from doing so, however, these are opportune moments for clients to establish their own agenda or introduce new material that is far more meaningful—as in the dialogue here. This type of process comment invites the client to approach more substantial material but, in line with our goal, without taking the impetus away from the client. The therapist has directly intervened by sharing her observation and asking about the client's perceptions of their current interaction, but she has still left the client an active participant in revitalizing the discussion. Routinely, clients produce far more significant material when they are invited to lead in this way rather than to follow. As always, try it out with your clients, assess their reactions, and see what works best for you.

Discovering More about the Problem   Important new information, central to understanding and changing the client's problems, is often identified when the therapist intervenes without taking the locus for change away from the client. Imagine the following situation: A 19-year-old client in a college counseling center had been telling his therapist about his attempts in early adolescence to observe his mother undressing. The client had been detailing his voyeuristic efforts at great lengths, but the therapist did not feel that this was genuinely of much concern to the client. Unless the therapist could find the relevant meaning for the client, he wanted to move on to other, more salient issues in the client's current life. In the following dialogue, the therapist uses a process comment to refocus the client in a way that places the locus for change more fully with the client and, in the process, reveals a new and more salient treatment issue.

THERAPIST: Does this feel like an important topic that you want to discuss with me? As I listen to you, I don't get the feeling that you are really very interested in what you are telling me. Is that the case, or am I not understanding the meaning this holds for you?

CLIENT: I thought therapists were interested in this oedipal stuff. I figured you would want to hear about it.

THERAPIST: I'm struck by the fact that you are telling me what you think I want to hear, rather than working on what is most important to you. Maybe that's something we should talk about. I wonder if you find yourself doing this in other relationships as well—trying to sense other people's needs or responding to unspoken demands at the expense of expressing your own interests or concerns?

Here again, the therapist has used a process comment to revitalize treatment without taking the impetus away from the client. In this vignette, the therapist has also used their current interaction to discover a more important aspect of the client's problems—compliance issues. Because this new issue arose from their joint interaction and not from the therapist's own agenda, most clients will be highly motivated to explore it. Thus, the therapist has effectively focused the client internally, identified a key concern to be explored, and helped the client become a more active participant in shaping the course of treatment.

As we see here, adopting an internal focus reveals new aspects of their problems and clarifies related issues that are shaping and contributing to their problems. As the client begins to work with these salient new issues, and with the themes and patterns that have been identified in the way the therapist and client interact together, important progress occurs on the client's presenting problems. How does this occur? The therapist makes links between understanding and changing what is going on between them in the therapeutic relationship, and making the same types of interpersonal or behavioral changes with others in their lives. If changes occur with the therapist but do not generalize to the client's everyday life, treatment has failed, of course. To illustrate this transfer, let's continue where we left off with the therapist and client in the previous dialogue. As we will see, the therapist is making connections between what is going on between them and other relationships in the client's life (such as his girlfriend) where the same maladaptive relational patterns are occurring.

CLIENT: (*sarcastically*) Well, duh! Of course—isn't that what everyone has to do with teachers and parents and therapists like you? Figure out what you want and go along with it?

THERAPIST: Well, it's what some people have learned they have to do with some teachers and some parents and some therapists. But it's not what you have to do with everyone, and it's not what you have to do with me. I like people who have their own mind.

CLIENT: (*sarcastically again*) Do you now.

THERAPIST: Yes, actually, I do. So, tell me, who in your life can you be honest with and talk about what you really care about—and who do you have to comply with and talk their talk?

CLIENT: Well, I guess I can be the way I want to be with my guitar teacher—he's cool.

THERAPIST: I'm glad you have him—you need good friendships in your life. Who's the most important person in your life that you have to go along with—like you thought you had to do with me?

CLIENT: Probably my mom.

THERAPIST: What happens if you to talk to her about what's real or what matters for you?

CLIENT: She's mad—like I embarrass her by being who I am.

THERAPIST: Ouch! I'm sorry for both of you. Have you ever tried talking to her about this? You just said it so clearly.

CLIENT: (*with disdain*) Of course not—do you believe in Santa Claus and the tooth fairy, too? She's not going to change!

THERAPIST: As you talk with me like that, especially in that contemptuous tone of voice, I feel insulted—as if you think I'm an idiot or a fool. Do you and your mom talk to each other like that?

CLIENT: (*pauses*) Yeah, probably. . . .

THERAPIST: I can see how things aren't going very well for either of you. I don't know if your mother can change with you or not, but I think it's worth testing the waters to see. Maybe she could respond better if you began talking to her differently—in a respectful and straight-forward way about what's really going on between you, rather than in the disparaging and sarcastic way you were just talking to me. That improves her chances of being able to respond better. And, even if she can't, I think the rest of your life would be better if you could change this way of responding with others. Want to work together on that?

CLIENT: Yeah, maybe, but how can things get better for me if she still needs me to be her way all the time?

THERAPIST: I have several ideas about what you could do differently with her—whether she changes or not—that could leave you feeling better about yourself.

CLIENT: How would things be better for me if it turns out to be hopeless with my mom? I think she'll just say what she always does—everything's my fault; I'm always doing it wrong.

THERAPIST: She might, but if you tried talking with her about this in a more respectful way, maybe you'd stop believing that all of the problem is always your fault. That would be a real change that could help you stop feeling so bad about yourself. And that might help you stop feeling so depressed and feeling like just getting stoned every day. Another thing is you wouldn't have to go along with everyone else—you know, talk about what you think they want—and then insult them because you're so angry about going along like that. So, other relationships in your life might go better too, even if it doesn't change with your mom.

CLIENT: Like with my girlfriend?

THERAPIST: Yes, exactly. Tell me how this goes with your girlfriend. . . .

In summary, the therapist in this dialogue is providing interpersonal feedback about the client's insulting manner and challenging the client to take more responsibility for causing and solving his problem. Carkhuff (1987) calls this "personalizing," and encourages therapists to respond in ways that help clients realize that they usually have some shared responsibility for creating and maintaining their problems. Therapists who avoid this challenge of personal responsibility generally lose credibility with the client and the working alliance is diminished. The therapeutic dialogue may be supportive and understanding, but it doesn't go very far—it isn't enough to lead to change for a client such as this. Thus, therapists are encouraged to respond in this personalizing way and say, for example, "You're upset because they took advantage of you, and because you didn't stand up for yourself when it happened."

Enlist Clients in Solving Their Own Problems

In this section, we bring the internal focus one step further into the therapeutic relationship. Once clients have begun to look within and share their inner world, the therapist can further clients' sense of efficacy by engaging them in understanding and resolving their own problems. In other words, intervening in ways that give clients more ownership of the change process so they can learn from their mistakes and gain from their successes. A common misconception about therapy is that the therapist takes responsibility for figuring out what is wrong and prescribes what clients should do. New therapists often hold unrealistic expectations that they must be experts who possess insightful solutions to their clients' problems; they then suffer under these performance demands with painful feelings of inadequacy. Here again, this misconception places the impetus for change in the therapist's lap and gives clients the role of passive recipients, waiting to be cured. Rather than adopting this hierarchical model, a more effective approach is to extend the collaborative alliance into the problem-solving phase of therapy as well.

Therapists will be more effective when they elicit clients' active involvement in resolving their own problems. For example:

- What have you tried in the past that has worked for you?
- Looking back, what have others done to make things better — or worse — for you when you were in this situation before?
- What could you and I do together to try and help with this?

In contrast, therapists too often establish expectations that set the stage for the therapist to tell clients how to lead their lives — by explaining what it all means, giving advice, and prescribing solutions. Instead, when clients participate in a collaborative working alliance, it is empowering and gives clients the opportunity to become more capable of managing their own independent lives. Thus, rather than just providing answers, therapists can intervene in ways that help clients learn how to think about and explore their own problems, and generate their own solutions and alternatives. Real change has occurred when therapy has not only resolved clients' presenting problems but has fostered their sense of self-efficacy in this way.

To illustrate this, let's look at two different ways of responding to a client's dream. Although the interpretation of the dream remains the same, the therapeutic approaches differ and enact very different interpersonal processes with the client.

Reenacting Clients' Conflicts

Anna, a 24-year-old client, lives at home with her embittered and chronically embattled parents. For several years, Anna has been struggling with the developmental transition of emancipation from her family of origin. She cannot establish her own adult life. Anna has few friendships, dates little, and has no serious educational or career involvements. Anna's "martyr-ish" mother and alcoholic father bicker constantly, and Anna feels that it is her responsibility to stay home and help her mother cope with her failed marriage. Anna entered therapy complaining of depression.

After several sessions, Anna recounted to her therapist a dream that was of great importance to her. As the dream began, Anna was riding a beautiful horse across an open savannah. She felt as one with this graceful animal as they glided effortlessly across the broad grasslands. They sped toward a distant mountain, past sunlit rivers, birds in flight, and herds of grazing elk. Anna felt strong and free as she urged the tireless animal onward.

The distant mountains held the promise of new life amid green meadows and tall trees. Anna felt their promise quicken inside her as she urged the horse onward. But as the mountains drew near, Anna and the horse began to slow. The horse's legs became her own and grew heavier with each step. Anna desperately tried to will them on, but their footing became unsure and they began to stumble. At that moment, she was surrounded on all sides by menacing riders, ready to overtake and capture her. As she awoke, Anna choked back a scream.

"What do you think my dream means?" Anna asked.

Her bright, concerned young therapist offered a lengthy and insightful explanation. The interpretation focused on Anna's guilt over leaving home. The therapist suggested that, if Anna went on with her life and pursued her own interests, she would feel powerful and alive—just as she had in the dream. But before Anna could reach her goal and experience the satisfactions of having her own adult life, she would have to free herself from the binding ties of responsibility that she felt for her mother and her parents' marriage. Her loyalty to her mother threatened to entrap her in guilt and prevent her from being able to live her own adult life.

ANNA: (*enthusiastically*) Yes, you're right, I do feel like I'm doing something wrong whenever I leave my mother and do what I want. Is that why I dreamed that?

THERAPIST: We've been talking about whether you are going to move into an apartment next month. I think the dream reflects your guilt over taking this big step on your own.

ANNA: I want to move out, but I can't leave my mother with my father. She says she will divorce him if I move out, and it'll be my fault. What should I do?

THERAPIST: I can't make that decision for you. It's important for you to be responsible for your own decisions.

ANNA: But I don't know what to do, and you always know what's best. You're so much smarter than me. I thought about that dream a lot, and I didn't know that's what it meant.

Throughout the rest of the session, Anna continued to plead for advice and expressed her discouragement about being able to resolve her own problems. The well-intended therapist did not want her to become dependent on therapy and kept refusing to tell her what to do—giving her a lengthy explanation about autonomy, independence, and the need for Anna to find her own solutions. At the end of the hour, Anna felt agitated and depressed, and the therapist was still trying to explain the need for Anna to make her own decisions.

Although the therapist was astute in linking Anna's dream to her separation guilt and the current manifestation of her problem about leaving home, this was an unproductive session. How did their therapeutic process go awry?

In this session, Anna experienced a relational pattern with her therapist similar to the pattern she was struggling with at home. In her family, Anna was indeed trained to be dependent and believe that she didn't have the wherewithal to live successfully on her own. Furthermore, she was made to feel guilty whenever she did act independently or on her own behalf. In the beginning of their session when the therapist interpreted the dream so accurately, the therapist acted as the knowing parent who gave all the necessary answers to the needing child. This interaction behaviorally communicated the message that Anna will be dependent on the therapist's superior understanding. On another level, however, this message was contradicted by the therapist's verbal message about independence. This mixed message from the well-intentioned therapist (being told what the dream meant and then being told to act independently) immobilized Anna. The therapist's direct interpretation could have worked well for some clients, but (recalling the concept of client response specificity) such an approach was problematic for Anna.

Anna reacted so strongly because this interaction cut right to the quick of her problem. On the one hand, she needed and greatly wanted permission from the therapist to become more independent but, on the other, expected that the therapist—deep down—really needed her to remain dependent—just as her mother did. When the problematic relational pattern of having to remain dependent on the authority was reenacted with her therapist, Anna's faulty belief that she was incapable of making her own decisions and having her own life was confirmed. Anna remained depressed until the therapist successfully reestablished a more collaborative interpersonal process with her. With the help of his supervisor, the therapist did this in the next session by soliciting Anna's ideas about something they were discussing. The therapist expressed his genuine pleasure in watching her be so insightful and, following this, shared some related ideas of his own. This time, Anna readily picked up on his ideas and used them to further her own thinking—it was a partnership.

Providing a Corrective Emotional Experience

Recalling "ruptures and repair," Anna and her therapist soon corrected their reenactment and restored their working alliance, and treatment successfully progressed. However, let's examine a more effective way of responding to Anna's question, "What do you think my dream means?" Interventions that invite a collaborative partnership—engage clients in exploring issues with the therapist—will usually be more effective than explanations or guidance provided by the therapist, no matter how accurate. Therapists can choose to respond in many different ways and can engage clients in exploring their own dreams (and any other material they present) as follows:

- It sounds like a very important dream to me, too. Let's work on it together. Where should we begin?

- What was the primary feeling you were left with from the dream? Can you connect that feeling to anything going on in your life right now?
- What was the most important image in the dream for you? What does that image suggest to you?
- Let's exchange ideas. I'll give you a possibility I'm thinking about, and you share one with me. It'll be interesting to see what we can come up with together. Who should go first?

Each of these varying responses encourages the client to participate actively with the therapist. Ultimately, the therapist may want to give the same interpretation of the dream. Even though the therapist provides an explanation, however, the process will be significantly different and far more enlivening for clients if it evolves out of their joint efforts and is integrated into their continuing collaboration. This approach gives clients a relationship in which they are not one down or dependent on the therapist but are encouraged to exercise their own abilities. When this occurs, clients feel more ownership of the therapeutic process, their motivation to work in treatment increases, and their self-efficacy is enhanced. Facilitating such self-efficacy provides a corrective emotional experience for clients like Anna, who get the permission they need to grow out of their dependence and act capably in their relationship with the therapist—something they may not have experienced in the past. To illustrate this more concretely, let's return to the same dialogue and see how quickly the therapist can make it more collaborative and go better:

ANNA: (*enthusiastically*) Yes, you're right, I do feel like I'm doing something wrong whenever I leave my mother or do what I want. Is that why I dreamt that?

THERAPIST: We've been talking about whether you are going to move into an apartment next month. I'm wondering if the dream has something to do with you feeling guilty about taking this big step on your own. What do you think?

ANNA: I want to move out, but I can't leave my mother with my father. She says she will divorce him if I move out, and it'll be my fault. What should I do?

THERAPIST: Tell me what you think as I slowly reflect back what you just said: It's your fault if your mother leaves your father.

ANNA: (*pauses*) It sounds crazy . . . it's making me crazy.

THERAPIST: That's right—it makes you feel crazy, and it's not true. Children are not responsible for their parent's marriages, and it's not fair when they are pulled into marital conflict like this—it keeps you from having your own life.

ANNA: I feel better when you say that . . . but what should I do?

THERAPIST: I'm wondering if you could talk to your mother about this. What would it be like for you to tell her you love her, you're sorry her marriage has been so disappointing, but you don't want to feel like you are responsible for it anymore?

ANNA: Wow, that would change everything, but I could never say anything like that.

THERAPIST: What's the threat or danger if you told your mother you didn't want to feel responsible for her marriage anymore? What's going to go wrong?

ANNA: It would hurt her so much—I'm all she has.

THERAPIST: This is such a hard problem. I can see that you really don't want to hurt her, but you don't want to give up your own life anymore either. It's been so hard for you to live with this dilemma.

ANNA: You understand this. You really do.

THERAPIST: Right now, I think we're figuring it out together and both appreciating the dilemma you've been stuck in for so long.

ANNA: You're right—this isn't fair. I do want to speak up, but I just can't say what you did.

THERAPIST: OK, that's too much. So let's find a smaller step that you could take—what could you say to her now to begin setting some limits?

ANNA: I don't want her to complain to me about my dad anymore. I don't want to hear what's wrong with him. I don't want to have to take sides.

THERAPIST: Yes, that's great. Let's start there and help you say that. Should we practice how the conversation might go? You know, anticipate the trouble spots. I could role-play you, and you could be your mother and say the things she will probably say when you talk to her about this. . . .

Tracking Clients' Anxiety

"Tracking clients' anxiety" is a clinical skill that will help therapists determine when and where to focus clients inward (Sullivan, 1970). Although the subjective experience of anxiety is uncomfortable and may even be painful, it is still the therapists' ally. Anxiety serves as a signpost that points you to the heart of the problem—a signal that the real threat or danger is at hand. That is, tracking what makes this client anxious will help therapists identify the client's key concern or generic conflict. Therapists do not fully understand clients' problems—or what needs to occur in counseling to resolve them—until they understand what makes each client anxious. Why do certain situations or interactions make a particular client anxious, and how has the client learned to cope with this threat or insecurity? Therapists are encouraged to begin formulating working hypotheses to these questions in order to clarify treatment plans and interventions. As we will see, focusing clients inward on their anxieties—exploring together what activates or triggers them, will lead the therapist and the client to the clients' key concerns.

Therapists can use a four-step sequence to track clients' anxiety and better understand, at core, what's really wrong for them. In this sequence, the therapist

1. identifies manifest and covert signs of client anxiety;
2. approaches signs of client anxiety;
3. notes the topic presently under discussion and considers the interpersonal process currently transpiring that may have precipitated the client's anxiety; and
4. focuses the client inward to explore the meaning of the client's anxiety.

Let's look at these four steps more closely.

Observing Signs and Symptoms of Anxiety

Clients will become anxious in the session when the issues they are discussing cut close to home and touch on their key concerns. Clients develop certain interpersonal coping strategies, which they employ over and over again in different situations, to ward off this anxiety that has been triggered. The therapist is trying to identify what makes this particular client anxious (for example, needing to ask for help and expecting the other to resent or feel burdened by their need; not knowing how to fix or do something and expecting the other to be impatient or disdainful; and so forth). When therapists can do this, they are better prepared to recognize the unifying themes or repetitive patterns that have triggered the client's anxiety. Thus, throughout treatment, therapists are continually asking themselves, "What makes this client anxious? When does the client feel threatened or insecure? Where does safety lie, and where does danger lie?"

To identify the patterns and themes that are causing problems in clients' lives, the therapist tracks their anxiety and remains alert for signs indicating that something they are experiencing is making them anxious right now. Clients may express their anxiety in a thousand different ways—nervous laughter, nail biting, hand gesticulation, agitated movement, stuttering, hair pulling, speech blockage, and so on. These "anxiety equivalents" will be expressed in endlessly varied ways across clients, and therapists will have to learn how each particular client tends to express their anxiety. (This insight will be especially important when working with clients from cultures in which certain behaviors—such as gaze aversion—may be normative rather than a signal of anxiety.) When they become anxious, most clients utilize the same interpersonal coping strategies in a repetitive or characteristic way. Routinely, these strategies include pleasing others, becoming critical or controlling, acting helpless or confused, withdrawing from people or situations, and so forth. We will examine these interpersonal coping strategies closely in Chapter 7. For present purposes, it is enough to note that the first step in tracking the client's anxiety is to observe and mentally note when clients become anxious (for instance, when they have to assume a leadership role, when they succeed, when they have to disagree with or challenge others, and so forth).

Approach Anxiety Directly

In Chapter 3, I suggested that therapists ask about and work with the feelings and concerns that clients may have around asking for help, having a problem, or entering treatment (resistance). Further along in treatment, therapists similarly are encouraged to approach clients' anxiety about an issue that they are discussing or about what may be occurring with the therapist at that moment. Therapists can focus clients on their immediate experience in an open-ended way (for example, by asking, "What are you feeling, right now, as you talk about this with me?"). Therapists can also label the anxiety more directly. For example:

THERAPIST: Something seems to be making you uncomfortable right now. Any idea about what that may be?

CLIENT: No, I'm not sure.

THERAPIST: What's the first thing that comes to mind when you ask yourself, "I'm afraid that _____ ."

CLIENT: You're mad. I think I'm afraid that you're mad or upset with me somehow. . . .

By focusing clients inward on their anxiety as they are experiencing it, and helping them explore it further and discern or name it more precisely, the therapist is leading clients closer to the source of their problems. As a result, some clients may be ambivalent about the therapist's request to look within and explore their anxiety further. On the one hand, clients may welcome the opportunity to share their anxiety with the therapist—and the sadness, shame, or pain that often follows close behind. On the other hand, clients may also recognize that approaching their anxiety will bring them closer to faulty beliefs about themselves ("I'm too demanding or needy") and unwanted expectations of others ("They don't really want me") that, yet again, evoke the same unwanted feelings.

Thus, though at times a client may experience the therapist's focus on his anxiety as validating and reassuring, at other times it may intensify anxiety and even arouse resistance. As discussed in Chapter 3, therapists need to respond to the client's resistance when it emerges by exploring why it is threatening or does not feel safe to approach this anxiety-arousing topic. This approach will be far more effective than ignoring the client's reluctance as if nothing significant is occurring, or just pushing through it and trying to persuade the reluctant client to talk further about the difficult issue. Before the therapist and client can address the problem itself, they can explore the resistance and together learn why it doesn't feel safe for the client to discuss this further:

THERAPIST: OK, don't talk anymore about this with me but, instead, help me understand what's the threat or danger for you if you did talk to me about that. What's going to go wrong between us—or how are you going to end up getting hurt again—if you did?

Perhaps the client fears that the therapist will not like or respect the client anymore if he talks further about this; he will start crying and won't be able to stop, and so forth. As I have already emphasized, the therapist's intention is to respond in a way that honors the client's resistance. Recall from Chapter 3 that the therapist does this by (1) trying to understand the original, aversive experiences that led the client to behave in this particular way; (2) helping the client appreciate how this "resistant" or "defensive" response was once a necessary and adaptive coping strategy; and (3) providing the client with a different and more satisfying response than the client has come to expect, as in the following example:

THERAPIST: No, I don't find you "needy" or "demanding" for asking me if I would call you and touch base for a minute before you make this big presentation on Tuesday. It would be easy for me to take a few minutes to do that, and I'm honored that you are willing to risk sharing this with me.

Observing What Triggers the Anxiety

As we have seen, the therapist first observes when the client is anxious, and then helps the client focus inward to explore and try to understand what she is experiencing. At the same time, the therapist has a third task: to try to recognize (1) what issue was just being discussed (content) or (2) what type of interaction between the therapist and the client (process) may have precipitated the client's anxiety. If the therapist can identify what it was that made the client anxious, the therapist will be better able to identify the faulty belief, interpersonal pattern, or difficult feeling that may have generated the anxiety. If therapists can discern the precipitating or triggering event, it will go a long way toward helping them understand what's really wrong or troubling for this client.

The client will be able to discuss many different issues comfortably with the therapist. However, when the client becomes anxious, the therapist's aim is to try to identify the central issue or concern that has just precipitated the anxiety. What was the client just talking about? Was the topic about sex, death, intimacy, money, vulnerability, success, divorce, inadequacy? *The answer will highlight the key concerns that are most central to the client's problems.* As we know, therapists can help the client explore this more effectively if they have already formulated working hypotheses about the possible issues that may be generating the anxiety. Keeping process notes after each session, as suggested in Appendix A, will help therapists formulate their working hypotheses more effectively.

In addition, the client's anxiety is often a signal that the interpersonal process between the therapist and client is reenacting a developmental problem or maladaptive relational pattern for the client. For example, the therapist might observe client anxiety in the following circumstances:

- As the therapist was expressing confusion about what was going on in the session at that point, the client became anxious, perhaps because her alcoholic father would start demeaning her whenever he felt uncertain or inadequate.
- The client had just made an important insight and the therapist acknowledged his achievement. The client then became anxious, perhaps because he believed that he would always have to be so insightful and perform so well, as a parent had always expected of him.
- The therapist had just been supportive, and perhaps the client felt anxious because strings had always been attached to what she had been given in the past. Following her schemas, the client may have become concerned that the therapist would make her pay for his support in some unwanted way, just as a caregiver used to do.

The therapist can generate working hypotheses such as these about how the current therapist–client interaction may have triggered a developmental problem or unwittingly reenacted a maladaptive relational pattern for the client. These hypotheses can then be used to help therapists and clients explore what evoked their anxiety, understand what this threat or danger really means, and find more effective ways to respond in their lives.

Focusing Clients Inward to Explore Their Anxiety

As we have seen, the therapist is alert for signs of client anxiety and approaches it whenever it seems to be occurring. Simultaneously, the therapist tries to identify what precipitated the client's anxiety—especially what has just transpired between them—and begins to generate working hypotheses about the relational patterns, faulty beliefs, or difficult feelings that may have triggered the anxiety. Finally, the therapist also focuses the client inward—to explore more specifically what the threat or danger seems to be for the client right now. If the therapist can help clarify the thoughts and feelings that are associated with the anxiety, the client's vague discomfort will usually become more specific. That is, once the client's basic concern is highlighted, the client can address the anxiety-arousing problem more directly. (Client: "I guess I'm afraid they won't like me if I speak up and disagree—but maybe that's a consequence I can live with.") The following dialogue illustrates this process:

> THERAPIST: It seems like something's going on for you right now. I wonder what's happening?
>
> CLIENT: You know, this doesn't sound very nice to say, but I don't think my mother really wants me to change very much. I don't think she's a mean person or anything, but I do think I'm getting better in therapy, and she's not completely happy about that.
>
> THERAPIST: How so?
>
> CLIENT: Well, I haven't been depressed for a while now, and I've actually been feeling pretty good the last month or so. Maybe it's just coincidence, but it seems that, as I've gotten better, my mother has withdrawn and been harder to talk to. And I think this has gone on between us before.
>
> THERAPIST: You know, I think you're right. You have been doing very good work in here and getting better, and that may be kind of hard for your mother sometimes.
>
> CLIENT: (fidgets, begins picking at her nail, and then starts talking about another topic)
>
> THERAPIST: Maybe something just made you feel a bit uncomfortable?
>
> CLIENT: I don't know. How much time is left?
>
> THERAPIST: I'm wondering if something we're talking about, or something that might be going on between us, could be making you uncomfortable right now?
>
> CLIENT: (pauses) Well, I don't really know why I'm saying this, but maybe I'm afraid that you're going to go away or something?
>
> THERAPIST: That sure makes sense to me. We were just talking about how your mother seems to act hurt and withdraws when you are feeling stronger, and I just told you that I thought you were doing very good work in here.
>
> CLIENT: Well, I guess so, but that just makes things worse. You're not going to stay with me either if I get better—isn't that the point of this whole therapy thing anyway?

THERAPIST: That sounded important, but I didn't understand it as well as I wanted to. Can you say that again or say it differently?

CLIENT: If I get better, then we stop—just like with my mother. Then I won't have my mother, or you, or anybody. Don't you see? It's hopeless.

THERAPIST: You're saying this so clearly. I didn't understand at first, but you're right—the issue we are struggling with right now is like what's gone wrong before. To feel better and act stronger has meant that you have had to be alone. In the past, you have had to be sad and depressed and needy in order to be close or connected to others. I think you've been struggling with this dilemma most of your life, and, right now, it feels like it has to be the same old story again for you and me.

CLIENT: Yeah, it does.

THERAPIST: But I wonder if you and I can do this differently for once. Can we work together and try to find a way to make it come out better this time?

CLIENT: It's not working very well so far.

THERAPIST: Well, one difference I'm thinking about is that we're talking about it—naming it and sharing it—and that hasn't happened before.

CLIENT: What difference does that make?

THERAPIST: Maybe it's different because I can see what's happening for you and get how discouraging it's been for you to be undermined in this way so many times. And right now, I'm still feeling connected to you—still for you—as you are changing this and acting stronger in here with me.

CLIENT: Well, that's probably true, but these don't really seem like huge differences to me.

THERAPIST: OK. I see these differences as more significant than you do right now, but let's keep talking about this.

CLIENT: Why? What's the point?

THERAPIST: If you can stop holding yourself back in here with me, and find that everybody doesn't have to respond the same way your mother does, it will help you act stronger with others in your life. Like your boyfriend, and your professor, and others we've been talking about.

CLIENT: Do you really think that could happen?

THERAPIST: Yes, I do. How do you think your boyfriend will respond to you if you let yourself be as smart with him as you are in here with me? Would he be threatened, like your mother has been and some others in your life will be, or would he enjoy that nice part of you, as I do and some others would?

In this way, tracking the client's anxiety can help the therapist and the client focus inward and clarify the client's key concerns. In particular, *it will often reveal how the same problem that the client is having with others has been activated in the therapeutic relationship.* By clarifying how aspects of the client's problems with others seem to be reoccurring between the therapist and client, the therapist has the opportunity to intervene in two important ways. First, therapists can talk about and change their interaction with the client to ensure that the therapeutic relationship does not repeat the problematic patterns that have occurred with others. Second, as we have just seen, they can link the positive new behavior with the therapist to others in the client's life. We will return to these basic, orienting constructs in the chapters ahead and explore them further.

Closing

The basic theme in the interpersonal process approach is that therapists from every theoretical orientation can better help their clients change by living out or enacting solutions to their problems in their real-life relationship with the therapist. Helping clients adopt an internal focus for change is one important way in which therapists provide clients with a different type of relationship than the problematic patterns they have experienced and come to expect with others. By codetermining what issues and goals will be pursued in treatment, and by actively participating in finding solutions to their problems, clients are far more likely to change, and gain an increasing sense of self-efficacy in the process. Perhaps Bowlby (1988) has captured this interpersonal process most succinctly by saying that the basic therapeutic stance toward the client is not "I know; I'll tell you" but "You know; you tell me."

As clients begin to reflect on their own reactions and become less preoccupied with the problematic behavior of others, not just their anxiety but their entire affective world opens up and becomes more accessible. This affective unfolding is a pivotal step in the change process. However, the strong feelings aroused by pursuing an internal focus can be distressing to clients and uncomfortable for therapists as well. The next chapter explores how therapists can help clients come to terms with the significant feelings activated by looking within.

Suggestions for Further Reading

1. As we are seeing, therapists in training are trying to develop new interpersonal skills that are not easy to acquire. To help develop these important but often challenging skills, a self-assessment measure is provided in Chapter 4 of the Student Workbook. Completing this self-assessment will help readers evaluate their current level of skill along several important therapeutic dimensions.

2. Several microskills texts are available that teach new therapists how to use "immediacy" interventions and provide interpersonal feedback. See Poorman (2003,

Chapters 2–6), Egan (2002, Chapters 11 and 12), and Hill (2004, Chapters 13–16).

3. Albert Bandura argues convincingly that perceived self-efficacy is the underlying mechanism of change across all treatment approaches; see his classic article "Self Efficacy: Toward a Unifying Theory of Behavioral Change" (1977) or his more recent book *Self-Efficacy: The Exercise of Control* (1997). He emphasizes that little enduring and generalized change results from verbal suggestion or persuasion. Instead, clients change when they have authentic experiences of "enactive

mastery"—that is, when they have the experience of performing adaptive new coping responses with the therapist.

4. Alan Wheelis (1974) discusses the internal process of change in his book *How People Change*. In particular, see the chapter "Freedom and Necessity," which uses the language of responsibility to clarify how clients must focus on themselves rather than others in order to change. Therapists can learn much from other existential therapists as well. In particular, see Irvin Yalom's (2002) highly engaging book, *The Gift of Therapy*, which helps therapists work with clients in the here-and-now.

5. As therapists work with diverse clients from varying cultural contests, it is helpful to become more aware of normative verbal and nonverbal behavior. In particular, readers may wish to examine Chapters 1 and 2 of *Counseling the Culturally Different* (Sue & Sue, 1999), which address language differences among cultural groups. For a discussion of how to assess anxiety in Asian clients by focusing on nonverbal behavior, see the case study "Rape Trauma Syndrome" in *Casebook in Child and Adolescent Treatment Cultural and Familial Contexts* (McClure & Teyber, 2003).

CHAPTER FIVE

Responding to Painful Feelings

Jennie, a 1st-year practicum student, was off to a good start with her client, Sue. At their third session, however, Sue entered in true crisis: "I have breast cancer!" she cried. "My doctor says I have breast cancer!" With undisguised anguish, she went right to her biggest fear: "My girls may not have a mother to help them grow up!" Taken aback sharply by the intensity of this raw emotion, Jennie didn't know what to do at first. Urgently thinking of what she could say, Jennie jumped in and began doing what she had always done best: getting the facts, making plans and solving problems: "Have you got a second opinion . . . ?" More questions and answers followed: "Have they talked to you about your treatment options yet? They can do so much more now than they used to; chemo has become so much more effective. . . ."

In supervision the next day, her supervisor tried to help Jennie see that she had moved right in to a problem-solving and reassuring mode without responding very directly to Sue's profound fear: "Such big feelings, Jennie, such vulnerability. It sounds like you wanted to help by helping Sue figure out what to do. Instead, what do you think would have happened if you had just tried to stay with her feelings longer?" Jennie was rather quiet, and vaguely suggested that she wasn't sure how to do that. Wanting to help, her supervisor tried to be more specific: "I'm thinking of several ways you might be able to do that in the future—like expressing more directly your own compassion or concern for her. Or perhaps helping her clarify or name her feelings that seem so overwhelming, or maybe just affirming how frightening all of this is right now? I think the sequence is important here. If you respond to her feelings first with this type of understanding or empathy, I think that then she could better use your suggestions about what to do."

As their discussion unfolded, Jennie disagreed and expressed that "the client wasn't ready to go that deep yet—we haven't been working together long enough to push her into feelings like that." Jennie looked puzzled as the supervisor replied,

"But you weren't 'pushing' her to 'go deeper' than she wanted. Sue initiated this—she brought these feelings to you."

It was too new and too much for Jennie to absorb now, but, by the end of her practicum year, Jennie had became better at letting her clients feel what they were feeling.

Conceptual Overview

Painful feelings lie at the heart of enduring problems. Change in therapy is a process of affective relearning, in part, and this change occurs when the therapist responds effectively to clients' emotions. When the therapist focuses clients on themselves and they look within, clients feel more intensely, and express more directly, the difficult feelings that accompany their problems.

This affective unfolding is a pivotal point in treatment because it reveals the emotional basis of clients' problems. On the one hand, clients will welcome the promise of having therapists make contact with them on a deeper, more personal level. On the other hand, however, clients may also want to avoid feelings that are too painful, shameful, or just seem hopelessly unsolvable. Thus, the therapist's response to the challenging feelings that clients bring will have a pivotal impact on the outcome of treatment.

Therapists facilitate change by providing a more helpful response to the client's feelings than he has come to expect from others (that is, the therapist is validating and takes them seriously rather than minimizing them or offering superficial reassurances). Furthermore, when therapists avoid or do not respond to the client's feelings, the therapeutic relationship loses its vitality and meaning—treatment is reduced to an intellectual pursuit. Thus, the purpose of this chapter is to help therapists respond effectively to the feelings evoked when clients look within.

Before starting, however, I want to begin this chapter with a note of encouragement and an eye toward realistic expectations for new therapists. The material presented in this chapter is more personally evocative and challenging for many therapists than the material presented in other chapters. The issues discussed here often evoke personal concerns for therapists in training, and you may not be able to apply these principles as readily with clients. Understanding these processes intellectually is a realistic goal for first-year therapists; being able to utilize these interventions effectively in ambiguous therapeutic interactions is not. Be patient: In a year or two, many of these interventions will be your own. Like so many things, it comes with practice.

Responding to Clients' Feelings

Many of the problems that clients present can be resolved by trying out new coping strategies or behavioral alternatives, questioning faulty beliefs, or reframing problems in a new perspective. These brief interventions are sufficient to help with many of the problems clients present, and they are an important aspect of change for most problems. However, for the more enduring and pervasive problems that clients often present, the conflicted emotions that accompany their problems need to be addressed as well.

At times, clients may be reluctant to explore more fully the difficult feelings that emerge in treatment. On the one hand, clients hope that the therapist will be able to help with their emotional reactions; on the other, they may want to avoid the difficult feelings that they have not been able to resolve on their own. Clients may be reluctant to share certain feelings in the beginning for many different reasons. To the client, it may seem pointless

- to share the same painful feelings again when—based on what has happened before—they have scant hope of receiving help;
- to suffer shame by revealing secrets, failure experiences, or feelings of inadequacy;
- to risk unwanted but expected judgment from the therapist for having feelings that have always seemed unacceptable; or
- to struggle with their own fear of "losing control."

As a result of these and other concerns, clients sometimes resist the therapist's attempts to touch the emotional core of their experience. At the same time, however, they often long for this empathic understanding as well.

Once the therapist recognizes the client's relational schemas and understands how significant others have characteristically responded to the client's vulnerability in the past, the fear or shame that motivates her resistance will make sense. As outlined in Appendix B, therapists can generate working hypotheses and try to anticipate the clients' potential resistance to conflicted feelings and attempt to work with those feelings rather than avoiding them. If not, treatment will be reduced to just an intellectual exercise that will not be potent enough to propel meaningful and enduring change. The therapist can appreciate the client's reluctance to approaching certain feelings yet also keep in mind the client's simultaneous wish to be responded to in this deeply personal way.

Although difficult emotions are central to most clients' problems, the therapist's affirming response to these feelings also provides the avenue to resolution and change. The corrective emotional experience that the therapist provides to clients—the experience of sharing important feelings with another who remains connected and validating—loosens the hold of faulty beliefs and expectations. As rigid schemas become more flexible and expand, clients are empowered to make the changes they desire. At this critical juncture when clients experience a reparative or corrective response to such significant feelings, the treatment process springs forward. As we will explore, clients become more receptive to a wide range of interventions from varying theoretical orientations at this point.

In the next section, we explore intervention guidelines to help therapists respond effectively to their clients' feelings and set this change process in motion.

Approaching Clients' Feelings

With each turn of the conversation, therapists are challenged again to choose how they want to respond right now to what the client has just said or done. For example, the therapist might respond by seeking more information, clarifying what the client has just said, making connections between this issue and other material the client has presented, or linking it to what is occurring between the therapist and client at that moment. As a guiding principle, however, *the most productive intervention is to approach the most salient feeling in the material that the client is talking about right now.* In other words, the therapist's first priority is to acknowledge the affective component—the feeling or emotion, in the client's response.

Imagine that the therapist and client have just sat down to begin their first session:

> THERAPIST: Tell me, Mike, what is the difficulty that brings you to therapy?
>
> CLIENT: I'm having a lot of problems with my 15-year-old son. We disagree about everything and can't seem to talk to each other anymore. He doesn't do what I ask him to, and I don't like his values. I guess I'm pretty angry at him. His mother and I are divorced, and I'm thinking it may be time for him to go live with her. Do you think it's OK for a teenaged boy to live with his mother?
>
> THERAPIST: I don't think either of us understands what's going on between you two well enough to make any decisions yet. You said you were pretty angry at him. Tell me more about your anger.

The client quickly has presented many different issues that the therapist could have chosen to pursue. Following the general guideline, however, the therapist approached the primary affect or feeling that the client presented. The therapist could have responded differently and inquired further about the issues that the father and son disagreed about, worked on values clarification or communication skills, obtained more background information about the father–son relationship or the divorce, provided research findings about the effects of mother custody versus father custody on boys, and so forth. These and many other responses will often work well. However, *responding to the central or most salient feeling that the client expresses will usually produce the most meaningful information and intensify the interaction.*

In this dialogue, the therapist responded right away when the client spoke directly about his anger. Additionally, therapists need to respond to the *covert* or unverbalized feelings that clients experience sometimes. If it seems to be present, for instance, the therapist in our example might have chosen to respond instead to a salient but more covert affect. For example, the therapist could do this by tentatively wondering aloud, "As I listen, it sounds like you're feeling really discouraged about being a dad right now—maybe so discouraged that you just feel like giving up sometimes?"

When clients begin talking about what matters most to them, important feelings usually come up. Such nonverbal affective signs as tearing, grimacing, or blushing are signals to the therapist that the client has entered a significant topic that holds real meaning. Responding to these nonverbal cues, the therapist invites the client to share the feeling or deeper meaning that this situation holds for the client. For example, imagine that the client has been discussing her marital problems:

> CLIENT: I don't know if I should stay married or not. I haven't been happy with him for a long time, but I can see how hard he's trying to make our relationship work. And our 4-year-old son would be devastated if we broke up. I don't know what to do, and I just have to make the right decision.

> THERAPIST: As you speak about this, your face tightens. I'm wondering what you might be feeling right now?

When the therapist responds to the client's feelings in this *open-ended* way, it often serves to clarify the client's key concern. For example, clients may respond by expressing more specific concerns. Continuing with the previous dialogue, the client may respond:

- I'm so sad about hurting the people I love.
- I'm afraid that everyone will think I'm selfish for leaving, and they'll blame me for the divorce. They'll think I haven't been a good mother.
- I don't want to be alone. I don't want to be married to him anymore, but I'm afraid of leaving and trying to make it on my own.
- I'm furious that I'm the one who has to make this decision. I'm responsible for every decision we make, and I always pay for it in the end. It's not fair!

Throughout each session, clients have emotional reactions to the issues and concerns that trouble them most. Sometimes the client's affect will be presented forthrightly; at other times, it will be subtle or more covert. Cultural, class, and gender factors will further shape how each client displays affect (Sue & Sue, 1999). For example, some clients may overtly state their reactions, whereas those from other cultural or familial backgrounds may simply divert their eyes or become noticeably quieter when certain feelings are evoked. The therapist's abiding intention, however, is to keep giving invitations or open-ended bids to explore the feelings that seem most important to the client right now. In other words, we are trying to reflect or highlight the most powerful or immediate feeling that holds the most intensity at this moment. As the therapist joins the client in sharing and understanding her feelings more clearly—as in the examples just listed—it will

- facilitate further disclosure from the client,
- highlight her key concerns, and
- provide a clearer direction or focus for treatment.

To emphasize, the therapist wants to find *collaborative* ways to invite and explore the client's affect. We don't want to "push" for feelings, of course, or become invested in the client having sad, angry, or any other feelings during the session. For most therapists in training, however, the problem usually goes the other way. The client introduces a feeling and the therapist does not "hear" or register it, keeping their interaction on a more superficial level. Instead, we want to invite whatever feelings the client seems to be experiencing by trying to reflect, clarify, and affirm them.

All of this may sound easy enough to do, but it is not. Researchers find that even experienced therapists often do not approach sensitive feelings and concerns very effectively (Hill & O'Brien, 1999). Recalling the discussion of "rupture and repair" in Chapter 2, we have already seen that clients routinely hide or withhold their negative feelings and reactions toward the therapist (Regan & Hill, 1992), and many clients leave one or more important things unsaid during sessions (Hill et al., 1993). Clients reported leaving important things unsaid because (1) their own emotions felt "overwhelming" to them, (2) they wanted to avoid dealing with the therapist about the disclosure, and (3) they feared that their therapist "would not understand." In another study, about one-half of the clients sampled had major life secrets that they had not told their therapists, even in long-term therapy (Kelly, 1998). They withheld these secrets for two reasons. First, returning to our old friend, shame, they reported feeling embarrassed and ashamed. Second, these clients believed that either they themselves would be "overwhelmed" by their emotions, or that their therapists "would not understand" or could not handle the disclosure.

These findings tell us a great deal about approaching clients' feelings. Recalling Jennie and her client with breast cancer at the beginning of this chapter, therapists often shy away from strong feelings and keep things on the surface by not reflecting the most significant feeling in what the client just said. Actively intending to "hear" and approach the feeling that seems to be the most sensitive, to hold the most meaning, or is the most intense is a new way of responding for most therapists in training. For example:

> THERAPIST: You're saying so much here. Out of all of this, what's the most important feeling you're left with right now?

It often feels awkward to approach feelings so directly because it breaks the social norms that most of us grew up with—we are not used to talking to friends and family like this in everyday interactions. However, with help from supervisors and practice with colleagues, new therapists can learn how to approach their clients' feelings more directly and respond effectively to the sensitive material that they choose to share.

Expand and Elaborate the Client's Affect

Open-ended Questions   Let's look at "open-ended" questions and explore further how the therapist can best respond to clients' feelings. At times, it can be helpful to suggest or tentatively name a client's feeling (for example, by saying, "It sounds like you were just furious" or "I'm wondering if you're feeling frightened right now?"). However, as a general guideline, a more effective response is to give clients a more *open-ended* invitation to explore the feeling further:

- What are you feeling right now?
- Tell me more about that feeling?

- How do you feel now about what happened there?
- Help me understand what you're experiencing right now as you tell me this.

These open-ended invitations encourage clients to explore whatever it is that they feel is central in their affective experience, rather than what the therapist thinks is most important. In particular, these open-ended responses also bring more *immediacy* to the therapeutic relationship. That is, the client's feelings are dealt with in the here-and-now with the therapist, which usually intensifies them.

In contrast, we don't want to ask too many "closed questions." They lead us to get all the facts and details about a problem and away from the feelings and what this event really means to the client. For example, suppose the client says, "I got my test back." Much more is gained if the therapist responds with the open-ended bid "Tell me your thoughts about it" than from the closed question "What did you get on it?" This takes practice, of course, because most of us have not been socially trained to respond in this open-ended, exploratory way that "unpacks" or expands the personal meaning this holds.

An open-ended invitation also is more effective than trying to label the client's feeling in an either-or fashion—for example, by saying, "Were you feeling ____ or ____ in that situation?" Such a choice may restrict the client rather than invite a free range of responses. Clients often experience something that the therapist has not anticipated (no matter how much experience we have, clients always seem to surprise us), and the client may have several different feelings about the same situation. Thus, the open-ended queries are more effective in drawing out exactly what the client is feeling and does so in a more collaborative manner.

Additionally, open-ended responses usually are more effective than asking clients "why" they are experiencing a particular feeling. For example:

THERAPIST: Why were you crying?

CLIENT: I don't know.

Although *why* questions will be effective at times, as a rule of thumb, therapists are advised to minimize them. Clients often don't really understand why they feel, think, or do something. Asking them why may only make them feel "on the spot" or defensive. The therapist will usually elicit more information and collaboration from clients by offering an open-ended invitation—"Help me understand your anger better" or "Tell me about your anger"—than by asking, "Why were you angry about that?" By inviting clients to express more fully whatever feeling they are experiencing, therapists are giving the message that they are interested in clients' own subjective experience and are comfortable with whatever clients say. As we have seen, clients frequently withhold affect-laden material from their therapists. Most clients have not had permission to share their emotions so genuinely in other relationships, which is why these open-ended invitations to explore whatever they are feeling right now is one of the most important responses therapists can offer. As they discover that they can be authentic with the therapist and that it is safe to feel whatever they are feeling, clients become more willing to take risks, try out new behavior, and invest further in the therapeutic relationship.

Clarifying Clients' Subjective Feelings   Giving clients an open-ended invitation to explore their feelings will also help them clarify what each particular feeling means to them. Too often, therapists assume that a particular affective word, such as *angry* or *sad*, means the same thing to the client as it does to them. Therapists should not assume that they understand a particular affective word without clarifying what it means to this particular client. Although this may be especially apt when there are cultural differences between therapist and client, it applies with every client. In this process of *mutual exploration*, just as we did with empathy in Chapter 2, therapists share their understanding of what the client meant and then encourage clients to correct or further clarify the personal meaning that a feeling holds for them. This clarification and mutual exploration is especially important if clients repeatedly use the same ("compacted") affective word to describe themselves, such as boring, overwhelming, and so forth. When this occurs, often several different and important meanings are contained or compacted in that single affective word. For example,

> THERAPIST: Here again, you're describing yourself as feeling "erased" and "dismissed." It's sounding to me like you feel totally insignificant when he does that—you are so completely unimportant that nothing you say or do matters. Am I saying that right? Can you help me capture it better?

Having the client elaborate her experience in this way—*clarify it more specifically*—will also safeguard therapists against becoming overidentified with the client and misperceiving the client's situation as being the same as their own, as so often occurs.

Not listened to or invalidated in many relationships, many male and female clients do not know very well what they feel. In many families, children grew up learning that it was not acceptable to feel a certain feeling—such as being sad, angry, or even happy. As a result, many clients are uncomfortable with, or unaware of, what they are often feeling. As a result, many clients need to clarify the broad, "undifferentiated" feeling states that they often experience (Client: "I don't know, guess I'm just mad all the time"). Therapists can help clients learn more about themselves, their own inner life, and the emotional reactions that are linked to their problems by asking questions such as these:

- Can you bring that feeling to life for me and help me understand what it's like for you when you are feeling that?
- Do you have an image that captures that feeling or goes along with it?
- Is there a particular place in your body where you experience that feeling?
- Is this a familiar or old feeling? When is the first time you can remember having it? Where were you? Who were you with? How did the other person respond to you?
- How old do you feel when you experience that emotion? Can you attach an age to it, such as 7 years old or 13 years old?

Some of these exploratory questions will work well with a particular client and not at all with another. Following client response specificity, therapists will need to assess the client's reactions to each way of responding and find what works best for each client. In general, however, these responses are all ways of approaching clients' emotions, entering their subjective worldview, and clarifying the subjective meaning this feeling holds for them. These types of responses also tell clients that the therapist is concerned about their feelings—who they are and what is most important to them. Furthermore, they behaviorally inform clients that the therapist is different from many other people they have known. The therapist is comfortable sharing whatever emotions the client has and, in addition, wants to know the client's authentic experience more fully—as good or as bad as it actually is for them.

When the therapist works with clients to clarify their emotions, it creates an opportunity for the therapist to see and respond to the clients' most personal experience. An important aspect of most clients' problems is that significant others have repeatedly invalidated, ignored, or disdained certain feelings that the client had, such as being sad, mad, or afraid. This occurs, for example, when caregivers reply:

CAREGIVER: Oh, come on sweetheart, you're not really mad about anything.

OR

CAREGIVER: One more look like that, and I'll give you something to be sad about.

As old expectations of judgment, invalidation, and so forth, are disconfirmed by the therapist's empathy and validation, the client receives an important, real life experience of change—a corrective emotional experience.

Experiencing Feelings versus Talking about Feelings   On the one hand, our treatment goal is not to just help clients "get their feelings out." Of itself, "venting" does not usually lead to much behavior change. On the other hand, therapists want to do much more than just cognitively label, explain, or interpret a feeling—they want to help clients experience it more fully in the moment, and then respond in a corrective way that provides validation and containment. Little change will occur for most clients until they are able to stop talking about their emotions in an intellectualized manner and actually experience or feel their feelings in the presence of the therapist (Gendlin, 1998; Carkhuff, 1987). With consistent validation and accurate empathy, the therapeutic alliance continues to develop, and clients feel more *safety*. In turn, this interpersonal safety allows clients to be able to risk "feeling what they feel" and "knowing what they know"—something that often has been too threatening or anxiety arousing in the past. The exploratory responses that we have been discussing, especially the interventions that create immediacy, will intensify clients' experience and allow them to enter their own experience more fully.

Therapists may also draw out the client's affect by addressing any "incongruence" that they perceive between the client's narrative and the accompanying affect. That is, in relaying a story that the therapist finds poignant or disturbing, the client may display no feeling at all or an incongruent affect, such as laughter. Therapists are subtly but powerfully "pulled" by social norms to match the client's incongruent affect. New therapists, in particular, often feel they are violating unspoken social rules if they do not respond in kind and laugh—even though the story seemed sad to them. However, clients are usually relieved when the therapist risks breaking the social rules and makes a process comment by acknowledging this discrepancy between what was said and the feeling that accompanied it. By working with the process dimension and making this discrepancy overt, the therapist makes a strong intervention: she begins to invite the client's true feelings or congruent experience. For example:

THERAPIST: I'm a little confused. What you are saying seems so sad to me, yet you are almost smiling as you tell me this. What are you thinking as I share this with you?

OR

THERAPIST: You're telling me how angry you are about this, Lee, but your voice is so pleasant as you speak, so soft and quiet. What you're saying doesn't seem to match the way you are saying it. I'm wondering if this discrepancy is telling us something?

Our treatment goal here is to provide *interpersonal feedback* that helps clients register the feelings that accompany their actual experience—to become more congruent. So many symptoms and problems stem from clients' inability to feel what they feel: to be sad about what was lost or missed, afraid of what really was once frightening, or angry about what actually was hurtful. No small intervention, this is about identity, personal power, and the ability to be who you really are. *Simply allowing clients to be able to have the feelings that are commensurate with what actually happened to them is a lifelong gift to many.* Recalling the concept of invalidation, it allows clients to claim their own experience and be empowered to know their own mind. Working within different theoretical traditions, many forerunners in the field have written eloquently about this outcome: "congruence" for Carl Rogers, "having your own voice" for Carol Gilligan, "authenticity" for Irvin Yalom and other existentialists, "identity" for Erik Erikson, and others.

The best opportunity to provide a corrective emotional experience and help clients change occurs at the moment when they are experiencing the full emotional impact of their problem. If therapists can communicate their accurate understanding and genuine compassion to clients in the moment, as they are experiencing difficult emotions, it will help them resolve the affective component of their problems. Clients move forward and make behavioral changes with others as their schemas expand as a result of these real-life experiences of change occurring with the therapist. In particular, this in vivo relearning changes their expectations of how some others in their lives can respond to them, and it changes their selective biases in filtering and distorting how others are actually responding to them. However, as the following case example illustrates, clients cannot make such progress when they are psychologically alone with their feelings or can't let their true feelings emerge.

During one session, Jean had been telling her therapist about her hopelessness. She described her life as a merry-go-round of endless ups and downs that always returned to the same discouraging spot. This past week she had dropped out of school, as she had done "so many times before," and had given in to pressure from her self-centered and undermining boyfriend to let him move back into her apartment again. Jean described her hopelessness about ever being able to do or have anything for herself.

The therapist could feel Jean's despair and was moved by it. The therapist tried to work with Jean's feelings by acknowledging her despondency and validating her experience.

> THERAPIST: I can see your hopelessness right now, and how painful it is to feel so defeated. You haven't been able to follow through and do what you wanted to do for yourself again, and that feels so discouraging right now. As you've done so many times before, you felt that you had to meet someone else's needs rather than say no and do what you wanted for yourself.

As in all of their previous sessions, however, Jean was not able to let the therapist be emotionally connected to her while she was experiencing this. She talked about these feelings, but, both agreed, she held the therapist at arm's distance. Because this emotional connection to the therapist was missing, there was no secure base for Jean—even though she was talking to the therapist, she was still alone with her feelings and hopelessly alone in her life.

The setbacks of the previous week had intensified Jean's distress. The therapist recognized that this crisis provided an opportunity to be more connected to Jean in her distress—as she was feeling her feelings, for the first time:

> THERAPIST: Right now, it just seems hopeless.
>
> JEAN: (*long pause, nods*) Yes, it seems hopeless. I'm just hopeless.
>
> THERAPIST: You don't want to try anymore—you don't want to feel so discouraged anymore. You just want to stop feeling that you are so hopeless, and it doesn't seem like there is any way to stop hurting so much. Am I saying that right?
>
> JEAN: (*nods*)
>
> THERAPIST: You are so sad right now. Is this familiar? Do you feel this way a lot?
>
> JEAN: Yeah, it's not just with my boyfriend and school, you know. I've kind of always felt this way.
>
> THERAPIST: Uh huh, you've felt this way a lot before, and I can see how painful this is for you right now.
>
> JEAN: (*looks away, puts hands over eyes*)
>
> THERAPIST: I'm feeling with you in your sadness right now. I can't take it away, but you don't have to be alone with it any more.
>
> JEAN: But don't you think I'm just pathetic? Come on, isn't this just so weak?
>
> THERAPIST: Oh, no, not at all. I see your sadness right now and how much all of this has hurt you. I respect your courage to risk sharing all of this with me.

JEAN: (*looks at therapist and slowly begins to cry*) Well, maybe I'm not completely hopeless. . . .

For the first time, Jean was able to let the therapist be emotionally connected with her while she experienced her despondency. The therapist acknowledged the risk that Jean had taken, and they went on to debrief and discuss how it felt for her to share her sadness so openly. Jean said that, to her surprise, she felt like the therapist still respected her and that she felt "lighter." At their next session, Jean said that her depression had been better that week and, almost as an aside, mentioned that she had told her boyfriend she wasn't ready to live together yet.

Identifying the Predominant Affect

Clients often feel confused by their emotions; their feelings seem to occur for no reason and often are stronger than, or different from, what the situation calls for. Clients' inability to understand their own emotional reactions undermines their self-confidence and engenders doubt. Helping clients make sense of their emotional reactions is an important pathway to their greater self-efficacy. One way that therapists can do this is by working with the client's *predominant affect*. As we will see, clients' experience often revolves around a central feeling state, such as being shame-prone or guilt-prone, and therapists help when they can identify and punctuate this predominant feeling.

Clients often enter therapy in response to one of two situations. First, a current life crisis echoes an emotional problem that originally occurred years before—the current stressor taps into an *old wound*. Second, clients may enter therapy in response to feeling overwhelmed by *too many stressors* in a short period. These multiple stressors have overwhelmed the clients' usual coping strategies, and symptoms such as anxiety, depression, sleep disturbances, and interpersonal conflict develop.

An Old Wound

Many clients enter therapy because of an "old wound." For example, a client who has suffered a painful loss through divorce or death will find it more difficult to cope with the stress of that loss if it simultaneously evokes painful feelings from previous losses in the client's life ("compacted grief"). The current loss may be associated with a predominant feeling—perhaps regret, loneliness, or sadness. *The feeling becomes understandable, and far more manageable, when it can be linked to similar feelings that have been aroused by other losses in the client's life.* Thus, the therapist needs to identify the core or primary affect that the current crisis has evoked and link it to the original wound. Connecting this primary affect to the way clients felt in the past enables them to make sense of their seemingly irrational feelings. When this occurs clients change by feeling more accepting or forgiving of themselves. They are better able to appreciate the bigger problem they are actually dealing with—a story playing in two times—and gain a greater sense of self-efficacy. Although this is important in all treatment modalities, it is one of the most significant interventions that therapists can make in crisis intervention or short-term therapy.

Multiple Stressors

When his company downsized, Jim was laid off work. Unable to find a new job during the economic downturn, Jim felt like a failure—but his problems were just beginning. As his self-esteem decreased and his financial worries increased, he became "negative" and "impossible" for his wife to put up with. His marriage had never been the best, but living all day, every day, with this short-tempered, pessimistic complainer was too much for his wife. Six months after being laid off, she had had enough and moved out. Because his wife had always created their social life, and with no more coworkers to visit with at his office, Jim became even more isolated and lonely. Eating and drinking too much to cope with his anxiety and depression, Jim's blood pressure rose markedly. At his doctor's strong prompting, Jim reluctantly entered therapy, describing himself as a "loser."

Most individuals have adaptive coping mechanisms that allow them to manage a single stressful event, unless it taps into the type of preexisting vulnerability or "old wound" described here. However, if a second or third stressful event follows from the first, as it did for Jim and many others, it becomes much harder for individuals to cope. The client's usual coping strategies fail and the cumulative stressors precipitate symptoms. In this way, most problems are not caused by simple, isolated events. They usually develop over time out of a sequence of events or interaction cycles—as they did for Jim.

When subjected to multiple stressors, clients may describe themselves as "overwhelmed," "burned out," or even "broken." They tend to repeat a *compacted phrase* that encapsulates their primary emotional reaction to the stressful events:

- I want to go away and just be alone.
- It's too much for me; I can't stand it.
- I don't care anymore; there's no point in trying.

As noted in Chapter 2, therapists are looking for *recurrent affective themes* and trying to discern the most central or unifying feelings that are encapsulated in these compacted sentences. Underlying each compacted sentence are one or two core feelings that keep coming up and capture the fundamental impact these stressors have had on the client. Generally, the various stressors are arrayed around one or two overriding feelings, like spokes around the hub of a wheel. For example, in response to losing too many people or things, the client may have an overriding feeling of being alone or powerless, or the client may be afraid of any further losses, changes, or new commitments. When therapists are able to articulate and capture the underlying feeling that links the impact of several different crisis events, clients often feel deeply reassured. For example:

THERAPIST: As I listen to what happened over the weekend, it seems that in each case—with the soccer coach, the telephone call from your boss, and then the argument with your husband—you were not heard. It seems like you became so upset because you were never really being listened to you weren't being taken seriously.

When the therapist has been accurately empathic, and able to identify and articulate the integrating theme, clients feel seen and understood. In turn, they feel more in control of themselves and invest further in the working alliance. To succeed in short-term therapy, in particular, therapists need to develop this skill in order to establish credibility quickly, make an empathic connection, and establish a working alliance early in treatment.

A Characterological Affect

As clients relay their narratives, it is important for therapists to respond to the most important or meaningful feeling the client is experiencing. They can also listen for a recurrent feeling that keeps coming up in the client's life—*a characterological affect.*

As treatment proceeds, the therapist can often identify a predominant feeling that captures the client's current conflict and also characterizes her life more broadly—the client experiences this familiar feeling as part of the fabric of her life. That is, if therapists are accurately empathic, they can often identify a central feeling that comes up repeatedly—a familiar feeling that pervades the client's life or expectedly returns when she is in crisis. Clients often have one or two of these core feelings that have been continuous throughout much of their lives. For example, when the therapist can identify these core feelings that come up again and again in different situations, clients often respond emphatically:

- Yes, that's how it has always been for me.
- That's what it's like to be me.
- That's who I really am.

Clients may describe these core feelings as their "fate," because it feels as if they have always been there and seems like they always will be. One of the most significant interventions therapists can make is to accurately identify these characterological feelings and reflect them to the client. For example, after listening to the client for a while, and grasping the integrating affective theme, the therapist may respond:

- It seems as if you've always felt *burdened* by all the demands you feel you have to meet.
- I'm getting a sense that you've always been *afraid* that people are going to find you out and point out your true inadequacies.
- Maybe you're saying that you've always felt *resentful* that, no matter how much you do, it's never enough.
- I'm getting a sense that you've always felt *wary* that others are trying to put you down or take advantage of you.
- I'm wondering if you're *worried* that if people really knew the true you, they couldn't possibly love you?

Therapists are prepared to respond to all of the varying emotions clients present in the course of treatment. However, therapists will be most effective when they can identify the repetitive feelings that have recurred throughout a client's life and that the client considers to be central to her sense of self. There isn't much more than this that therapists can do to establish their credibility with clients and strengthen the working alliance.

Affective Constellations

It is deeply rewarding but sometimes challenging to work with the painful feelings that clients often bring. Some clients want to avoid their emotions; others may talk at length about the same feelings without any change resulting. When clients are not progressing in treatment, one reason may be that therapists have only responded to the client's single presenting affect. In this section, we will see what to do about this by learning about the constellation or sequence of feelings (the "firing order") that clients often present.

If the therapist acknowledges the client's current affect and invites the client to explore it further, *a sequence of interrelated feelings will often occur together in a predictable way.* For example, the client feels anger, which occurs in response to feeling hurt or sad about something that happened. However, the hurt or sad feeling is unacceptable to the client and is shunned or disavowed because it seems "weak" to this client and evokes shame. To ward off the unwanted feeling of shame, the client "restores" his sense of self and personal power by returning to the "stronger" or safer feeling of anger again, and he is perceived by others as being an "angry person." This constellation of emotional reactions is central to the client's problems and each successive affect in the sequence needs to be addressed. Although different sequences occur for each client, the therapist can often identify a triad of interrelated feelings that cycle repeatedly when clients are distressed. Making a similar point, Lazarus (1989) writes informatively about tracking a client's "firing order." The process of change involves coming to terms with each of the feelings in this affective triad. In this section, we examine two affective constellations that commonly occur. We will see that therapists help clients change by responding to the entire sequence of feelings in their affective constellation, rather than just responding to the first feeling in the sequence—the client's primary presenting affect.

Anger-Sadness-Shame

The first affective constellation that we will examine consists of anger, sadness, and shame. The predominant or characteristic feeling state for some clients is anger. These clients often experience and readily express angry feelings, including irritation, impatience, criticism, and cynicism. However, the client's anger is usually *reactive*; it is a secondary feeling that occurs in response to an original experience of sadness or pain.

The therapist needs to acknowledge the client's anger, clarify more specifically who and what the client is really angry about, and help the client find appropriate ways to express it. For example, becoming more appropriately assertive, setting limits, or telling someone what it is that you don't like rather than losing your temper, yelling, or demeaning them. However, this approach is not enough. If the therapist stops at this point, the client may still remain stuck in an angry feeling state and more substantial change will not result. Instead, *the therapist can help the client identify and stay with the primary emotional response of sadness or hurt that led to the reactive feeling of anger.* This primary feeling is more threatening or unacceptable for the client than the anger that it prompts.

To reach the original affect, the therapist might wait for clients to ventilate their anger spontaneously and immediately afterward invite them to examine what they are feeling. Following the expression of the reactive feeling, there is often an open window to the primary feeling of hurt or sadness. For example,

> THERAPIST: OK, you've just expressed how angry this made you. Can you stop here for a moment and check in on what you are feeling right now?

Often, the original feeling (for example, vulnerability, embarrassment, helplessness) will often surface next. By joining the client in this feeling and working together to accurately name it, the therapist can often dislodge the client from the familiar (but safer and nonproductive) reactive feeling of anger.

If this approach does not work, the therapist can inquire directly about the original experience (such as her husband's criticizing her in front of friends, and so forth) that led to the client's reactive or secondary feeling of anger.

> THERAPIST: Something must have hurt you very much for you to remain so angry. Tell me how you felt when that first happened.

If the client again begins to express anger or indignation about what occurred ("It really pissed me off when he put me down like that . . ."), the therapist focuses the client back on the original experience:

> THERAPIST: Yes, I can see why it made you so mad, but how did it feel right then when he was putting you down in front of everybody?
>
> CLIENT: Well, I guess maybe it hurt my feelings a little.
>
> THERAPIST: Yes, it hurt your feelings. Of course it did. I'm so glad you can share that hurtful feeling with me. You've just made a big change, you know: you've never been able to risk sharing the hurt part before. How does it feel to take that big step with me?
>
> CLIENT: (*emphatically*) I hate it—I just hate feeling so weak that he can hurt me like that. . . .

Thus, in response to this query, the client's original feeling of "hurt" was expressed for the first time, rather than going on repetitively about how angry she is. The therapist responds affirmingly as the client begins to experience and share the internal aspect of her problem—that is, the feeling of hurt or sadness that is so unwanted or unacceptable for her. Recalling our internal focus for change, *it is this internal aspect of her problem—her sense of herself as "weak" or shame-worthy because he has been able to hurt her—that has kept leading her back to her safer, reactive feeling of being angry all the time.* This is a complex sequence for new therapists to follow in the beginning; let's keep going and try to make it clearer.

As we have just seen, when this client experiences anger, hurt or sad feelings about what happened to her often follow. Although her anger is easily experienced and readily expressed, the client has learned that it is not safe to experience or share the original hurt or vulnerability. To avoid the original painful feeling, the client defensively returns to the reactive feeling of anger in a repetitive manner. This defensive coping pattern often leads others to regard this client as an "angry person"—and contributes to the problems she keeps having with others at work and at home.

The goal of the therapist's intervention is to help such clients experience the sadness, hurt, or vulnerability that precipitates their reactive/repetitive anger. However, as soon as the original hurt is activated, the client feels ashamed and will defend against this unwanted feeling by reflexively returning to the safer expression of anger that, in her belief, is "stronger." Thus, the therapist needs to explore the client's resistance to the original feeling of being hurt. Often, clients avoid the original painful affect because it arouses a third, aversive feeling of anxiety, guilt, or—most frequently—shame, as we saw here.

For example, as the therapist inquires about clients' reluctance to feel the original hurt, they will often reveal faulty beliefs and say something like this:

- If I let the pain be there, it's admitting that he really can hurt me.
- If I acknowledge that it hurts, it means he won.
- If I feel sad, it shows that I really am just a baby feeling sorry for myself.

Thus, if the therapist draws out the original pain that underlies the reactive anger, a third feeling of shame will also be aroused. To defend against the original sadness and the shame associated with being sad, the client has learned to automatically return to the "stronger" feeling of anger that is safer or more acceptable for this client to reveal.

Although shame is a common reaction, other clients will experience anxiety or guilt in response to feeling the original hurt. For example, some clients will say that if they let themselves feel sad, hurt, or vulnerable, then "no one will be there," "others will go away," or "I will be alone." These clients feel painful separation anxieties upon experiencing their hurt or pain and avoid this anxiety by reflexively returning to their anger. Still other clients will say that, if they let their sadness be real, they would be admitting that they really do have a need, which would make them feel selfish or demanding. For these clients, guilt is the third element of their affective constellation.

This triad of feelings exists for many clients: *Frequent anger defends against unexpressed sadness, which, in turn, is associated with shame, guilt, or anxiety.* Enduring change results when therapists can help clients better come to terms with each feeling in the triad. More specifically, clients resolve such conflicted emotions when they can

1. allow themselves to fully experience or feel each feeling;
2. share or be able to disclose the feeling with the therapist so they are no longer alone with it or hiding it; and

3. with the therapist's support, be able to stay with the difficult feeling for a moment or two (for example, tolerate, hold, or contain it) rather than having to move away from it, as they have in the past.

This is what it means to integrate or come to terms with unresolved feelings.

Thus, clients integrate their conflicted emotions when they can progress through these three steps for each feeling in their affective constellation; they are internally resolving their conflict. With this resolution, clients no longer need the outdated coping strategies or defenses that they have employed in the past. Furthermore, at this point clients are prepared to adopt new, more adaptive responses with others in their lives that they were previously unable to incorporate. For example, rather than always get mad and alienate others, they can opt to respond differently sometimes:

> CLIENT: (*to husband*) I'm feeling criticized right now and I don't like it—I'm asking you to find a better way to talk about this or wait until we are alone.
>
> HUSBAND: (*contemptuously*) I don't care what you think. Why don't you just shut up!
>
> CLIENT: You're not listening to what I asked or being respectful. I can't make you stop, but I'm not going to participate in it. Either apologize right now, or I'm leaving the room. It's your choice.

Clearly, this affective integration is a pivotal experience in the process of change.

Before going on to examine another common affective constellation, let's pause to highlight further the special role of shame. Although much has been written about anxiety, depression, and guilt, shame has long been eschewed in clinical training. Correspondingly, clients could be anxious, guilty, or depressed with their therapists, but not ashamed. Thankfully, shame is no longer the taboo topic it has been; indeed, this most difficult and miserable feeling state is now regarded by many theorists as the "master emotion" (Nathanson, 1987; Scheff, 1995). The cardinal role of shame in most resistance and in so many of the symptoms and problems that clients present was first illuminated by the groundbreaking work of Helen Block Lewis (1971) and later developed by many others (Kaufman, 1989; Karen, 1992).

As we began to explore in Chapter 3, new realms of understanding will open up to therapists as they allow themselves to begin hearing the family of shame-based emotions that pervade their clients' narratives. To feel shame is to feel fundamentally bad as a person—inadequate or worthless in the essence of your being (Kaufman, 1989). Clients directly reveal their shame-based sense of self by repeatedly using words such as *bad, self-conscious, sheepish, embarrassed, stupid, low self-esteem, humiliated*, and so forth. Less direct indications of shame may include the repetitive use of words such as *should, ought to, must*, and *perfect*. On an interpersonal level, clients defend against their own shame, and defensively attempt to "restore" their sense of self by inducing shame in others. Routinely, clients try to "disprove" their feelings of shame through rage and violent temper outbursts ("You can't tell me what to do. I'll show you!"); showing contempt for others ("You are stupid and disgusting"); and repetitive, harsh blaming of the other—especially in chronic marital conflict ("It's all your fault"). Clients also defensively try to restore themselves and deny their feelings of shame by rigidly striving for perfection or power, internal withdrawal, and other shame-anxiety and shame-rage cycles (Balcom et al., 1995).

When therapists are able to break cultural taboos and begin approaching shame reactions sensitively but directly, significant progress follows readily. However, most therapists have been socialized strongly to avoid or rescue clients from these shame-based affects (that is, ignore, make light of, or change the topic). As a result of these social norms, new therapists will need the support and guidance of their supervisors, and additional reading to begin responding effectively (see Suggestions for Further Reading at the end of this chapter). In a nutshell, however, therapists help clients resolve their shame when they demonstrate compassion for the client's feelings of inadequacy or incompetency, harsh or contemptuous judgments toward themselves, and feelings of worthlessness. Often, these shame-based emotions are referred to euphemistically as "low self-esteem," masking the far-reaching emotional wound that is really being addressed. Therapists help clients resolve such shame-based reactions by first providing or modeling, and then helping clients adopt, a more accepting or forgiving stance toward themselves and their own humanness. It is very difficult for most of us to bear witness to our client's shame, especially as we begin our training. With supportive supervision and role models, however, therapists can soon learn how to offer clients this gift.

Sadness-Anger-Guilt

Whereas some clients lead with their anger and defend against their hurt, others characteristically lead with their sadness and avoid their anger. This type of client may present with a general ("undifferentiated") feeling state of sadness, helplessness, vulnerability, or depression. These clients do not experience or express anger, they avoid interpersonal conflict, and tend to respond to others' needs at the expense of their own. To illustrate this affective constellation, we return to the case of Jean.

Earlier in this chapter, we examined a dialogue in which, for the first time, Jean was able to let the therapist respond to her sadness and be emotionally connected to her while she was experiencing it. Subsequently, the therapist explored why it had always been so difficult for Jean to share her vulnerable feelings with others. Jean responded that her painful feelings were all sickeningly more real if someone else saw them. She had always experienced these feelings, but they had always been ignored or dismissed in her family of origin. Over time, Jean had taken on her family's taboo against unhappy and angry feelings, and she now disdained her own feelings in just the same way as her caretakers once did. As part of this, she denied to herself the reality of the emotional deprivation and disparagement that she had suffered throughout her childhood. Following her schemas, Jean continued in adulthood to reenact the affective themes and relational patterns of her childhood in the problematic love relationships that she kept reconstructing.

Recall that when Jean returned to therapy the following week, she had made a significant change: she had told her boyfriend she wasn't ready to live with him yet. Two weeks later, when the boyfriend borrowed her car and arrived late to pick her up from work, she expressed her indignation for the first time. Jean told him that she deserved to be treated with more respect and that she had the right to have her own needs considered. Jean had remained in this relationship because she was "afraid to be alone." This time, however, she realized that this relationship was no good for her and was able to end it—a strong new change for her. Once the therapist was able to touch Jean's sadness and hurt, her anger was activated and subsequently expressed in an appropriate manner. Jean had remained locked in her sad and helpless victim role, in part, because the direct expression of anger and the assertive expression of her own needs were unacceptable. As often happens, Jean began to claim her own personal power when her feelings were expressed and she received affirming responses to them. However, we must look more carefully at the sequence of feelings that unfolded for Jean.

Although Jean began the next therapy session by sharing her good news about ending the problematic relationship, she had difficulty sustaining this stronger stance. In the course of the session, Jean progressively retreated from her indignation at how badly this man had treated her, began to feel concerned about how she may have "hurt" him, and wondered whether she had been "selfish" to end the relationship because he needed her so much. However, the therapist's working hypotheses had prepared her for the possibility that Jean would feel guilty upon experiencing her anger, setting limits, or responding to her own needs. With this awareness, the therapist was able to quickly recognize Jean's guilt and engage her in questioning its validity.

As with the anger-sadness-shame constellation, Jean's affective constellation included a third, aversive feeling—guilt. Jean defensively avoided her anger because it aroused guilt, which served to reflexively return her to her characteristic affect of sadness and interpersonal presentation of helplessness. As the therapist helped Jean identify and consider all three of these interrelated feelings, she was able to relinquish her victim stance for increasing periods.

With the therapist's support, Jean could acknowledge the reality of the painful deprivation and disempowering invalidation that she had originally experienced in her family. As she stopped avoiding the reality of these painful experiences, she stopped reenacting them in her current relationships; she no longer recreated these familiar relational patterns that kept evoking the same disheartening emotions. With the therapist, Jean began to protest the hurtful ways she had been treated—both as a child and in her current adult relationships. The therapist affirmed how unfair this treatment had been, and Jean began to assert her own limits and preferences in the presence of others' competing needs or expectations, without becoming immobilized by guilt. Coming to terms with each component of this affective constellation increasingly left Jean feeling stronger and more in control of her own life than she had ever been. Once again, significant life changes followed from Jean's internal resolution of her conflicted emotions. By responding to each feeling in clients' affective constellations, therapists help clients feel better about themselves and change how they can respond to others.

This is a lot of material for most new therapists to absorb—let's pause and try to tie things together a bit. Stepping back for a moment, let's see how the previous topic, the predominant characterological affect, relates to the client's affective constellation. The predominant characterological affect will typically contain the leading edge of the affective constellation or triad. For example, clients who present affectively a predominant sense of helplessness can easily access and express sadness, which is first in their triad. Often, *the first emotion in the triad is the one that was allowed in the client's family of origin*; other feelings became subsumed under, or covered over by, this emotion. Oftentimes, the second affect in the triad was considered too threatening to a caregiver—and therefore was not allowed to be experienced or expressed. Over time, the client also came to see these legitimate feelings as unacceptable, and the client's own authentic voice was muted. The third affect in the sequence, shame or guilt in the two examples here, is often avoided because it is too disruptive or aversive for the client. Hence, the client repetitively returns to the first or leading affect that is safer and more familiar.

The therapist's intention is to validate the legitimacy of each feeling in the affective triad and remain empathic or emotionally connected to clients as they are experiencing them. This gives clients the experience of expressing themselves authentically without the expected yet unwanted responses resulting (on the one hand, for example, the parent/therapist withdrawing, rejecting, or criticizing the child/client, or on the other hand, the parent/therapist feeling burdened, immobilized, or overwhelmed by the child/client).

Meaningful changes are set in motion for many clients as they find that their previously unacceptable feelings can be shared without the familiar but unwanted consequences occurring. In particular, many begin to feel that they can center in themselves and "live inside their own skin," rather than being so anxiously preoccupied with how others see them. Thus, outgrowing these limiting familial rules about affective experience brings clients into a more cohesive personal identity. Feeling safer and stronger, they are able to risk revealing other aspects of themselves, are better able to make use of a variety of therapeutic interventions, and continue making changes with others in their lives.

Holding the Client's Pain

Having drawn out the full range of the client's feelings, the therapist's next concern is how to respond most effectively. Attachment theory—and its illuminating concepts of a secure base, attuned responsiveness, and a holding environment to provide affective containment—offers clear guidelines. Attachment-informed therapists respond effectively to clients' painful feelings by providing a "holding environment" to "contain" the client's feelings. How do therapists provide such emotional containment? In essence, *therapists do so by maintaining a steady presence in the face of the client's distress.* The therapist's nonverbal communication is "I can encompass or tolerate these difficult feelings and, together, we can get through it." Let's explore these attachment concepts further—they teach us so much about how to work with clients' pain and distress.

As we saw in Chapter 2, therapists are trying to empathize with the client and communicate that accurate understanding to the client. The therapists' intention also is to communicate their genuine compassion for the client's distress, overtly with their words and by their emotional presence. To provide this attuned responsiveness in the face of strong feelings, the therapist strives to be a "participant/observer." That is, the therapist is an involved and responsive "participant" who is empathically connected and feels with the client. At the same time, however, the therapist is also an "observer" who simultaneously maintains an observing self who thinks about, understands, and tracks what is going on (Sullivan, 1968). As a "participant/observer" in Sullivan's interpersonal terms, or an attachment figure who provides a secure base in attachment terms, the therapist provides a secure holding environment by conveying three essential messages to the client:

- these emotions will not overwhelm the therapist,
- the intensity of the pain will decrease, and
- the upsetting event is not catastrophic but can be endured.

When the therapist can provide such a holding environment, it provides clients with a new and more satisfying response to difficult feelings than they have received in other relationships. The therapist's intention is to be accurately empathic and register what the client is actually feeling—without exaggerating the problem and overreacting to it, on the one hand, or downplaying the affect by minimizing it (reassuring, explaining or dismissing), on the other hand. As an emotionally responsive and affirming therapist continues to provide these corrective experiences, clients' feelings are contained, and they readily restore their own emotional equilibrium. Real comfort comes from being seen and psychologically "held" in this way, which leaves clients feeling secure and empowered. They are not dependent, as some sometimes fear; in sharp contrast, they are becoming independent as they learn how to "self-regulate" their emotional reactions (Greenberg, 2002). In the moments following such understanding and containment, clients often make meaningful insights, risk trying out new behavior with the therapist, and begin acting in new and more adaptive ways with others in their lives.

Clients Resist Feelings to Avoid Interpersonal Consequences

In Chapter 3, we saw that both therapists and clients may be reluctant to address the client's resistance. Similarly, therapists and clients alike may steer away from the client's difficult feelings, as Jennie did at the beginning of this chapter. Thus, treatment doesn't progress very far—it soon doesn't hold much meaning—when strong feelings are not dealt with. Let's look further now at how clients can be resistant and protect themselves against difficult feelings at times. As with resistance to treatment, the therapist does not simply want to push through these defenses in order to reach clients' sadness or other significant feelings. Again, that would be winning the battle but losing the war—we don't want to press clients to feel or disclose anything they don't want to share. So what do we do instead—just wait and try to be patient? No, as before, *therapists can join collaboratively with the client in trying to understand the threat or danger if they shared a certain feeling*, rather than press for it. Let's see how therapists can take this approach effectively.

Resistance and defense against feelings are more about expected but unwanted *interpersonal consequences* (Client: "You wouldn't respect me if you really knew how afraid of that I am") than the more traditional notion of *intrapsychic dynamics* (Client: "It hurts too much; I can't stand it"). That is, clients often avoid difficult feelings because they expect to receive *familiar but unwanted responses* from the therapist and others that evoke fear, shame, or guilt. Therapists create a more affirming interpersonal process, with greater safety for clients, if they enlist clients' collaboration in trying to understand the very valid reasons why it has been threatening to experience or share certain feelings. For example, rather than pressing the client to disclose a difficult feeling, the therapist may use responses such as these to explore the relational threat:

- OK, let's not talk about how you feel about this. Let's talk instead about what might go wrong if we did.
- If you did cry or let me see your vulnerability, what might go on between us? Is there something that has happened with others that you don't want to happen again here with me?
- What is the danger for you if you let me see your sadness? Help me understand how you have been hurt in the past.
- How did your parents respond to you when you were feeling _____? What did they say and do? Can you recall the look on their faces or what they were feeling toward you?

In these ways, the therapist does not push for the content—the specific feeling—but works with the client to clarify the threat or danger—the reality-based reasons for why it has not been safe to experience or share the feeling. As in Chapter 3, the therapist is honoring the client's resistance by collaborating with the client in trying to understand why this reluctance or defense was once a necessary and adaptive coping mechanism. The therapist's basic orienting assumption is that the client's resistance to certain feelings makes sense in terms of what has transpired in other relationships, although it is no longer necessary or adaptive in many current relationships. *If the therapist and client can clarify how the client's defense was necessary and adaptive at another point in time, and in some current relationships as well, the therapist can make it clear that she (and later, that some others in the client's life) will not respond in these familiar but unwanted ways.* With this new understanding, it will soon be safe enough for the client's conflicted feelings to emerge.

Let's look further at the *interpersonal reasons* why clients often want to avoid difficult feelings—this is a big conceptual shift for many new therapists. Though clients may tell therapists that they don't want to explore their underlying feelings because they are "too painful," "overwhelming," or it "doesn't do any good anyway," this is not the whole story. Therapists can begin to generate working hypotheses about the unwanted *interpersonal consequences* that clients may be expecting if they express certain feelings. Clients usually avoid difficult feelings because of their expectations that certain unwanted interpersonal consequences will result if they share with the therapist feelings that have been unacceptable to significant others.

For example, on the basis of their repeated experiences with others, clients commonly have expectations of being ridiculed or judged, being invalidated and having their feelings dismissed as unimportant or unrealistic, being ignored or not taken seriously, seeing that others are hurt or burdened by their feelings, having others overreact or dramatize their feelings, and so forth. Such interpersonal threats are usually far more important than their own subjective discomfort. Although it is very difficult for most beginning therapists to grasp this, *clients also hold these same expectations toward the therapist—even though the therapist has never responded in any of these problematic ways!* In sum, the discerning therapist can usually find very good reasons why clients learned that certain feelings were unsafe to experience or share. Therapists can also help clients differentiate then from now, and make overt how they would like to respond differently than others have in the past. For example:

THERAPIST: I'm wondering if it has ever seemed like I was responding to your feelings in that kind of judgmental way?

OR

THERAPIST: How do you think I am going to respond if you risked sharing that feeling and letting me see that part of you?

Of special interest, one interpersonal purpose of clients' defenses against their feelings may be to protect their caregivers. Many clients have learned to avoid or deny painful feelings in order to protect their caregivers from seeing the hurtful impact they are having. However, this loyalty to the caregiver is carried out at the clients' own expense. As adults, these clients are still complying with familial rules and colluding with their caregivers in denying the impact of hurtful parental actions. By so doing, these clients are preserving insecure attachment ties ("Everything's fine—no problem") at the cost of not acknowledging their own painful feelings. This protects the (internalized) caregivers from seeing what they did that hurt the client (such as embarrassing or ridiculing him, ignoring him or favoring a sibling, needing the child to be "perfect" or fulfill the family role of "hero," being afraid of getting hit or witnessing domestic violence, and so forth). If clients participate in this denial and dismiss the problems that actually were occurring, homeostatic family rules are maintained. The client remains attached to internalized parents who often remain idealized or "good" while, in this splitting defense, the client remains "bad" and feels that something is wrong with her. However, if painful feelings are permitted expression in therapy, and *the therapist can affirm the clients' experience without blaming or rejecting the parent*, chronic symptoms such as lifelong dysthymia or generalized anxiety often change. For example:

> THERAPIST: Yes, of course you feel so anxious and worried all the time. It must have been frightening to watch your parents fight like that. Somebody might get really hurt—and you couldn't stop them. That's a heartbreaking spot for a little girl to be in.

To sum up, the therapist's intention is not to help clients "get their feeling out." As noted earlier, helping clients "vent" does not usually help very much. Instead, the therapist's goal is to help clients understand the interpersonal themes that give rise to their difficulties with certain feelings. By clarifying that these unwanted interpersonal consequences have indeed occurred in other important relationships, but are not going to occur in the therapeutic relationship, the therapist provides the interpersonal safety that clients need. With this understanding and safety, therapists can help clients sort through the relationships in their current lives where these same problematic scenarios are recurring, and other relationships where better responses are occurring or could be developed.

Providing a Holding Environment

In attachment terms, the best way for therapists to help clients resolve challenging feelings is to provide a holding environment that contains their distress. However, when stressful life situations activate clients' most problematic feelings, their schemas will shape faulty expectations and distort their perceptions of the therapist. That is, based on the early maladaptive schemas that dominate in these affect-laden moments, *clients anticipate that they are going to receive the same problematic response from the therapist that they have received from others in the past*. It is difficult for most beginning therapists to grasp that clients believe this, often with great certainty, because the therapist has never responded to the client in these hurtful or unwanted ways. As we have emphasized, a corrective emotional experience occurs when the therapist responds in a new and safer way that resolves, rather than metaphorically reenacts old relational patterns. However, *unless the therapist clarifies or makes overt what the therapist is thinking and feeling as clients experience these conflicted feelings, clients routinely misperceive the therapist's response to fit their old schemas*. For example, by working in the here-and-now with what is going on between them, the therapist can clarify or correct these expectable misunderstandings or transference distortions:

> THERAPIST: Right now, as you are feeling so upset about this, what do you think is going on for me—what am I thinking about you?
>
> CLIENT: Well, since you're a therapist, you have to act nice and supportive so you'd probably never say anything. But you're probably thinking I'm overreacting

again. You know, acting like a child, and I should just stop feeling sorry for my-self and grow up. Like my husband says.

THERAPIST: Oh, no, I'm not thinking that at all. Actually, I was feeling how sad it is that this has hurt you so much for so long.

Without such exploration and clarification, clients routinely hold the inaccurate belief that the therapist is privately disappointed, frustrated, burdened, and so forth, just as significant others have been so many times before. With these considerations in mind, let's examine further how therapists can provide a holding environment to contain difficult feelings.

While growing up, the client had certain developmental experiences repeat over and over again that were painful, aggrandizing, disempowering, and so forth. Oftentimes, enduring problems arose because the adult caregivers were unable to provide the affirmation and understanding the child needed to learn how to manage or self-regulate their own emotional arousal. Caregivers provide such containment when the child is psychologically held in a close interpersonal envelope of empathic caring (Brisch, 2002; Cassidy & Shaver, 2002). Behaviorally, this means that the child's concerns are heard, taken seriously, and understood by a caregiver who remains emotionally present and responsive. In other words, the child's distress registers with the caregiver and is taken seriously, without being diminished or exaggerated—the caregiver is accurately empathic. The child feels secure that the caregiver will remain available to help manage the problem, if only by expressing interest and concern. That is, *the securely attached child is secure in the expectation of getting help or being responded to* when she is distressed or in need (it doesn't mean that the child is indulged or spoiled in any way or that the caretaker can always solve the problem). Before children can develop the capacity to manage feelings on their own and be more independent, someone must first provide this holding context for them. Too often, however, this essential developmental experience of affect regulation is not provided. For example, suppose the child is sad or hurt. In some families, the child's sadness

- arouses the parent's own sadness, and the parent withdraws, which leaves the child emotionally alone;
- makes the parent feel guilty, and the parent denies the child's sad feelings and tries to cheer the child up, which leaves the child confused and alienated from her own authentic experience; or
- makes the parent feel inadequate in the parenting role, and the parent responds punitively or derisively toward the child's sadness, which leaves the child ashamed of his vulnerability.

In such commonly occurring scenarios, the child's sadness cannot be heard and responded to; as a result, these feelings cannot run their natural course and come to their own natural resolution. Young children cannot contain a strong, painful affect by themselves without support from an emotionally available caregiver. As a result, the child will have to find ways to deny or avoid this painful feeling. Years later, when this child is an adult client and risks sharing the unacceptable or threatening feeling with the therapist, the therapist's intention is to provide the holding environment that was missed developmentally. Working within this attachment framework, the therapist is providing a safe harbor for a while—the secure base that the client missed developmentally. We have already seen two examples of this in Chapter 2: Marsha's second therapist (pages 60–61) heard and responded to her experience of emptiness, and the therapist of the depressed young mother sensitively articulated the mother's own unanswered cries (pages 80–81). Both therapists successfully provided a holding environment that allowed their clients to make important changes in the next few sessions. Turning to specifics, we'll look first at what therapists should not do; then we'll examine further what therapists can do to help clients contain and resolve difficult feelings.

Too often, therapists avoid or move away from clients' painful emotions. In so doing, they reenact the clients' developmental problem and leave clients disconnected from others while they are experiencing their emotions. The most common reason for this countertransference is because therapists assume too much responsibility both for causing the client's pain and for alleviating it. New therapists may think, "I got the client into this, so now it's up to me to get him out of it." Both parts of this belief are false. By responding in the ways suggested here, therapists reveal feelings that are already present in the client; they do not cause the pain. Moreover, it is patronizing for therapists to think they can "fix" clients or "get them back together again." In actuality, the therapist never has the power to manipulate another's feelings in this way, and attempts to do so often recapitulate the caretakers' controlling or dependency fostering stance with the client.

When therapists take on responsibility for the client's feelings and place on themselves the unrealistic expectation that they have to fix or solve the problem, the usual outcome is for therapists to feel inadequate. In addition, therapists may be uncomfortable with their own sadness, shame, or other feelings that have been evoked by the client—that is, countertransference. In such situations, therapists are likely to respond to the client's feelings in ineffective ways:

- Interpreting what the feelings mean and intellectually distance themselves
- Becoming directive and telling the client what to do
- Reassuring the client and explaining that everything will be all right
- Becoming anxious and changing the topic
- Falling silent and emotionally withdrawing
- Self-disclosing or moving into their own feelings
- Diminishing the client by trying to rescue him
- Overidentifying with the client and pressing the client to make some decision or take some action to truncate the feeling

When these ineffective responses occur, there is no holding environment. Therapists' avoidance, as reflected in these responses, often replays the client's developmental predicament. For example, by trying to talk her out of her feelings, Marsha's first therapist reenacted her experiences in her family. Marsha's first therapist went even further to avoid Marsha's feelings by, in effect, sending her away: When Marsha got near her sad, empty feelings, the therapist suggested that she join a group and talked to her about seeing a psychiatrist for antidepressant medications. Such referrals are certainly necessary at times, yet in this case they were used to distance the therapist from the client's feelings. Unfortunately, this scenario is a common occurrence in treatment.

If these distancing and controlling responses are ineffective, what should the therapist do instead? The therapist's aim is to welcome the client's feelings and approach them directly. For example, the following responses would be more effective ways to respond to Marsha's strong feelings.

> THERAPIST: I'm glad you're taking the risk of telling me about these sad and empty feelings. Let's keep talking, I want to understand them better.

> OR

> THERAPIST: I know talking this way and being together with such painful feelings breaks the social norms, but I appreciate being able to share all of this with you.

Therapists also want to find genuine ways to express their care and concern. Therapists can do this in words, of course, but it is equally important to communicate nonverbally in tone and manner. In particular, therapists want to be emotionally present with clients while they are experiencing their feelings. There are many ways to do this, of course, and much of this communication is nonverbal or involves only a few words. We'll now consider some of the elements found in effective responses that create a successful holding environment. Having assimilated these guidelines, however, therapists will want to tailor their responses to fit each particular client and to make them congruent with their own personal styles.

Initially, the therapist's primary goal is to stay emotionally connected or present with clients as they are experiencing painful feelings. As discussed in Chapter 2, the therapist may also want to reflect what clients are feeling and to affirm the reality of their experience. This can be done as simply as by acknowledging, "You're very sad right now." With some clients, it may be helpful to further validate their experience by saying, "Of course you are feeling mad right now. It hurt you very much when he did that." An important part of developing their own professional identities, therapists in training are encouraged to explore and try out their own ways of expressing their compassion or concern. Role models are important, yet instead of trying to be like your favorite supervisor or therapist, new therapists are encouraged to find their own style or natural way of being with people in these special moments:

> THERAPIST: I can see how much you miss your dad and how sad you are about his loss. I feel close to you right now and honored that you choose to share such a special part of you with me.

To provide a secure holding environment, therapists also want to demonstrate behaviorally that they can tolerate the full intensity of clients' feelings. In other words, they must communicate by their manner—and overtly in words with some clients—that they are in no way hurt, burdened, or undone by clients' feelings.

> THERAPIST: I'm privileged that you choose to share this with me. I'm with you right now, and we're going to stay together in this until we have worked our way through it.

To undergo a new or reparative experience, clients need to see that the therapist is not responding in the problematic ways that others have and that they expect from the therapist. It needs to be clear that the therapist is not threatened by their feelings or judgmental of them, does not need to move away from them in any way, and is still fully committed to the relationship. As emphasized before, new therapists usually fail to appreciate how clients' early maladaptive schemas are activated in these affect-laden moments, take precedence over the current reality, and distort their perceptions of the therapist.

As the client regathers his equilibrium from a strong emotion, the therapist may also want to clarify what caregivers and others have done when the client felt this way. The therapist may want to discuss how their current relationship is different from the client's relationships with others (for example, the therapist remains emotionally present and affirming rather than becoming impatient or withdrawing) and emphasize that the therapist is accepting of this aspect of the client that has been revealed. Thus it is important to *debrief* clients by asking them how it has been to share these feelings or personal disclosures with the therapist. Too often, therapists miss this important opportunity. Genuinely touched by the client, they fail to appreciate that it is at these intense, special moments that the client is most likely to misperceive the therapist's caring response in terms of earlier schemas. To correct these expectable misunderstandings, it is necessary to ask clients what they think the therapist was thinking, feeling, or going through while they were feeling sad, angry, and so forth. Early in their training, it can be awkward for new therapists to broach this. However, long-standing transference distortions, which also cause problems and disrupt relationships with others, can be identified and resolved at these moments. For example, the therapist might clarify the client's distortion by means of a self-involving comment:

> THERAPIST: No, I wasn't thinking that you "looked silly" and "sounded weird" in any way. In fact, I was touched by how alone you felt in that awkward, embarrassing situation. Your feelings sure made sense to me.

When offered sincerely, simple human responses of validation and respect mean much to clients and help to restore their dignity. Unfortunately, new therapists often believe that they have to do much more than this to respond adequately to the client's pain. These concerns may reflect the therapist's own unrealistic expectations that there is an absolute right or wrong way to respond, or that they have to respond in some precise, clinical way. In reality, the client does not need the therapist to provide eloquent, crafted responses in a calmly self-assured manner. Such perfectionism is as unnecessary as it is unrealistic. It often serves only to make the therapist more preoccupied with evaluating her own performance and less focused on just being present with the client and grasping what he is really saying. As Winnicott (1965) says, the child needs only "good enough mothering." Therapists provide a "good enough" holding environment when they find their own genuine ways to enact these guidelines. Clients will be appreciative of sincere efforts and find them helpful, even if they are not always expressed articulately.

Change from the Inside Out

When therapists provide a holding environment to contain feelings that clients have not been able to come to terms with previously, most clients will return to sessions and begin reporting that meaningful changes are occurring in their lives. How does this come about? A corrective emotional experience occurs when the therapist can be affirming and remain emotionally present with clients without superficially explaining, reassuring, or otherwise moving away from their feelings.

Many clients will be profoundly relieved to disconfirm their faulty beliefs and find that neither they nor others need be hurt by their feelings. How does this translate into behavior change? When the therapist compassionately accepts clients' most threatening emotions, and it is made clear to the client that the unwanted consequences they expected did not ensue this time, clients forge a more authentic connection to the therapist—and to themselves. In other words, it is safe this time to be seen and known as they truly are. This reparative experience is empowering. It allows clients to change their faulty expectations of others and pathogenic beliefs about themselves, and to begin responding in new and more adaptive ways with others in their lives. Rigid schemas expand and clients now have greater freedom to pursue their own interests and goals—choosing more actively how they want to live and shape their lives. Let's explore further how the experience of change with the therapist leads to change with others.

In this supportive context, clients' feelings can run their course and come to a natural close, usually in just a few moments. Clients then recover, by themselves, and readily reestablish their own equilibrium. As clients continue to receive these types of corrective emotional experiences, clients disconfirm pathogenic beliefs about themselves and learn that their feelings—and they themselves—are not disgusting, overwhelming, and so forth. As they find that their previously unacceptable feelings do not rupture this relationship, their feelings no longer compel them to withdraw and be isolated, feel ashamed and strive to be perfect, or fall back on other interpersonal coping strategies that just bring more symptoms and problems. Thus, the feelings that previously had to be sequestered away can now be experienced, understood, and integrated; clients no longer need to employ these faulty coping strategies that engender so many problems with others. Furthermore, when the therapist can hold clients' threatening feelings and allow them to tolerate experiencing and sharing it, they, in turn, will become more able to contain these feelings on their own, which greatly changes how they can respond to others. *This is change from the inside out.* This interpersonal process often propels a far-reaching trajectory of enduring behavioral change, and long-standing symptoms often abate at this point. Ironically, when clients risk exposing their pain, vulnerability, or shame, and the therapist responds with compassion, they feel more powerful. One of the most satisfying aspects of being a therapist is to facilitate such self-acceptance and self-efficacy. To illustrate this approach, let's look at a brief case study.

Sherry, a 1st-year practicum student, was struggling to assimilate all the new ways of thinking about therapy that she had been learning. It was an exciting but difficult year as her entire worldview was shifting. Part of the difficulty was that her practicum instructor and her individual supervisor were proposing very different ways to approach her clients. Having grown up with combative, divorced parents, she had conflicts about trying to please "both sides." This unwanted old scenario was being activated in this situation, which made the already complex challenge of clinical training even more difficult. At times, Sherry was almost immobilized as she tried to sift through these competing ideas, find what made sense to her, and come up with something that she thought would be useful to say to her clients. Two-thirds of the way through her first practicum, she felt more self-conscious and inhibited about responding to people than she had been when she started the program!

Despite her insecurity, Sherry actually was learning a lot and had been responding quite helpfully to her client, Maria. Near the end of one of their weekly sessions, Maria began to disclose a painful memory. With tears in her eyes, Maria recounted the sickening experience of a date rape 15 years earlier, when she first went to college. Listening to Maria's story, Sherry felt sad and angry. However, as she began to think of all the things she should and should not say, and the different things her instructor and supervisor would probably want her to say, she felt frozen and could not find the words to say anything at all. Although it was much less than a minute, the silence grew painfully awkward as Maria finished speaking and waited for Sherry to respond. Unable to speak, Sherry just sat there. Deeply moved by Maria's painful story, but immobilized by thinking of all the different ways she "should" respond, she couldn't find any words.

After an excruciating pause, Maria tried to take care of Sherry and started talking again, but this did not last long. After a few minutes, Maria chided, "Don't you understand? He hurt me a lot when he did that!" Sherry's inability to respond had ruptured their alliance. Although Sherry felt she had made a "terrible mistake," she was able to restore by validating Maria's anger and talking through what just happened between them.

SHERRY: Yes. He did hurt you very much—it was frightening and painful. I'm very sorry that happened. And I understand why you are angry at me for not responding. I was touched by what you said, but I just couldn't find the right words for a minute.

MARIA: You couldn't?

SHERRY: No, I wanted to, because I felt for you. But I just couldn't find the words I wanted to say, and the harder I searched for them, the worse it got. I'm sorry this happened; it must have been hurtful.

MARIA: It's OK. I don't know what to say either sometimes. . . . (*pauses*) I don't know why, but this makes me think of my daughter when she was little. If I would take the time to stop and pick her up and really listen to her when she needed me, we would get through her problem pretty quick. My husband called it "collecting her." But if I was in a hurry and tried to ignore her or speed her up, she would escalate and it would go on for a longer time.

SHERRY: You're a very good mom. I think you're remembering that right now because that's what you needed from me—to be attended to, cared about, collected. You needed me to give you what you gave your daughter.

MARIA: Yeah, I think that's just exactly what I needed. (*tearing*) You know, I never dreamed of telling my mother what happened to me—you know, about the date rape thing. She wanted me to be "happy" all the time; she didn't really want to hear about problems. . . .

Obviously, Sherry recovered from her "mistake" very effectively, and began providing the holding environment that the client herself had articulated so well with her own young daughter. As in this example, being nondefensive and validating the client's dissatisfaction often leads the client to act stronger in the next few minutes and make progress in treatment. For example, clients may bring up related issues in other relationships—as Maria did; achieve meaningful insights that lead to behavior change; bring up significant new material—such as other problems they have been having with the therapist or others; and so forth. Moreover, by recovering and affirming Maria's feelings at the second attempt, Sherry gave Maria a corrective emotional experience that helped her progress in treatment.

As Sherry is teaching us, we're too afraid of making mistakes—misunderstandings can be resolved. If the therapist is willing to take the risk of making problems overt and talking them through with the client, the expectable ruptures that will occur in every therapeutic relationship can be restored (Safran & Muran, 2003; Hovarth, 2000).

To sum up, when difficult feelings emerge, therapists have the best opportunity to help clients change. The most important corrective emotional experiences usually occur around the therapist's new, more satisfying response to the client's emotions. On the other hand, however, the client's maladaptive relational patterns are most likely to be reenacted with the therapist in these affect-laden moments as well. Why? Vulnerability activates our schemas and leads clients to expect the familiar but unwanted response from the therapist they have received in the past. Per Beck's "hot cognitions," clients are most apt to distort or misperceive the therapist's response and slot it to fit old expectations, when strong feelings have been triggered. Compounding this situation, the therapist's own personal issues or countertransference reactions are also most likely to be evoked in these sensitive and intense interactions. Thus, we need to think about our own countertransference issues and turn now to factors in the therapist's own life that make it hard to respond to clients feelings.

Countertransference Prevents Therapists from Responding to Feelings

Just as it is easier sometimes for clients to externalize and focus on others rather than themselves, it can also be easier for therapists to focus on the client than it is to look within at their own issues and dynamics. If therapists are not willing to work with their own emotional reactions to the material clients present, however, they will tend to avoid clients' feelings or to engage them only on an intellectual level. The purpose of this section is to help therapists responsibly manage their own reactions to the evocative material that clients present by learning about their own countertransference propensities.

Neither a working alliance nor a secure holding environment will occur unless therapists are accurately empathic and can communicate their respect and compassion for the emotional conflicts clients are struggling with. Realistically, however, it is difficult for therapists to respond to all of their clients' feelings, because all therapists bring their own developmental histories and current life stressors to the counseling session. Let's explore how these personal factors can keep every therapist from responding effectively to clients' feelings at times.

The Therapist's Need to Be Liked

Most trainees who are drawn to clinical practice are generous people, genuinely concerned about others, and ready to give of themselves. Often nurturing individuals by nature, they can readily empathize with and respond to others' hurt and pain. At the same time, however, many therapists also have strong needs to be liked. As a result, many therapists have difficulty responding nondefensively to the problematic relational patterns that some clients have been enacting with others in their lives (that is, they are confrontational, controlling, competitive, and so forth) and begin to replay these faulty interpersonal patterns with the therapist as well. In particular, researchers find that clients often act in the same angry, critical, demanding, or distrustful ways toward the therapist that they have been with others and that most therapists respond poorly to this hostility with their own countertransference reactions (Strupp & Hadley, 1979; Najavits & Strupp, 1994). More specifically, Binder and Strupp (1997) report that therapists of every theoretical persuasion routinely respond to the clients' negativity personally, with their own emotional reactions of anger, withdrawal and, most frequently, a pejorative attitude toward the client! Most clinical training programs do too little to help new therapists learn how to respond nondefensively to this negativity, which is especially problematic for therapists who need their clients to like them.

Other therapists have strong needs to nurture others who are dependent on them. These therapists enter clinical practice, in part, to fulfill these needs, and they will have difficulty supporting the client's healthy individuation at times. Continuing, other therapists grew up fulfilling the role of "caretaker" in their families of origin. They are drawn to clinical practice, in part, by the often unrecognized need to rescue their caregiver, and later others, from alcoholism, an unhappy marriage, chronic depression, and so on. This countertransference issue of "parentification," in particular, can make it harder for therapists to address clients' vulnerable feelings of sadness or hurt. Those feelings can evoke for therapists the familiar predicament of needing to protect or take care of their caregiver. Because, in reality, this is always an impossible task for parentified children to actually succeed with, it threatens them once again with the painful prospect of disappointing their caregiver (and now their client) and feeling inadequate.

For these and a thousand other countertransference issues, all therapists need to assess their own motivations for becoming clinicians. We need to reflect on the personal needs of our own, and the roles we played in our families of origin, that are fulfilled by becoming a therapist. The issue is not whether therapists have their own needs and countertransference tendencies, but how they deal with them. Every therapist has certain countertransference reactions—it's just human—and self-examination is one way to manage them effectively.

New therapists may be taken aback when their well-intended attempts to approach the client's affect occasionally meet with negativity: anger, distrust, defensiveness, and so forth. Too often, beginning therapists absorb the client's rebuff, comply with the defensive side of the client's feelings, and stop approaching the client's feelings. This response is most likely from therapists whose needs to nurture or to be liked have been frustrated. To be effective, however, therapists want to continue by working with the client's negative feelings that have engendered distance or discouraged others in their lives from responding as well. Thus the therapist's aim is to be *flexible* and look for other ways to approach the client's feelings and, thereby, to begin changing the maladaptive relational pattern that usually ensues (Safran & Muran, 1995, 2003). That is, rather than responding automatically with the usual reaction that clients get in social situations (that is, hostility, defensiveness, or withdrawal), the therapist tries to be thoughtful about the feelings or reactions the client usually elicits, and consider other ways of responding that do not repeat the familiar, problematic pattern (observing this aloud with a process comment, such as "I'm wondering if you're aware of how angry you sound right there. Has anybody ever given you this kind of feedback before?").

To respond more flexibly and nondefensively, therapists can adopt the strategies for working with the client's resistance that were presented in Chapter 3. That is, therapists should not press clients to express a feeling they are reluctant to share but, instead, explore what the threat or danger may be for them if they do so:

• What does it say or mean about you if you feel ____?
• What could happen between us, or what could go wrong for you, if you let yourself feel ____?
• How have people responded to you in the past when you felt ____?

Therapists do not want to win the client's approval by trying to be a "nice" person who does not approach difficult feelings or talk about awkward interactions. On the other hand, therapists do not want to demand that clients enter into and share, or stay with, a certain affect. Instead, they repeatedly offer invitations or bids and allow clients to enter and leave difficult emotions as they wish. Letting clients have it their own way avoids the issues of coercion and compliance that so many clients had in their families of origin and that people who have been marginalized often have experienced in the dominant culture. Therapists provide an effective middle ground when they work with the client to understand the resistance by exploring together the threat or danger that once existed. *With understanding comes safety*, and clients soon will be able to explore feelings that needed to be held away in the past.

To sum up, all therapists are going to have difficulty responding to certain feelings at times, and therapists are especially likely to have difficulty working with the client's critical, angry, or negative feelings toward them. For many, learning to respond nondefensively to the client's negativity (anger, irritability, competitiveness, criticism, control, and so forth) is the most challenging task in clinical training. New therapists need practice with their instructors and support from supervisors to keep from personalizing this negativity and responding defensively with their own countertransference. Instead, therapists can learn much by

- looking for "negativity" in clients' maladaptive relational patterns with the therapist and with others;
- exploring collaboratively how this negativity that is occurring with the therapist is likely to play out and cause problems in clients' relationships with others; and
- making process comments to intervene and begin changing this problematic pattern in their current interaction.

For example:

THERAPIST: When you speak to me like that, Joe, I feel criticized. Maybe it's just me, but it seems like that happens pretty often. You've been talking about problems communicating with your wife and getting along with your son. Do you think they are feeling criticized or put down by you as I often do?

Therapists Misperceive Their Responsibility

Too often therapists believe that, if they respond to a client's feelings, they are somehow responsible for causing the feeling or for taking the client's pain away. Assuming responsibility for the client's feelings is unrealistic and causes problems both for the therapist and the client. Let's look at two aspects of this problem.

First, some therapists avoid their clients' feelings because they do not want to see them hurt or feel that they have made their clients feel bad. These therapists mistakenly believe that responding to a painful feeling in the client is the same as causing that feeling to exist. Therapists can keep in mind that they did not cause the client to have this feeling; it was present before the client met the therapist. The therapist has simply responded to the client in a way that reveals the feeling. The misconception that the therapist has caused the client's painful feeling often leads therapists to feel guilty and avoid sensitive feelings.

A second problem arises when therapists labor under the impossible burden of having to "fix" the hurt they believe they have "caused" the client to feel. When therapists assume an unrealistic responsibility for causing the client's feeling, they also mistakenly assume responsibility for doing something to make the client feel better. This is an unrealistic responsibility; therapists can never take away or fix clients' pain. In sharp contrast, however, it is empowering for the client when therapists can share the client's experience and respond affirmingly with empathic understanding.

Therapists who assume responsibility for clients' feelings are apt to avoid those feelings. Typically, they do this by telling clients what to do, reassuring them, and explaining or interpreting their emotions before they can fully experience or share them. These approaches often prove ineffective.

If therapists do not shoulder this inappropriate responsibility for causing or for fixing the client's feelings, what should they do? The therapist's response to the client's feelings is threefold: to identify, to join, and to affirm. First, the therapist tries to help clients identify and express more fully whatever feelings they are having. Second, the therapist's intention is to be empathic or emotionally responsive to clients, so that clients can share their feelings with a concerned other rather than having to experience them alone, as they often have in the past. Third, the therapist wants to affirm or validate the clients' feelings, by helping clients understand their feelings and make sense of why they are experiencing this feeling in a particular situation. If the therapist accepts this threefold challenge, but does not assume responsibility for causing or changing clients' feelings, the therapist will be free to respond effectively to whatever feelings clients present.

Even therapists who are well aware of the pitfalls of assuming responsibility for clients' feelings are likely to struggle with this issue when the client's affect becomes intense. For example, a graduate student therapist was treating a young boy who was in the midst of a potentially life-threatening battle with leukemia. The therapist would sit with the boy in the hospital, and, among other topics, they would sometimes talk about the possibility that he could die. In response to the child's questions about death, the therapist would try to discern the central or key concern for him: "What would be the scariest thing about dying? What are you most afraid of?" As the therapist approached the feelings the child was expressing, the boy responded, "I don't want to die! I'm afraid to die!" Seeing the fear in this young child's eyes and feeling totally inadequate, the therapist questioned what she was doing and felt "cruel" to be evoking such anguish.

But the child went on to say, "I don't want to be alone. I'll be all alone if I die!" Inviting the child to explore his feeling more fully revealed his key concern that he would be "all alone." Hearing the boy's intense separation anxiety, the therapist now understood the specific meaning this fear held for him and was able to talk with him in more helpful ways. For example, with this clarification, the therapist was able to intervene by helping the child's mother understand his basic fear and help him with it. With the therapist acting as "coach," his mother was able to reassure him that, if he were to die someday, their love for each other would last forever and that he would always be close with her inside. By talking with the mother and son together about his fear of being left alone, the therapist helped them to become closer during his illness. The felt sense of security that came from this shared understanding tangibly alleviated the child's distress.

When working with such profound emotions, therapists almost inevitably feel to some extent that they are responsible for causing the client's suffering. As a result, they tend to avoid these intense feelings. To prevent this, therapists need to consult with a supervisor or colleague to help them manage their own exaggerated feelings of responsibility at times. Otherwise, these countertransference reactions will stop therapists from responding effectively and giving clients the holding environment they need. In the example of the boy with leukemia, the therapist could not have stayed with the intensity of the boy's fear, or the mother's sadness, without regularly scheduled consultations with a supervisor to help her manage her own emotional reactions.

Family Rules

Therapists may not be able to respond effectively to certain feelings that the client presents because of family rules about emotional expression that they learned in their own families of origin. Families have unspoken rules that govern emotional expression: A rule-bound "homeostatic" system prescribes which feelings can be expressed and which are unacceptable. This system also prescribes when, how, and to whom certain feelings can be expressed. The permitted intensity or degree of emotional expression is also controlled by homeostatic family mechanisms. For example, if too much conflict, independence, or closeness is expressed between certain family members, a corrective homeostatic mechanism is activated to bring the level of emotional expression back within acceptable limits. The following example illustrates how homeostatic mechanisms operate to keep family emotional expression within prescribed limits.

Suppose that in the Smith family a moderate degree of anger can exist between the older sister and the younger brother. A lesser degree of conflict can be expressed between the father and both children. However, anger or even modest disagreement is rarely expressed between any family member and the mother, and is never expressed between the father and the mother. What is the homeostatic mechanism that serves to keep this system operating within these boundaries? Whenever a child begins to express angry feelings or have direct conflict with the mother, she stops the child's protest by taking the role of martyr and looking sad and hurt. If the child's anger continues, the father intervenes and says, "Don't talk to your mother that way," or else the mother employs the father as a regulatory influence by saying, "I'm going to tell your father about this when he comes home."

In contrast, if anger or differences of opinion begin to escalate between the mother and father, the children serve as homeostatic regulators and respond in predictable, patterned ways to terminate the parental conflict. For example, the oldest child, 8-year-old Mary, serves as a go-between in the parental relationship. She tries to make peace between her parents by carrying messages back and forth between them. This mechanism usually succeeds in terminating parental conflict, except in periods of high family stress. If the parental conflict continues, the role of the younger may be to provide a diversion to the parental conflict by acting out in some predictable way—by breaking something or having an accident. When the son makes himself a problem that mother and father must jointly address, their rule-breaking level of conflict is terminated, and the family's emotional equilibrium returns to acceptable levels.

Fifteen years later, Mary, the oldest child in the Smith family who used to serve as the mediator to assuage parental conflict, is now a graduate student in clinical training. The family rules that Mary has learned in her family of origin will influence her ability to respond to her clients' emotions—just as the family rules that all therapists have learned in their families will influence how they respond to their clients.

For example, Ann, a depressed and irritable client whom Mary was seeing at her internship site, became frustrated with Mary because of her continuing depression. Just as she accused her husband and others of always letting her down, Ann angrily criticized Mary: "I keep coming here every week, but you're not doing anything to make me better. Sometimes you just make me feel worse!"

Mary was taken aback by Ann's angry charge, even though it was short-lived. Mary was especially surprised because she thought they had been working together more effectively in recent weeks. Throughout the rest of the session, Mary backed away from Ann's anger and frustration by offering reassurances and trying to convince Ann that she would get better if she would continue to come to therapy and work with the difficult feelings they had been sharing. Mary felt "terrible" that Ann had been so critical of her and anxiously sought help from her supervisor the minute the session was over.

Mary's supervisor was responsive to her distress over Ann's anger and was able to help Mary sort through what went on for her during the session. In reviewing a videotape recording of the session, the supervisor helped Mary recognize how she had tried to assuage Ann's criticism with reassurances, rather than acknowledging Ann's anger and inviting her to discuss her dissatisfaction more fully. Supportive but also challenging, the supervisor was encouraging Mary to join Ann in figuring out what wasn't working and collaboratively explore what they could change to make treatment more helpful. She suggested that Mary say, for example, "Your depression isn't getting better, and it doesn't feel like I'm helping you. I'm glad you're bringing this up. Let's talk about that and see if we can come up with some other ways of working together that might be better."

With her supervisor's help, Mary quickly realized that she had been unable to tolerate Ann's anger. She had indeed tried to assuage it, just as she had always done with her parents and others. Mary's role in her family was to take care of the conflict, rather than to allow others to talk about their dissatisfaction and sort through what they could do to resolve the problem. In addition to clarifying Mary's old response pattern and modeling other responses she could try out in the future, the supervisor helped Mary generate working hypotheses about what this critical response may have meant for Ann. The supervisor observed that Ann had expressed her anger just after sharing her hurt, and hypothesized that the anger may be a part of her affective constellation. In other words, Ann needed the experience of being able to express her anger at someone who would take her concerns seriously but also who would not be run over or intimidated by her—as she had learned to expect. Thus, Ann may have been "testing" to see whether she could be angry or critical of Mary and still remain in a relationship with her—which Ann had not been able to do in other important relationships.

Without this helpful consultation, Mary and Ann may have become stuck at this point in their relationship and treatment might not have progressed. Ann could not move forward and make progress with her depression, and its interpersonal fallout, as long as Mary could not tolerate the anger component in Ann's affective constellation. In their next session, however, Mary was able to recover and restore their working alliance by inviting Ann to discuss her frustration. Although Mary still felt an anxious compulsion to reassure Ann and move on, she was able to contain her discomfort this time and discuss Ann's concerns more fully. Through this frank discussion, Mary learned about things that weren't working for Ann in their relationship, and together they came up with changes that each thought could help them work more effectively. In particular, Ann wanted Mary to be more active in sessions, "say more" about what she thought was going on, and make suggestions about what Ann could do differently at home each week or try out between sessions.

Most clients will not be able to resolve long-standing problems unless the therapist can respond to each component in their affective constellation. *Treatment often stalls at the point where the therapist cannot respond to an important feeling the client is experiencing.* With this in mind, all therapists are encouraged to think about how emotions were dealt with in their families and reflect on how they are likely to bring their own familial history to their clinical work. For many therapists, their own therapy will be the best way to learn about and change limiting family rules and roles that still govern how they act with clients.

Situational Problems in the Therapist's Own Life

Situational stressors in therapists' own lives may also prevent them from responding effectively to clients' feelings. For example, when personal problems in therapists' own lives are similar to those their clients are experiencing (such as marital conflict, parenting problems, or health concerns), therapists may have difficulty approaching their clients' affect or will not be able to allow clients to experience fully the feelings they are struggling with.

Therapy is a very human interchange. Although the therapist's personal qualities and emotional responsiveness are the greatest asset in helping clients change, they will also be an obstacle at times. When the therapist repeatedly fails to be accurately empathic, avoids rather than approaches the client's feelings, or, most importantly, becomes personally *invested* in the choices clients make (for example, whether to leave a relationship, file a complaint, or have an operation), the therapist's own countertransference is operating.

Countertransference reactions occur for every therapist at times. The issue is not whether they will occur but how they are dealt with (Bergin, 1997). When countertransference reactions occur, therapists want to consult with a colleague or supervisor to better understand and manage their own personal reactions. When therapists find that they are repeatedly having difficulty with the same type of affect, or when supervision does not free them up to respond effectively with more neutrality, therapists are encouraged to seek treatment for themselves. *Because change is predicated on the relationship therapists create with clients, therapists are encouraged to make a lifelong commitment to working on themselves and their own personal development.* Without this willingness to work on their own personal reactions to clients, the help therapists can provide will be limited (Robbins & Jolkovski, 1987). Furthermore, therapists who are unwilling to acknowledge or work on persistent countertransference reactions are those most likely to have a negative therapeutic impact on their clients (for instance, "It's the client's problem—it doesn't have anything to do with me"). It is a privilege to be able to help people change, and therapists honor that privilege by acknowledging their own limitations and personal involvement in the therapeutic process.

Closing

As we saw in Chapter 4, clients' feelings often do not emerge unless the therapist focuses clients inward. An externalizing focus will lead therapists into a superficial problem-solving and advice-giving mode that precludes clients' affective exploration. As clients talk about the problem out there with others, the therapist's complementary social role is to give advice, reassure, or self-disclose what the therapist has done to solve similar problems. Although these responses can be helpful at times, limited change occurs if they characterize the ongoing course of treatment.

But if the therapist does not tell clients what to do, what does the therapist really have to offer? *An essential component of facilitating change is to help clients resolve difficult emotions by integrating or coming to terms with feelings that have been too painful, shameful, or unacceptable to tolerate in the past.* Many clients have experienced tragedies in their lives, and painful feelings will be entwined with the symptoms and problems they present. The therapist helps clients resolve their problems by providing a holding environment for feelings that others have not been able to understand or accept. Perhaps the most important relearning occurs when clients risk sharing certain feelings and find that, this time, they do not receive the same problematic response they have come to expect. Thus, a corrective emotional experience occurs as old expectations and unwanted relational patterns are disconfirmed. Yet, therapists also need to be aware that the biggest obstacle to providing clients with a relationship that can produce change will be their own discomfort with certain feelings. Thus, therapists are invited to make an ongoing commitment to working with their own personal issues and how they influence their work with clients.

What's ahead? In the next three chapters, we will see how therapists can clarify the client's problem and better understand what's really wrong and needs to change. We will learn more about how to conceptualize clients and make intervention plans on the basis of their conflicted feelings, faulty beliefs and expectations, and the maladaptive relational patterns they have been enacting with others and now begin to play out with the therapist. For now, this chapter has presented a great deal of challenging information that often takes several years to integrate and make your own. This is hard work—be patient with yourself.

Suggestions for Further Reading

1. New therapists need help identifying their countertransference propensities. In Chapter 5 of the Student Workbook, a self-assessment scale is provided to help therapists become more aware of their own response tendencies.

2. For information on basic communication skills to help therapists approach or inquire further about clients' feelings, see Chapters 5 and 6 of Egan's (2002) *The Skilled Helper*. These two chapters also help therapists listen without judging and become more aware of their own cultural filters and biases.

3. Readers are also encouraged to examine Leslie Greenberg's (2002) *Emotion-Focused Therapy*. This informative book teaches therapists how to help clients express emotion in sensible ways that are appropriate to context. Also, *Working with*

Emotions in Psychotherapy (Greenberg & Paivio, 2003) helps therapists learn more about working with specific emotions, such as distinguishing between adaptive sadness and maladaptive depression. Both books assist therapists in understanding and responding more effectively to their clients' feelings.

4. One way for therapists to increase their effectiveness with clients is to explore their own motivations for becoming therapists and explore how their own personalities are expressed in their clinical work. It is just as important to focus on the personhood of the therapist as it is to focus on the client and interventions. One good source to help therapists explore these issues is J. Guy's (1987) *The Personal Life of the Psychotherapist*.

Conceptualizing Clients and Developing a Treatment Focus

CHAPTER SIX

Familial and Developmental Factors

Conceptual Overview

The previous chapters have focused intensively on the interaction between the therapist and the client. In these next three chapters in Part Three, we continue to track the interpersonal process. However, we begin to consider more fully how therapists conceptualize their clients' problems and develop a treatment focus. That is, we are trying to become clearer in our thinking about what's wrong and what to do. Therapists are far more effective when they can formulate a case conceptualization that leads to specific treatment plans and intervention strategies. To do this, therapists are trying to understand

- how the clients' problems originally developed or came about (Chapter 6),
- how they are being played out now and disrupting current relationships (Chapter 7), and
- how they are brought into the therapeutic relationship and shaping the way the therapist and client are interacting together (Chapter 8).

The more specifically therapists can clarify their thinking and understand their clients in these three ways, the easier it will be to formulate the *treatment focus* therapists need in order to help clients change. However, most new therapists have had little training or experience conceptualizing clients. Especially in the beginning, it is usually difficult to formulate a treatment focus to guide their interventions, and to write the treatment plans and case reports that third-party insurers and treatment agencies require. To help with this, guidelines to meet these professional demands will be provided in the next three chapters.

In particular, we will learn how to identify interpersonal patterns and organizing themes that help therapists understand and respond to their clients' symptoms and problems.

As listed, the first way to help therapists conceptualize their clients and formulate treatment plans is to recognize the familial and developmental antecedents of many clients' problems. Thus, this chapter introduces family systems concepts that clarify how many problems develop. One way to understand how enduring problems develop is to examine what occurs for offspring in families that function well. Once the basic familial processes that produce healthy offspring are identified, it will be easier to understand how normal development can go awry and lead to the enduring problems clients present. Following this discussion about how problems develop, the next two chapters will examine the other two areas for conceptualizing clients listed above.

Structural Family Relations

This chapter examines three fundamental dimensions of family life: the "structure" of family relations and the nature of the parental coalition, the "separateness–relatedness" dialectic, and child-rearing practices. Let's begin with family systems concepts on the structure of family relations and see how individual therapists can use these ideas to intervene with their clients.

Structural family relations entail the relatively enduring patterns of alliances, coalitions, and loyalties that exist in the family. These relationships are the basis for family organization and define how the family operates as a social system (Minuchin, 1974; Minuchin & Nichols, 1998). Such family patterns are strongly influenced by cultural values and beliefs that shape norms and define acceptable behavior and family functioning. Thus, structural family relations, and the cultural context within which they are embedded, shape family communication patterns and the roles that family members adopt. Among the different structural family relationships that exist, the parental coalition is the basic axis of family life and shapes much of how the family functions (Teyber, 2001). In particular, the nature of the parental coalition will be pivotal in determining how well children adjust (Beavers & Hampson, 1993; Teyber, 1981).

Shifting Loyalties and Establishing a Marital Coalition

In nuclear families that function well, the marital relationship is the primary two-person relationship in the family. Each partner has a primary loyalty commitment to the other, and the marital coalition is stable and cannot be divided by grandparents, children, friends or employers. Parents will still have differences or problems between themselves, of course, and have substantive personal relationships with hers, yet the marital relationship remains a stable and secure alliance. Researchers find that its cohesiveness is a fundamental feature of healthy families (Beavers, 1982; Minuchin, 1984). In contrast, two-parent families in which the primary emotional bond or involvement is not between mother and father often have more conflict and are more likely to produce offspring with enduring behavioral problems. In these more symptomatic families, the primary alliance is between a grandparent and a parent, between a parent and a child, or within some other dyad. Cultures vary widely in how strongly they emphasize family loyalties, and extended family members may play a more important role in child care responsibilities in some cultures (for example, in Native American cultures). Within varying cultural contexts, however, family researchers find that a cohesive parental coalition is a mainstay in families that function well.

How does a primary parental coalition develop? Recalling Erikson's (1968) stages of development, we note that both parents have been able to "individuate" within their own families of origin. This means they were able to achieve an authentic sense of their own personal identity in late adolescence and early adulthood—a sense of self that is based on their own genuine feelings, interests, and values. In this developmental transition of "leaving home," establishing new commitments in love relationships, and forming their own marital coalition, they need to be able to establish a new balance of *family loyalties*. This complex juggling act involves maintaining ties and responsibilities toward the previous generation, while shifting their primary loyalties or emotional commitment to the new peer relationship with a partner or spouse. Successfully negotiating this developmental hurdle or "crisis" means that young adults have been able to transfer their primary loyalty commitment from their parents to a spouse. When negotiated successfully, this developmental transition is accompanied by increased self-efficacy and self-esteem, permitting these young adults to further define their own values and beliefs. This personal evolution does not mean losing their sense of responsibility to their own parents or connectedness to their family of origin. Instead, it signals increased flexibility in their relational capacities and involves transferring their primary relational commitment to a peer relationship characterized by mutual empathy and respect.

If young adult offspring have not been able to shift their primary loyalty commitment from their family of origin to their spouse, a primary marital coalition will not develop. The primary loyalty of one or both spouses will remain with their parent, grandparent, or other caregiver. When this occurs, the primary dyadic relationship in the family is a *cross-generational alliance* between a parent and the adult offspring, rather than a primary marital coalition between the husband and the wife. With the birth of children, this cross-generational alliance (for example, maternal grandmother–mother) usually repeats in the next generation and results in a primary parent–child coalition as well (for example, mother–oldest daughter). Before learning how these cross-generational alliances become so problematic for offspring and are expressed in treatment, we need to better understand the important developmental transitions that occur while forming a primary parental coalition. To do this, let's examine the wedding ceremony as a cultural rite of passage.

The wedding ceremony is a familiar ritual; its central purpose is to publicly mark a transition in loyalty bonds from the parents and family of origin to the new spouse. How does this occur? In the traditional Christian ceremony, once family, friends, and clergy are all assembled, the father walks the bride down the center aisle to the front of the church. In front of all, the father symbolically gives the bride away by placing her hand in the hand of the waiting groom. The father then leaves the couple by stepping back from the front of the church and sitting down beside his own wife. The bride and groom, who have now been demarcated physically from the parental generation, turn away and step forward to be married. In this ceremony, the bride and groom shift their primary loyalty commitments and publicly define themselves to family and friends as an enduring marital couple.

This shifting of primary loyalties from the previous generation to the new spouse is a major developmental task. According to Erikson (1968), this transition is a normative developmental *crisis* that arouses difficult feelings of loss and requires a complex redefinition of individual identity, as well as familial roles, membership, and boundaries—even in healthy, well-functioning families. This developmental challenge becomes far more conflict laden, and often leads to symptoms, when the young adult offspring are trying to emancipate from families that have not had a primary marital coalition. The question of who comes first, parent or spouse, may be painfully enacted in conflicts that surround planning for the wedding day. Offspring who grew up embroiled in cross-generational alliances often find that their wedding day, and the preparations for it, are a wrenching battle between conducting the ceremony as the couple being married would wish and doing it the parents' way. Usually, the underlying conflict is over the shift of loyalties from the previous generation to the spouse, but it is played out in arguments over who will be invited to the wedding, where it will be held, how it will be conducted, who is financially responsible, and so on. In contrast, young adults who enjoy their wedding day usually have been allowed to expand and shift their primary loyalty from parents to each other. These couples are not being pulled apart by competing loyalty ties, with the associated feelings of *guilt*.

How does this model of a primary parental coalition apply to single-parent families? Are they inherently problematic because there is not a primary marital coalition? Of course not. However, this structural model can still help us understand why some single-parent families function very effectively, whereas others experience more problems.

In single-parent families that function well, as in nuclear families that function well, clearly defined *intergenerational boundaries* separate adult business from child business. By necessity, children in single-parent families usually need to take on more household duties, and often more emotional sharing occurs between parent and child than in two-parent families. Even so, in healthy single-parent families, roles and responsibilities are clearly differentiated between adults and children. This means that the single parent still performs "executive ego functions" for the family, such as providing an organized household with predictable daily routines for children, making decisions and plans for the family, and setting limits and enforcing rules. Furthermore, *although parents have legitimate needs for companionship and support, these emotional needs are met primarily in same-generational peer relationships rather than through the children.*

In contrast, single-parent families (and intact, nuclear families as well) will have more problems when adult and child roles are not clearly distinguished, when intergenerational boundaries are blurred, and when too many adult needs are met through the children. With this general framework in mind, we now examine how the parental coalition influences child development and why primary marital coalitions are more effective than cross-generational alliances.

The Parental Coalition Shapes Child Adjustment

Why is the nature of the parental coalition so crucial for family functioning and child adjustment? To answer this question, we look at four common problems of adjustment among children from families in which the primary coalition is not between mother and father.

1. *Problems with limits and authority*. When mother and father have a stable alliance, children cannot come between their parents and play one parent against the other. For example, if a primary coalition exists between the father and an 8-year-old son, the mother may have trouble setting and enforcing limits with her son. If the mother tries to put her son to bed at 9 P.M. or to have him pick up his room, he can often defy these attempts at discipline by appealing to his father. The son learns that he does not have to do what his mother says because his father is likely to take his side. When children can play one parent against the other in this way, they learn that rules do not apply to them. Having learned that they do not have to conform to adult rules in the home, they often will disobey teachers and try to manipulate other authority figures in their lives as well. These problems with limits and authority will carry over into adulthood and are especially likely to lead to problems in the work sphere, where adults need to be able to conform to rules and cooperate with others. A tragic illustration of this principle is found in Woodward's (1984) biography of the late comedian John Belushi.

When Belushi was an early adolescent, his authoritarian father would work long hours in the restaurant business. Belushi was supposed to stay home and do chores for his mother. By making up humorous but hostile imitations of his father and performing these routines for his mother, he could get her to laugh at his father and manipulate her out of enforcing household duties. With his mother and siblings in stitches, Belushi would escape his chores and head out the door to meet with his friends. Among other problematic messages here, Belushi learned that, if he could successfully play one parent against the other, limits and rules did not apply to him. Despite his talents, his life was a study in compulsively pushing limits and being out of control with food, money, sex, and especially drugs. After years of chronic drug use, he died from an overdose of cocaine and heroin.

Therapists who work in alcohol and substance abuse programs are likely to see large numbers of clients with this background. Clients who learn that they can *manipulate* others, and escape rules and limits by dividing the marital coalition, are more likely to develop impulsive acting-out symptoms. When counselors see them in treatment, these clients often will push limits, avoid taking personal responsibility, and be demanding of the therapist. For example, they will often arrive late for appointments, expect to be able to stay beyond the scheduled time, be delinquent in or try to avoid payment of fees, be insistent on calling the therapist at home, and so forth. These challenging clients receive a corrective emotional experience when therapists can tolerate the clients' disapproval, set firm limits, and not be manipulated out of their usual treatment parameters. At the same time, however, therapists want to do this in a nonpunitive way that also communicates their continuing commitment to the client and compassion for the insecurity and loneliness that has been engendered by these developmental experiences. When therapists have the interpersonal flexibility to combine firm limits with compassion, they create secure boundaries for these clients for the first time. This safety allows these clients to stop testing and manipulating others and acting out in self-destructive ways.

2. *Exaggerated self-importance.* Children in families with a primary marital coalition are more likely to gain a realistic sense of their own personal power and control than are children raised in cross-generational alliances. When children form the primary emotional bond with a parent, they become too important to the parent and can exert too much influence over the parent's well-being. As a result, these children gain an *exaggerated* sense of their own importance and a grandiose sense of their own ability to influence others. Children cannot gain a realistic sense of their own limits and capabilities when they are encouraged in the illusion that they can prop up a parent's sagging self-esteem, maintain their parent's emotional equilibrium, or make important decisions for the parent.

It is highly reinforcing for children to feel so powerful vis-à-vis their parent, and they will be reluctant to give up their special role. Following their schemas, these clients will presume this special status with the therapist and are often effective in reestablishing this relational pattern. In particular, such clients may offer the therapist a subtle but all-too-easily accepted invitation to establish a "mutual admiration society." Growing up, these sophisticated clients learned well how to make others feel better and are often adept at flattering the therapist and establishing an elite sense of mutual, shared superiority. For example:

CLIENT: No one has ever understood me like you do—this is amazing.

In this cozy relationship, the client has "antennae" out to grasp nuances in the therapist's mood, is skilled at making the therapist feel special, and, in turn, expects to be treated as special by the therapist. However, once this unspoken deal is struck, the therapist feels constrained because challenging or disagreeing with the client, or focusing the client inward on his participation in conflicts with others, feels like a betrayal of this special relationship. Of course, this type of reenactment in their interpersonal process will keep the therapist and client from addressing the client's real problems.

In parallel with their grandiosity, such clients also have a strong sense of inadequacy. This inadequacy is a pervasive source of anxiety that arises because, as children, they were never actually capable of meeting their caregiver's emotional needs. These clients often suffer from strong performance anxieties and feel inadequate to meet the exaggerated demands they now place on themselves.

As before, the therapist's task is to remain empathic toward both sides of their dilemma. On the one hand, these clients struggle both with feelings of inadequacy and anxiety over all of the things they must try to control but ultimately cannot (for example, their father's drinking or temper; their mother's unhappiness or depression). On the other, if they relinquish the demands they are struggling to meet, then clients surrender their family role and identity as someone who is special. In so doing, they dissolve the illusion of safety or secure attachment that this coping strategy has provided. Furthermore, to become healthier and relinquish these attempts to control or rescue caregivers also threatens to undo or betray their caregivers, who have, convincingly but covertly, communicated for decades that they cannot get along without them. Thus, therapists can anticipate the paradox that making progress in treatment and doing better in their lives will evoke two intense conflicts for these clients. On the one hand, these clients feel guilt over leaving their attachment figures to manage their own lives—these clients feel as if they are selfishly abandoning their caregivers who cannot cope without them. On the other hand, anxiety is evoked over disrupting their lifelong attempts to fashion a secure attachment by meeting the needs of their caregiver.

What intervention guidelines does this type of case conceptualization suggest? The therapist does not want to overreact to such clients' compliments, criticisms, or distress. The therapist needs to be emotionally responsive, of course. However, recalling client response specificity, these clients will backtrack in treatment if the therapist's own personal equilibrium can be too readily influenced. In the past, these clients were able to exert too much control over their parents' well-being, and, staying with our core concepts, the therapist does not want to reenact this maladaptive pattern in their interpersonal process with the client. In other words, these clients feel safer and are reassured when the therapist is responsive but without becoming over-reactive—even when they try to elicit over-reactions from the therapist.

3. *Emancipation conflicts.* Many children caught in cross-generational alliances have problems with individuation and emancipation. If the mother and father have a primary marital coalition or if a single parent has healthy peer relationships, children are free to grow up. As young adults, they can "leave home" successfully while still preserving a lifetime of close ties and shared involvement. On the one hand, they do not have to emotionally disengage or physically break off contact with the family in order to have their own independent life or, using a different strategy, engage in more risky or extreme behavior (such as drunk driving or teen pregnancy) to separate or "get away." This is possible because the caregivers do not need the child to remain dependent or centered in them in order to fulfill their own lives.

In contrast, if the primary coalition is between parent and child, these offspring may feel *separation guilt* over leaving their parent alone or forsaking their parent to the unfulfilling relationship with the other spouse (Weiss, 1993). As a result, they are especially likely to feel dissatisfied with their achievements, to be unable to establish fulfilling relationships with others, or to be chronically depressed. Guilt over emancipation frequently underlies academic failure in college, as well as many other symptoms and problems that students present in college counseling centers (Teyber, 2001). Thus, cross-generational alliances impose binding loyalty ties that make young adults feel guilty about leaving home, successfully pursuing their own career interests, and establishing satisfying love relationships. Although emancipation issues are expressed differently in varying cultures, symptomatic guilt and depression often ensue when offspring are not permitted culturally sanctioned avenues of individuation and coupling.

Many clients, like Anna in Chapter 4, struggle with binding separation guilt and survivor guilt. These clients may feel guilty about being happy, succeeding in life, or even getting better in therapy. It is important with all clients, but with these in particular, that *therapists enjoy clients' happiness and express pleasure in their success*. Some trainees are unsure of how to respond or what their role is when clients begin the session by saying happily, "I feel really good today; I don't have any problems to talk about right now." When that happens, clients with separation guilt are apt to assume, on the basis of their relational templates, that their happiness, success, or independence has somehow "hurt" the therapist. Other clients, with narcissistic or competitive caregivers, may be deeply concerned that the therapist feels envious of the happiness or success they have just communicated. The therapist "disconfirms" such clients' guilt or faulty expectations—and expands the healthier ways in which they can be connected to others—by welcoming the client's good feeling and enthusiastically responding, "That's great! What's the best thing going on for you today?" Without such unambiguous affirmation of the successful aspects of clients' lives, however, clients do not find a corrective emotional experience. Treatment will bog down and become repetitive as clients with these developmental experiences begin to retreat from their success and, in line with their old expectations, keep sharing sadness, fear, or other problems without any progress or resolution.

As we will explore in Chapter 10, conflicts over independence and success will be activated for many clients around the time of terminating therapy. In particular, clients who struggle with separation guilt will feel especially bad or "worried" about no longer needing the therapist and heading off successfully on their own. Therapists are looking for multiple ways to give these clients permission not to need them, to leave when they are ready to go, and to enjoy their own successful lives. Guilt over growing up and becoming stronger, succeeding in work and love, or perhaps even surpassing the parent or therapist with more happiness or success is a common issue, although few clients will be able to recognize it as such on their own. Fortunately, it is relatively easy to help many clients dispel these binding guilt-related beliefs by communicating in words and behavior that the therapist enjoys their success, takes pleasure in their competence, and is in no way hurt or threatened by their stronger stance.

Finally, cultural factors play an especially important role in these issues of individuation and family loyalty. Family connectedness or close family ties may be of special concern to clients from strong relational cultures—for example, Asian or Latino. Entering the subjective worldview of these clients, therapists can learn to appreciate the subtle balance between being loyal to family and culture yet still possessing their own authentic self. To become familiar with the culturally sanctioned avenues for individuation that exist, therapists will find it helpful to consult with others who are knowledgeable about the client's cultural context. With such consultation, therapists can help clients find ways to individuate that are acceptable within their own culture. Although these individuation and emancipation issues will be expressed differently in each familial and cultural context, therapists who are open to these differences can help clients find culturally acceptable outlets and role models within their culture for expressing their own individuality and forming love commitments.

4. *Parentification of children.* When the primary coalition is between a parent and a child rather than between the mother and father, the result is frequently the *parentification* of children. A role reversal occurs: Rather than the parent responding to the child's needs, the child takes on the role of meeting the parent's emotional needs. That is, when a parent's emotional needs are not met by his or her spouse or other (same-generation) adults, the parent inappropriately turns to one or more of the children to meet his or her own adult needs for affection and intimacy, for approval and reassurance, or for stability and control. Therapists should be alerted to the possibility of such parentification when they hear adults describe their children as their "best friend," "lifeline," or "confidant" (Teyber, 2001).

It is problematic for children when they become responsible for, or take care of, the emotional needs of their parent. The problem with this role reversal is that children must give to their parents rather than receive, and the children's age-appropriate dependency needs go unmet. These parentified children grow up to feel overly responsible for others, insecure about depending on others, and guilty about having their own needs met. As adults, parentified offspring often describe themselves as feeling empty or "having a hole inside" as a result of having given rather than received throughout their childhoods. Parentified children initially enjoy having such a special role vis-à-vis their parent. As adults, however, they often come to resent having been deprived of their own childhood. They also become angry and anxious in current adult relationships because (1) they do not trust that others will be there for them in times of distress, and (2) they are preoccupied about losing a relationship because they cannot meet the needs of the other.

Parentification is pervasive in the background of clients and therapists alike. Perhaps as many as one-half of all clients seeking outpatient treatment have been parentified to some extent. Because their basic relational orientation is to take care of others, parentified offspring select careers that fulfill this caretaking role—such as nurses and therapists. Although kind and capable, they often feel guilty about saying no, setting limits, and meeting their own needs. Because they do not draw boundaries well, they tend to become overidentified with others' problems (thinking, for example, "Her divorce was *exactly* like mine!") and are prone to experience burnout. Because they grew up having to take care of their parent, it is now threatening to relinquish this exaggerated sense of responsibility and need for control in their current lives. For example, it may be hard for them to let others share in meeting obligations, although they may also resent having to do everything themselves. These control issues may also be

evidenced in symptoms such as airplane phobias, for example, where these individuals must temporarily relinquish control to the pilot. These clients also
have problems in close personal relationships because it is too threatening to
relinquish the control necessary to be intimate with someone.

Note, however, that in some families—for economic, single-parent, or
other reasons—one child may temporarily assume a parental role with younger
siblings. This is not the same as parentification because it remains clear that the
parent is in charge when the parent is home. This child's role—to assist in the
functioning of the home in the parent's absence—is temporary or situational.
Furthermore, the child is caring for siblings, not for the emotional needs of the
parent. This responsible and helpful child, who still remains a child and is nur-
tured and given to without the role reversal that occurs for parentified children,
often grows up to be an especially resilient and capable person with many personality strengths.

Parentified clients will be highly sensitive and responsive to the therapist—
as they once were to their caregiver. It can feel great to work with these clients
in the short run, because they astutely discern what they need to be or do so that
the new therapist will feel competent or secure! However, if the therapist does
not collude in reenacting this relational pattern, clients can begin to explore the
far-reaching consequences of being parentified and having missed their own
childhoods. Simultaneously, it will be relieving for these clients—yet anxiety
arousing at times—when the therapist begins this exploration by making process comments that highlight this maladaptive relation pattern as it begins to
repeat in the therapeutic relationship:

THERAPIST: I really appreciate your genuine concern for me. But thinking about
the caretaking patterns we have been talking about, I wonder what happens to
your needs here?

To illustrate this important concept, let's consider a case study. Carol was
a highly regarded psychiatric nurse. An utterly dependable and take-charge person, she could seemingly handle every situation that arose. In a hospital emer-
gency setting where she had to deal with very disturbed patients in crisis, her
rapid and accurate assessments, good judgment, and compassion for highly disorganized patients had earned her the respect of the entire staff. Although considered a superstar at work, Carol sought therapy for her recurrent depression
and loneliness.

Carol was a bright and engaging client who was adept at getting the therapist to lead, talk about herself, and become involved in collegial discussions
about interesting clinical issues. However, the therapist had formulated working hypotheses about this potential reenactment and was usually effective in
recognizing Carol's strong "pulls" to repeat this pattern with her. Repeatedly,
she focused Carol inward by saying, for example:

THERAPIST: Help me understand what was going on for you when . . . ?

OR

THERAPIST: What were you afraid was going to happen if you said no and . . . ?

In response to these repeated invitations to consider what she was thinking, feeling, or wanting, instead of being preoccupied with what the therapist might be wanting, Carol began expressing that it was new for her to pay attention to herself and her inner experience in this way. And, although she enjoyed this very much, she also found that it made her feel "uncomfortable." The therapist did not press for this internalizing focus when Carol did not want it but instead asked Carol on occasion how it was for her to look within at her own experience. It soon became clear to both of them that guilt over being "selfish" and shame over being "the center of attention" were evoked by this internal focus and paying attention to herself. Thanks to the therapist's affirming responses to these reactions (for example, "Oh no, I don't think you're being selfish here at all . . ."), both soon agreed that important new material was emerging.

Four weeks into treatment, Carol disclosed something that she had never told anyone: Her alcoholic stepfather had sexually assaulted her in her early adolescence. Carol successfully fought him off, although she was scratched, had her blouse torn open, and suffered a bloody nose when she literally pushed him off her. Carol recounted her ordeal in detail, but without any emotion, and the therapist responded effectively in caring and validating ways. The therapist, herself a mother of two daughters, soon asked what was for her the burning question.

> THERAPIST: You have kept this awful secret for almost 20 years. I'm glad that you can share it with me now, but I'm sorry that you had to be alone with this for so long. What kept you from telling your mother?
>
> CAROL: I didn't want to put more on her. She couldn't have done much anyway, and I didn't want her to worry.

Carol's poignant response illustrates the plight of the seriously parentified child. As in most aspects of her relationship with her mother, Carol's own profound needs for protection and comfort were set aside in order to meet her parent's need.

To sum up, there are many ways in which children function better in families that have a primary marital coalition with clear boundaries between parent and children than in families with cross-generational alliances. Therapists will see the problematic consequences of these structural family relationships and blurred inter-generational boundaries operating with many of their clients. This is one reason why effective therapeutic relationships always honor treatment parameters and provide a corrective experience by maintaining clear boundaries between the therapist and the client. One of the best ways for new therapists to utilize supervision is to have their supervisors actively help them track this process dimension with their clients.

Next, we will see how the nature of the parental coalition also influences two other basic dimensions of family life: the separateness–relatedness dialectic and child-rearing practices.

Balancing the Continuum of Separateness-Relatedness

The second basic dimension of family life that informs therapists about clients' problems is the *separateness–relatedness dialectic*—an awkward term but an important concept drawn from the family systems literature. This family systems perspective refers, in part, to the family's ability to respond to the child's need for *both* relatedness (that is, secure attachments and family cohesiveness) and separateness (identity, autonomy, and personal authenticity). At times, the separateness side of this dialectic has been misunderstood. In this context, separateness does not imply cutoff, selfish, or uninvolved individuals who are insensitive to their responsibilities to others. Rather, separateness or individuation connotes being connected to our genuine self—our inner voice and our own feelings, needs, and wishes. It is the ability to make choices based on awareness of our own experience within our own relational contexts. Thus, the autonomy side of this dialectic connotes connection to our authentic self and the ability to respond purposively and choose, instead of merely complying with what others want us to do.

As children develop, each family faces the complex task of having to provide both secure attachments (a secure base with a sense of belonging and shared family loyalty), on the one hand, and support "differentiation" (independence and personal identity), on the other hand. In other words, children need to develop the ability to feel close and belong while still remaining a separate person "with her own mind" at the same time. Both of these behavioral systems—of attachment and of differentiation (curiosity, exploration, mastery)—are present at every stage of the life span and are interrelated. In this regard, attachment theorists believe that having a secure base as a child (or an internal, mental representation of a secure base later as an adult) is necessary to have the capacity to venture out confidently and explore the environment.

Although there will be varying expressions and timetables within different cultural contexts, the attachment system holds more importance during childhood than adolescence. Toddlers, for example, look for a parent or other important caregiver for reassurance when they find themselves in new situations or around strangers. If the parent is in close proximity during these times, children gain the confidence needed to explore the environment and begin to take steps away from the parent, trusting the parent will be available if something goes wrong. When a parent or other attachment figure is available to provide physical and/or emotional support, the attachment system provides a secure base from which infants and toddlers are free to engage in other activities, such as exploration and play. While the attachment system remains primary in young children, the exploration or differentiation system becomes primary as children mature and become adolescents (Colin, 1996).

As children's cognitive skills develop, they are able to form mental representations of attachment figures that allow them to interact independently in new environments, such as school and the neighborhood, without requiring a parent's physical presence in order to feel secure or attached. If a parent has provided a "good enough" secure base during early childhood, the school-age child is better able to meet the developmental crisis of "industry versus inferiority" (Erikson, 1968) by mastering a number of skills and engaging in cooperative relationships with peers. This, in turn, builds a foundation for forming a strong sense of personal identity during adolescence. Although older children and adolescents will continue to seek a secure base during moments of distress, moving into the larger circles of neighborhood, school, and the "world" takes on increasing importance. Teens thus take the initial steps toward eventually leaving home, yet they continue to need some reassurance that their parents will honor these movements, as well as be emotionally available to them when needed. Maintaining a proper balance between attachment and differentiation, one that honors the changing developmental needs of children and adolescents, requires great flexibility on the part of parents. Therefore, healthy families are characterized by *experiential ranging* between these two poles of intimacy and individuation (Farley, 1979). Effective caregivers have the flexibility to range back and forth along this separateness–relatedness continuum and meet children's changing needs both for closeness and for autonomy throughout their formative years. Easy to say, yet, even for committed and skillful parents, so challenging to do.

In contrast, less effective families do not have the interpersonal range to function well at both ends of the continuum. These less effective parents cannot meet the changing needs that children present at different developmental stages. As we will explore later, "enmeshed" or overinvolved families cannot support older children's growth and development toward individuation, whereas "disengaged" or emotionally cutoff families cannot provide a secure base for young children and meet their basic attachment needs.

This conceptualization provides a useful model for therapists, who similarly need a broad interpersonal range to respond flexibly to their clients' needs. This wide interpersonal range is a critical component of client response specificity. Depending on their developmental history, some clients will need help with issues related to separateness (becoming more assertive, choosing a career), whereas other clients have problems in relatedness (issues related to safety or trust in close personal relationships). Furthermore, these needs will change during the course of treatment. For example, as clients receive affirmation and understanding from the therapist for the hurt or vulnerable aspects of themselves, they often begin to improve in treatment. However, as they develop an increased sense of self-efficacy or personal power, they need a different kind of response from the therapist—unambiguous support for this new, stronger self that was not encouraged in other relationships. Keeping in mind this flexibility or wide interpersonal range for both the therapist and the client, let's see how the separateness–relatedness continuum relates to the parental coalition. The following example shows how families without a primary marital coalition tend to be more restricted or less flexible in their experiential range.

Suppose that both members of a young married couple have not made much progress with Erikson's developmental tasks of identity in late adolescence and intimacy in early adulthood (Solomon, 1973). As a result, they have not been able to separate psychologically from their families of origin sufficiently—they cannot shift their primary loyalty ties from their parents and establish a successful marital coalition with each other. Such newlyweds are likely to have a high degree of marital conflict that they are both unable to resolve. In many cases, the couple will attempt to deal with their on-going relationship problems by having children. As Bowen (1978), Beavers (1993), and other multigenerational family therapists illuminate, these "less differentiated" parents—who have not been successful in completing previous developmental tasks—are more apt to experience the children as extensions of themselves. Without being aware of this overidentification, they tend to project their own unresolved developmental conflicts and marital problems onto one or more of their children, in an attempt to distance and externalize those issues. Without being aware of it, such parents may select a certain child who matches their own schemas— perhaps an oldest or youngest child who matches their own sibling/birth order, or a boy or a girl like themselves—to carry or express certain unmet needs, unacceptable feelings, or unwanted aspects of themselves. Through this "family projection process," described by Murray Bowen (1978), Virginia Satir (1967; Satir & Bitter, 2000), Ivan Boszormenyi-Nagy (1991), and other multigenerational family therapists, children are scripted into roles, and restrictive family rules and myths are established to help parents defend against their own problems that have been activated by marriage and child rearing.

To illustrate, if the mother's need for closeness or intimacy cannot be met by the father, she may turn to a particular child to have her emotional needs met. This solution will work well in the short run, yet it establishes the basis for long-term family problems. In many cases, the mother could not emancipate psychologically from her own parents, and she is made anxious and/or guilty by individuation. As long as the child remains dependent on her, her need for closeness is met and her anxiety or guilt over individuation is managed. On the other hand, the father in this prototypical scenario is uncomfortable with closeness; it makes him anxious. The primary mother–child coalition allows him to gain greater distance from the anxiety-arousing demands of closeness from both the spouse and the child. In this way, the child's role is to hold the marriage together and, at the same time, to ensure distance between the couple.

This situation eventually poses a dilemma for the child's continuing development, however. The child's innate push for growth will destabilize the marital relationship, and this will arouse each parent's own internal conflicts. Thus, these parents need to undermine the child's increasing independence or competence in order to maintain the marital status quo. At the same time, the parents will express anger and frustration to the child about the child's refusal to grow up and act responsibly. This family's limited experiential range is most likely to come to a head during adolescence, when offspring face the developmental hurdle of "emancipation" or beginning to leave the family of origin. The mixed messages that the child has received about growing up and remaining dependent will often erupt in psychological symptoms and acting-out behavior at this time. Furthermore, these underlying issues have been overlearned as maladaptive relational patterns and often continue to be expressed in symptoms and problems with others throughout adulthood.

The separateness–relatedness dialectic is a useful construct to help therapists conceptualize basic familial and developmental processes. It also helps with some of the most common problems that new therapists face in their clinical training. As in this parenting model, therapists need to be *flexible* enough to respond to clients at both ends of the separateness–relatedness dialectic. Regarding the separateness side, for example, most beginning counselors find it far easier to be understanding and empathic than to challenge or disagree with their clients. Understandably, most therapists are reluctant to "confront" clients—it often connotes the unwanted hard edge of "exposing" clients, embarrassing them, or making them mad. Most people interested in clinical work would simply loathe to be called "pushy," and most new therapists—responsible and caring people—are worried about doing anything that might hurt their clients' feelings. However, without being confrontational or hostile in any way, it is often necessary to *challenge* clients by giving them honest feedback about the problematic ways they are effecting others or inviting them to take more personal responsibility by looking at their own participation in interpersonal problems. Therapists are not responding effectively at the "separateness" end of this dialectic when they cannot do the following:

- challenge clients at times or disagree with them and see things differently,
- address interpersonal conflict directly and talk through the inevitable misunderstandings and problems that come up between them,
- encourage clients to look more realistically at their own contribution to or participation in problems with others, or
- set firm limits with acting-out clients and maintain clear boundaries.

Therapists are encouraged to reflect on their own countertransference propensities in these ways and strive to expand their own experiential range during their clinical training. For example, with practice and rehearsal, new therapists can learn respectful ways to challenge clients, provide unwanted but necessary interpersonal feedback, and set limits:

> THERAPIST: It's OK to be angry with me, and I welcome talking about that, but you can't curse or swear at me like that. I don't want it.

In parallel, many therapists in training also struggle with issues on the relatedness side of this dialectic. In particular, new therapists often express concerns that they are "too sensitive" to their clients' feelings. For example, many caring and capable practicum students worry about overidentifying with their clients' feelings:

- What if I feel my client's sadness too much, you know, and just start to cry myself?
- Maybe I'll take on the client's feelings and feel them too much—you know, be so affected that I can't be objective about things.
- What if I can't stop thinking about their problems when I leave the session and go home?
- Some of these things really bother me. Maybe I'm too sensitive to be a good therapist!

The therapists' goal is to find a new middle ground for themselves along this separateness–relatedness dialectic. To help with this, recall Sullivan's (1968) concept of "participant-observer." The therapist's intention is to be empathic and resonate with the client's feelings yet, at the same time, also remain separate by being clear that these are the client's feelings—not our own—that we are trying to respond to and understand. Learning to become a participant-observer in this way is one of the most challenging but important tasks in clinical training; we will explore it closely in the chapters ahead.

We will return to the separateness–relatedness dialectic later in this chapter and see how therapists can use it to better understand the interpersonal process that is taking place between the therapist and the client. But first, let's turn to another dimension of family life that helps therapists understand their clients' problems: child-rearing practices.

Child-Rearing Practices

Child-rearing practices are the third dimension of family life that helps therapists understand how symptoms and problems develop. We will examine four different styles of child rearing that most families use and see how each is related to the problems clients present in treatment. Much of this discussion draws on research by Diana Baumrind (1983, 1991) on the long-term effects of different styles of parenting and discipline.

To help understand the different styles of child rearing that parents use, consider the two dimensions, Control and Affection, as diagrammed in Figure 6.1. As reflected on the horizontal axis, parents can vary along a continuum from firm discipline (high control/structure) to permissive or lax discipline (low control/structure). The vertical axis represents how parents' can vary on a continuum from much warmth, emotional responsiveness, and communication (high affection/support) to little approval, acceptance, or interest (low affection/support). Using the high and low points along these two dimensions, parents typically discipline their children using one of four approaches: *authoritarian*, *permissive*, *disengaged*, or *authoritative*.

Authoritarian Parenting

One of the most common but ineffective methods of discipline is the authoritarian approach, which is punitive, controlling, and unreasoning (see Quadrant II: high control, low affection). Authoritarian parents are strict disciplinarians— they give children clear expectations about what behavior is acceptable and unacceptable. Parental rules and expectations are clear and the consequences for violating them are consistently enforced. Authoritarian parents are demanding—they have high expectations for their children to behave in a responsible and mature manner. Children are expected to be competent and perform up to their abilities, and to be responsible and contributing family members. However, authoritarian parents do not have a wide or flexible experiential range.

FIGURE 6.1 | FOUR STYLES OF PARENTING

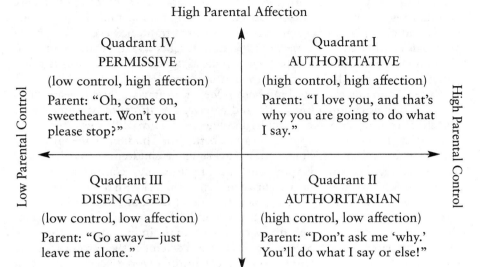

High Parental Affection

Quadrant IV
PERMISSIVE
(low control, high affection)
Parent: "Oh, come on, sweetheart. Won't you please stop?"

Quadrant I
AUTHORITATIVE
(high control, high affection)
Parent: "I love you, and that's why you are going to do what I say."

Low Parental Control

High Parental Control

Quadrant III
DISENGAGED
(low control, low affection)
Parent: "Go away—just leave me alone."

Quadrant II
AUTHORITARIAN
(high control, low affection)
Parent: "Don't ask me 'why.' You'll do what I say or else!"

Low Parental Affection

Along the separatedness–relatedness dialectic, they are limited to the separation and differentiation side of the continuum. Without the emotional support and affection so important to the development of a sense of belonging and security, especially during moments of distress, children of authoritarian parents learn to hide any signs of vulnerability from their parents and sadly, eventually from themselves as well. They develop an interpersonal strategy that in effect communicates to others, "I don't need you or anyone else." These children often are successful in school and later in their chosen careers, yet they tend to keep people at a distance and their own emotions under tight control. Experiencing vulnerability of any kind often activates a rejection schema, something they want to avoid at all costs. Children of authoritarian parents may have an avoidant attachment style.

In addition to a lack of emotional support, authoritarian parents often instill a fear of rejection in their children by their methods of discipline. Authoritarian parents do not give children reasons or explanations for the rules they set. The child is expected to obey without questioning or trying to understand why the parent has set these limits. Children cannot ask why a rule is set; they must simply obey. Children who grow up in an authoritarian household regularly hear their parents make statements such as "Don't ever ask me *why* you can't go out. I am your father, and you'll do what I say or else!" Furthermore, there is no room for compromise or verbal give-and-take between the parent and child. Children are not encouraged to suggest alternatives or to explain their side of the story. Authoritarian parents discipline their children strictly and expect much from them. They also may provide an "instrumental" form of love in the sense that they responsibly feed and clothe their children, help them with their homework, and even play games with them. Most important, however, *they do not give their children much warmth or affection*. Clients who grew up in authoritarian households often describe their parents as "good people" but usually experience them as cold, distant, and even intimidating.

There is no idle threat or bluff here—the authoritarian parent means business—and their children know this and behave. The threat of parental power and the fear of rejection keep children in line, especially while they are young. This strict, no-nonsense approach is far better for children than having no discipline at all, but it has major drawbacks. These children are obedient and achieving, yet they are also anxious and insecure; they comply out of fear (Wenar & Kerig, 2000). The insecurity that children feel with their parent often generalizes to teachers, coaches, principals, and other adults in their lives. This insecurity also carries over into adulthood and infuses many clients' relationships with anxiety. Although many children remain intimidated by their authoritarian parents, some become aggressive and defiant as they grow older and more verbal (or when marital conflict escalates or divorce occurs). As adult clients, however, they suffer low self-esteem and lack interpersonal confidence, even though they are often achieving and successful.

Another drawback to the authoritarian approach is that it limits the growth of children's intellectual abilities. When children are not given reasons to help them understand why parents have set certain rules and are not encouraged to suggest alternatives or compromises, they do not learn to exercise language and reasoning skills. Children of authoritarian parents score lower on verbal tests of intelligence than children who are given an opportunity to interact with parents over rules and directives.

In contrast, within healthy families, children learn to obey without sacrificing their own initiative and positive self-regard. Healthy children become self-controlled and self-reliant without losing their sense of being prized by their parents. However, for children of authoritarian parents, the trade-off between acting on their own wishes and maintaining parental approval is too severe. *Because authoritarian parents provide too little nurturance and affection, too much of their children's initiative and positive self-regard is lost in order to try and win parental approval*. By the time the child is of school age, these relational patterns with the authoritarian parent have been internalized as a cognitive schema: What was originally an interpersonal conflict becomes an internal conflict that shapes subsequent relationships with other adults, and with themselves, in problematic ways. Specifically, these well-behaved, insecure children become harsh, critical, and demanding toward themselves, just as their parents have been with them. Many will seek therapy as adults, presenting with symptoms of guilt, depression, unassertiveness, anxiety, and low self-esteem. These personal and emotional problems will be present, even though these clients typically are responsible, hardworking, and successful adults. Because so many clients and therapists come from authoritarian backgrounds, we will explore this topic further later in this chapter.

Permissive Parenting

Although authoritarian parents recognize that children need to know the rules and that the consequences for violating them will be enforced, their discipline is rigid and their parenting lacks empathy. In contrast to such strictness, other parents may err on the side of permissiveness. Although permissive parents provide more warmth than authoritarian parents, they are not able to take a firm stance, consistently follow through, and place appropriate controls on children's behavior (Quadrant IV: low control, high affection). This inability to take charge and discipline effectively may occur for many reasons. Permissive parents may falsely believe that if they are firm they are acting harshly and, in their own minds, being just like their own parents who were too harsh or intimidating. Or, sadly, some insecure parents fear their children will not love or be close to them if they say no and mean it. Still other parents, perhaps disempowered by their own authoritarian parents in childhood, don't really believe they have the right to be the one in charge and make their children behave or do what they want them to do. For these and other reasons, *the balance of power has tipped in permissive families, and children wield too much control in the parent–child relationship.* These parents often feel disempowered vis-à-vis their children and may be heard negotiating, bargaining, or even pleading with their children to behave ("Oh, come on, sweetheart, won't you please stop?"). Parent and child both begin to suffer when control shifts from the parent to the child in this way. Problems begin to develop for these children as they become bossy with peers and angry/demanding with adults. And, of course, these same relational patterns will be tested in the therapeutic relationship as these clients will try to manipulate, avoid consequences, and exert control over the therapist—as they have learned to do with their caregivers.

Lax disciplinarians, permissive parents often are indulgent with their children. Children of permissive parents do not know what behavior is expected of them or what will happen if they violate parental norms. Furthermore, permissive parents do not consistently enforce the few rules they set. As a result, children of permissive parents learn that they do not have to obey because parents will not *consistently* enforce the rules. Thus, permissive parents are often loving and communicative, but their children are not disciplined and are not expected to behave in a mature, responsible manner. Without parental expectations that they perform to the best of their abilities, children do not develop the skills or internalize the discipline necessary to succeed on their own. Dependent and demanding, they are often described as "spoiled" children.

Offspring of permissive parents will also have adjustment problems, but their symptoms and interpersonal problems are different than those from authoritarian families. These offspring may enter treatment with anxiety, depression, and other "internalizing" symptoms. Children and adolescents are right in knowing that they are not safe and cannot be protected by a parent who cannot say no to them, cannot tolerate their disapproval, or gives them too much power and control over the parent–child relationship. In a word, they will be insecure.

In addition, children of permissive parents are prone to develop acting-out or externalizing problems. For example, behavior problems may develop involving school authorities for truancy, with the police for reckless driving, or for involvement with drugs and alcohol. Children do not respect parents (or, later, therapists) whom they can manipulate or parents who cannot say no and discipline effectively because they need the children's approval. Such children do not learn to respect others or themselves. They often become demanding, selfish, and angry and are likely to fail with peers and friendships. Researchers describe these children as dependent, immature, demanding, and unhappy.

Later, as adults who enter treatment, they tend to be self-centered, demanding, and dependent in their relationships and less capable of making commitments and following through responsibly on obligations. Oftentimes, these clients will be mandated to treatment by judges and courts, and seen in alcohol and substance abuse treatment programs. Therapists working in Employee Assistance Programs will also work with this client. Typically, they will be referred for treatment by a supervisor at work who is dissatisfied with their inability to be a good team player and get along with others in their work group. When seeking therapy as adults, they tend to avoid taking responsibility for themselves and try to blame others for their problems. They also have learned that they can break the rules and escape the consequences of their own behavior by manipulating others. Of course, these same interpersonal themes that are causing problems with others will quickly appear in the interpersonal process they begin to play out with the therapist.

Disengaged/Dismissive Parenting

The third parenting style is *dismissive*. An especially problematic parenting style, these caregivers are disengaged from their children (Quadrant III: low control, low affection). Looking at Figure 6.1, these uninvolved parents are low on both axes—in other words, they are doing little for their children. Whether passively unresponsive or overtly rejecting, this neglectful parent says in word and deed, "Go away—just leave me alone." In addition to experiencing rejection, children of disengaged or dismissive parents also feel abandoned. Their very existence seems to annoy or disrupt their parents. In order to adapt to this painful situation, these children learn how to hide and not make waves, often limiting their ability to form a personal identity.

How does this disengaged/neglectful parenting style come about? Perhaps caught up in their own alcohol or drug problems, dismissive parents may be too self-absorbed to attend to their child's needs. Similarly, another type of disengaged parent may be chronically depressed and unresponsive to their child who grows up unnoticed—describing themselves in treatment years later as "invisible." Still other children, who may have grown up with a caregiver who had a personality disorder such as a narcissistic, borderline, or paranoid personality disorder, may be actively rejected or pushed away by an angry parent, or blamed for all of their parent's problems. Heartbreaking to observe, for these and other reasons, some children grow up without care. Unsupervised or unwanted, the disaffected child with dismissive parents becomes at risk for many different problems, including antisocial behavior and the peer influences of drugs, delinquency, and early sexual contact. These clients are often in the child welfare system, frequently being seen in group treatment homes and juvenile probation programs. These at-risk youth bring strongly held schemas for rejection and distrust to the therapeutic relationship, and they challenge the therapist to establish the working alliance they need but cannot trust. Their greatest need—and perhaps their worst fear—is for the therapist to "see" them, because often they have learned how to be invisible. In addition to communicating "I don't need anyone" (a characteristic they share in common with those who grew up in authoritarian families), these clients may also communicate "I don't really exist" and "I'm not worthy of your time and effort." As the therapist consistently passes the client's tests (doesn't respond with the rejection, judgmentalism, or disinterest that they expect and elicit), they will become better at establishing a working alliance based on trust. As this occurs, clients will feel more secure and typically will move in the direction of building a personal identity, with the therapist providing the kind of "benevolent guidance" that Erikson thought was so essential in helping a person achieve a viable sense of identity.

Authoritative Parenting

Finally, the *authoritative* child-rearing style produces the most well-adjusted children (Quadrant I: high control, high affection). The authoritative approach combines

- strict limits and reliably enforced rules,
- reasons and explanations for parental rules,
- high expectations for responsible and mature behavior, and
- parental warmth and overtly expressed affection.

These highly effective caregivers possess a wide range of parenting skills that allow them to combine firm discipline with nurturing child care. They are loving, consistent, and willing to listen to their children. In effect, authoritative parents are saying to their children, "I love you, and that's why you are going to do what I say."

The authoritative parent believes in strict discipline but, unlike the authoritarian parent, couples this with physical affection and spoken approval. For example, authoritative parents demand obedience yet also tell their children stories, roll with them on the floor, hold them in their lap, praise them when they do well, and look in their eyes and say, "I love you." Children more readily cooperate with requests from an affectionate parent than from one who is threatening or distant.

Although authoritative parents are firm about discipline, they also invite children's participation in the process. They encourage children to offer alternatives or compromise solutions. They also tell their children what they would like them to do and explain why certain behavior is encouraged or discouraged. In contrast, the authoritarian parent provides clearly defined and enforced limits but no room for compromises, alternatives, or explanations. Even very young children may be more willing to cooperate if they understand the reasons for the rules. Adult authority seems less arbitrary or unfair when children can participate in the discipline process.

Authoritative parents consistently enforce the rules that they set. Permissive parents may offer reasons and explanations to lessen their children's disapproval but, ultimately, they do not take a firm stance and convincingly enforce limits. Children are astute judges of how serious their parents are about enforcing rules. If parents enforce rules inconsistently, children will continually test and try to break them. Thus, whether or not the child has been able to understand or agree, authoritative parents follow through and enforce the rules that have been set.

Clearly, the authoritative parent exercises a wide range of parenting skills. We have seen that the authoritarian parent sets limits but does so harshly, the permissive parent cannot follow through or does not believe in establishing controls, and the disengaged parent has given up on attempts to manage children. Although many parents falsely believe that they have to be either strict or loving (that is, authoritarian or permissive), *authoritative parents are more effective because they have the flexibility to be both at the same time.* Despite the best of intentions, however, balancing these two domains is challenging for most parents. Indeed, researchers find that only about 10–12 percent of parents provide authoritative parenting. Most parents do not possess the wide experiential range necessary to be both firm and loving—just as many therapists find it difficult to be supportive or empathic with their clients and, at other times, forthright or challenging. Although it is not easy for caregivers to provide this authoritative child rearing, it produces the most well-adjusted children. Following them over time in longitudinal studies, researchers find that these children tend to be independent, self-controlled, successful with friends and to have a positive outlook.

As therapists listen to their clients' narratives and learn more about their developmental histories, they will hear relational patterns and themes derived from authoritarian, permissive, and disengaged parenting styles. What about children of authoritative parents—is this a "perfect" family without any problems? Parenting and development are always challenging, even in the best of circumstances, and conflict-free families and individuals are a myth. On average, however, children of authoritative parents are better adjusted and will be less likely to be seen in treatment for enduring problems. More likely, they will seek help in crisis situations (such as a child's major illness) or when negotiating developmental hurdles, such as seeking premarital counseling. And, because they have learned that some others can respond responsibly to their emotional needs, they are able to seek help and enter treatment when necessary.

In thinking about these four different styles of child rearing, we also need to recognize the enormous complexity of raising children and all of the different factors that contribute to family interaction. For example, birth order, gender, and temperament greatly influence how parents respond to children. Family functioning is fundamentally shaped by cultural values and beliefs as all child-rearing practices are embedded in a social context. Because of these and other influences, the four parenting styles we have discussed become far more complicated in everyday life. To illustrate, suppose a 10-year-old child grows up with an authoritarian father and a permissive mother. Not the best circumstances, perhaps, yet between his two parents, this child is benefiting from consistent discipline and affection in his life and probably adjusts well in this developmental situation. But let's play out just one of the thousands of different developmental possibilities that could ensue and that a subsequent therapist will need to understand.

What if the strict but cold father and the warm but permissive mother wrangle over their differing parenting styles and, because of this and other disagreements, eventually divorce? Pressured to "take sides" in the ongoing parental battle, the son chooses to live with his mother and feels disloyal to her when he visits his father. Fed up with infrequent and superficial "visits" with his son, the father remarries in 2 years, starts a new family, and, in effect, drops out of the son's life. Now the child is living with his permissive mother where, among other problems, he no longer has any effective limits. Angry at his father for leaving him and without discipline in his life, this boy begins to act out at home and at school. In particular, he becomes defiant and disrespectful toward his mother. Soon, she no longer feels like expressing the warmth and affection she once did to this son who has become demanding, bossy, and self-centered. An escalating negative interaction cycle ensues as the mother becomes increasingly frustrated with this angry boy that she can't control and who is "ruining my life." By age 13, the exasperated mother feels "defeated" by this "impossible" son and, transitioning to the "disengaged" parenting style, essentially gives up on this relationship. Three years later, a therapist enters the picture as the adolescent boy is court mandated to treatment for speeding and reckless driving while intoxicated.

In the next section, we will apply this parenting information and look more closely at how child-rearing practices in general, and authoritarian parenting in particular, are reflected in the problems clients present in treatment.

Other Child-Rearing and Attachment Issues
Authoritarian and Disengaged Parenting, Love Withdrawal, and Insecure Attachment

Many caregivers, especially authoritarian and disengaged/neglectful parents, often employ discipline techniques based on "love withdrawal." Instead of communicating that they disapprove of the child's behavior, parents respond with anger or rejection and communicate disapproval of the child's basic self. While disciplining, these parents withdraw their warmth and *emotional connection* to the child, engendering in the process the anxiety of an attachment disruption (even though the parent and child are not physically separated). This parental communication is often nonverbal; the withdrawal of love from the child is expressed as much by tone of voice and inflection, gesture, and facial expression as in words. In the child development literature, researchers call this "love withdrawal" disciplinary techniques. Central to his theory, Carl Rogers (1959, 1980) referred to this far-reaching developmental issue as "conditions of worth"—what children had to do to maintain their parent's approval.

Parents respond in this way to punish the child and communicate their anger—which too often is accompanied by the far more wounding affect of contempt or disgust by more severely authoritarian or disengaged parents. They may say, for example, "Get out of here! I don't even want to have to look at you. What's wrong with you anyway?" Better-functioning parents may respond in these hurtful ways occasionally, when they are tired or upset, and may later apologize or clarify that they have overreacted ("Daddy got too mad and shouldn't have said that. I'm sorry").

In contrast, for some parents, withdrawal of love and emotional disengagement occurs routinely, and the child's emotional ties to the parents are disrupted regularly. For example, the parent may say overtly, "I can't stand to be around you. Get away from me." In contrast to these overtly rejecting messages, caregivers may emotionally withdraw in more covert ways. For example, the martyrish parent may say nothing, sigh painfully, and turn her back and walk away from the child, silently shaking her head in disappointment or disgust. Of necessity, the child develops symptoms and defenses to cope with the painful separation anxieties and the shame-based sense of self that these attachment disruptions engender. For these children, self-schemas develop that leave them shame-prone throughout their lives as they come to believe "I'm a bad boy," "There's something wrong with me," or "Mom doesn't want me."

A Continuum of Love Withdrawal Love withdrawal techniques and "conditions of worth" occur on a continuum of severity. In some families, disruption of ties from love withdrawal may not be severe. And, as noted earlier, they may occur infrequently—only when caregivers are stressed or tired. If love withdrawal does not occur routinely, and if other opportunities for emotional connection are available, ties can soon be restored. Although the anxiety may still be painful, the child can fashion a reconnection through *compliance* and taking full responsibility for causing the disruption. For example, the child of a disengaged parent may adopt the interpersonal coping strategy of being quiet and "going away inside" so she needs nothing from her parent. Or, the child of an authoritarian parent may learn that by compulsively achieving or attempting to be perfect, this coping strategy will help them preserve as much as they can of their insecure attachment. As new therapists gain more clinical experience working with diverse clients, they will begin to recognize *the varied coping strategies their clients have adopted to use in their efforts to maintain ties to important others.* Such interpersonal coping strategies may sound like this:

It's my fault they don't like me. If I could just [play baseball better; be nicer to everyone all the time; do more for my mom; always get A's in every class; be thinner and look

prettier; disappear and not ask for anything], then I'll be
OK, and then they'll be happy with me.

In families that are more dysfunctional, however, the attachment disruption will occur more often and more severely. In daily interactions around discipline and control, *children in these families will be exposed regularly to experiences of interpersonal loss and emotional isolation, even though the parent and child remain in physical proximity.* As we move further along this continuum, ridicule and rejection occur more overtly. In dismissive/disengaged families, gross neglect, even actual abandonment may ensue. In highly authoritarian families, these behaviors may erupt into physical abuse. In these painful situations, the child experiences his parent's anger and contempt as assaults on his basic sense of self, leaving the child ashamed of who she is and psychologically alone. It is important for new therapists to know that they will see many clients with these developmental experiences, and that they will hold the pathogenic belief that they are justifiably to blame for their parent's neglect, rejection, physical domination, disgust, and so forth. In other words, these clients take responsibility or blame themselves for what their parents did and will tell their therapists:

- "If I hadn't fought with my brother so much, my dad would have stayed with my mom. It's my fault she cries all the time now."
- "If I had done all of my chores, like Mom asked me, she wouldn't have hit me so hard. I was bad."
- "If I hadn't dressed like that, he wouldn't have touched me down there. I feel dirty."

Routinely, when parents act irresponsibly in these ways, they overtly blame the child for their own inappropriate behavior. In addition, however, the attachment researchers teach us that this sense of shame and feeling blame-worthy are adaptive if these children are to experience some sense of psychological control in their lives. That is, clients hold onto such false beliefs tenaciously because the unwanted alternative would be to view the parents more realistically as rejecting and punitive, which would leave the child feeling even more powerless and unattached. In sum, when the child experiences this severely authoritarian or dismissive style of parenting, it is the genesis of a shame-based sense of self.

Moving further out on this continuum, some parents become so completely removed from themselves and the child that they become dissociated. For a few moments, usually while in the midst of their own shame–rage cycle or under the influence of alcohol or drugs, these parents lose all vestiges of self-awareness or connection to their own experience. More importantly, they lose any awareness of their child's experience and do not feel or register the frightening or humiliating impact that they are having on their child at that moment. It is terrifying for children when threatening caregivers become dissociated in this way; they may describe their parents in these moments as being "possessed," "not there," or "somebody else." Most clinical trainees will not be familiar with these *dissociative ego states* and, early in their training, will be understandably upset by the shame and terror these states engender in some of their clients. However, by learning about these highly dysfunctional families, also known as the Category D (Disorganized) attachment style in the attachment literature (Hesse & Main, 1999), therapists also learn important principles for working with clients from less troubling backgrounds.

Caregivers abuse or mistreat children for many different reasons. Some abusive parents were indulged by their permissive parents when they were children and experienced no consistent limits or consequences for their behavior. Whenever they did something wrong or got in trouble, their caregivers externalized all responsibility for the problem and blamed the teacher, coach, or another child for the problem. Others mistreat their own children in the same ways they were once mistreated. *This multigenerational reenactment is most likely when the parent–child relationship was highly ambivalent; that is, the parent who hurt them also was authentically caring or affirming at other times* (Rocklin & Levitt, 1987). Abuse of this sort tends to occur when parents are in dissociated ego states, which often have been facilitated by alcohol or other substance abuse. In those states, parents may reenact with their children abusive or traumatic events from their own childhoods. In many cases, these individuals can remember and talk about the physically or sexually abusive events that happened to them, yet the fear and shame that accompanied their experience are not available to them. That is, therapists will observe that these clients remember what happened and relay the story, but without emotion. By reenacting the abuse, rather than reexperiencing it, some individuals defend against the painful feelings evoked by their own mistreatment. Typically, *one of their own children comes to represent themselves or, more specifically, the unwanted or disowned aspects of themselves* (the emotional needs, vulnerability, or anger) that originally evoked the abuse. Which child is chosen will depend on age, gender, birth order, temperament, and other characteristics. However, most victims of abuse do not reenact what happened to them, although many live with the debilitating fear of being like their parent in any way, even though they are actually very different.

This disturbing reenactment has received different labels within different theoretical systems: projective identification (object relations theory); identification with the aggressor (psychoanalytic theory); the family projection process (family systems theory); turning passive into active (control/mastery theory). However labeled, it is a defense—an attempt to cope with trauma and manage or gain control of unwanted, painful feelings that once were intolerable. This intergenerational transmission of pathology stops if the abusive parent can not only remember the events but also tolerate experiencing the fear and shame they cut off from feeling in their own past (Eron & Huesmann, 1990). This psychological task is not easy to accomplish, however.

Let's review these complex dynamics again. Object relations theorists and family systems theorists inform us that the more conflicted or unintegrated we are, the more likely we are to instill or project unacceptable aspects of ourselves onto others—especially our children. As we discussed in Chapter 4, evoking or instilling our own conflicts in others is a way of seeking an external solution to an internal problem. Thus, in this type of abuse, the abusive parents' reenactment with their own child may be a psychological defense—often an attempt to gain some control over the painful feelings associated with their own past abuse (in particular, the shame instilled by the mistreatment). By externally reenacting some version of the trauma over and over again with their children, some caregivers who are abusive do not have to remember or experience internally their own painful feelings. That is, by evoking the same terrified or otherwise unacceptable feelings in their child that they were once made to feel, abusive parents do not have to experience the frightening trauma and shame-laden defeat of abuse as their own; their fear and shame is expressed vicariously through the child. For children in extremely authoritarian and rigidly "hierarchical" families, such abuse is a persistent threat that organizes their daily experience and, ultimately, their psychological adaptation to life. In dismissive/disengaged families, there may be less physical abuse and more neglect and abandonment, but the legacy of shame instilled in children raised in such families is still very real and debilitating.

Effects of Severe Love Withdrawal   Severe love withdrawal as a means of discipline occurs in many ways and in different types of families. For example, variations of this approach can be seen in parents who act like martyrs. These parents often communicate their hurt and disappointment nonverbally, by turning away and withdrawing emotionally or by providing a sigh or long-suffering look. Other parents demand perfection from the child, exert excessive control over the child in order to obtain it, and withdraw their warmth, approval, and emotional presence when the child does not fulfill their perfectionistic demands. In these moments of parental love withdrawal, however they occur, the child in effect loses the parent. The child's emotional connectedness to the parent is temporarily broken, and the child's attachment ties are situationally disrupted. As described earlier, this engenders separation anxieties and shame until the child can find a way to comply and restore the tie. Of course, all children's ties to their parents will be threatened at times. However, significant problems occur when disruption of such ties is so frequent as to characterize the relationship. Enduring problems also ensue when these rupturing interactions are disavowed by the parent—acting as if nothing significant happened—rather than acknowledged and resolved (for example, a parent saying, "I'm in a bad mood today and got too upset just now; that was my fault, not yours"). Thus therapists will find that many clients who grew up with these problems also lived with an *unspoken family rule* that hurtful interactions like these could not be talked about—named or made overt—as if the client could not say or even know what just transpired.

A constellation of significant emotional reactions occurs when parents cut off their emotional connection to children in anger or disgust. Even though the caregiver is physically present, the child is psychologically alone and suffers painful separation anxieties. This withdrawal also stifles the child's sense of self-efficacy and gives rise to feelings of helplessness and hopelessness because the child in this predicament cannot really win or earn the parent's love. This also leads to dysthymia and other forms of depression, and a shame-based sense of self as ineffectual or feeling unworthy of being loved. Thus, this child is made to feel bad and alone, desperately wants to restore the relationship and renew emotional ties, yet is ultimately helpless to do so until the parent chooses to reengage. Moreover, the child's inefficacy is further exacerbated by being made to feel responsible for the parent's withdrawal and deserving of it. In this immobilizing double-bind, children believe that their behavior (spilling a glass of milk, crying, feeling angry, needing help, and so forth) has caused the parent to go away, which is the response the child fears most because it threatens already insecure ties.

The child is angry at being abandoned and wants to protest, of course, but this reaction would only elicit further domination or intimidation from the authoritarian parent. Through power assertion, the authoritarian parent usually does not allow the child to disagree, let alone find appropriate means of expressing anger. The child may be told, for example, "I'm your father. You are never angry at me. Do you understand that? Look at me and say, 'Yes, sir.'" Thus the child cannot protest behaviorally, or even experience anger internally, because such reactions will further threaten already tenuous ties to the parent. As a result, this child often becomes intrapunitive by turning the anger inward; it is expressed through self-deprecation, dysthymia, and having a shame-prone self. This tendency toward self-blame is exacerbated as children come to identify with the parent and adopt the same critical or contemptuous attitude toward themselves that the parent originally held toward them. Just as this type of parent loses touch with the child's feelings or experience in these angry moments, the child, in turn, loses the clarity and authenticity of his or her own internal experience; such children lose touch with important aspects of themselves. Denial, idealization, splitting defenses, and disassociative tendencies result when these processes have been intense and pervasive. Poignantly, when they enter treatment as adults, such clients often describe their parents in idealized, conflict-free terms ("My family was great—no problems"). However, adult attachment researchers find that their descriptions tend to be abstract and global, lacking in specificity about real-life experiences that occurred, fearing that such detail would evoke unwanted, painful memories (Hesse & Main, 1999). The therapist, in turn, while listening to this client describe her idealized childhood in such abstract terms, may find it difficult to emotionally connect with her.

Low self-esteem, internalized anger, inefficacy, and loss of the caregiver's love constitute a prescription for depression. They also engender anxiety symptoms and the control issues found in eating disorders. In addition, these offspring develop identity conflicts because their feelings have been so pervasively invalidated. Because highly authoritarian parents are so rigidly demanding of conformity and obedience, and dismissive/disengaged parents are so invalidating of their children's feelings and needs, children soon lose touch with their own internal experience. That is, they may not know what they like and dislike, and may even be uncertain about what does or does not feel good to them. Their own subjective experience can be so completely overridden that they have no basis for later developing their own belief systems, clarifying their own values, or formulating occupational interests in late adolescence and young adulthood. In other words, they haven't been allowed to develop an identity or, more basically, a self and "have their own mind" or "find their own voice." New therapists often will find themselves working with the emotional problems and family dynamics described here. With these clients, the therapist's initial goals are

1. to validate their subjective experience which has been so pervasively invalidated;
2. to encourage their initiative which has been undermined—for example, by following their lead in treatment and supporting their own initiative with others; and
3. to provide a treatment focus by helping them clarify their own interests and act on their own goals whenever possible.

To better understand these issues, let's look further at the developmental experiences of children in highly authoritarian and dismissive/disengaged families when emotional ties are disrupted by severe love withdrawal.

Interpersonal Strategies to Cope with Insecurity   There is already too little affection in both authoritarian and dismissive families, and these thin threads of connection are repeatedly disrupted when angry or demanding parents emotionally disconnect from children. When young children are unable to maintain parental affection, they will not grow up to feel love-worthy or secure. These children have missed the essential experience of constancy in their attachment bonds. This developmental deficit has two aspects.

First, these children missed the experience of someone actively reaching out and choosing them. In many cases, they do not feel loved or, especially in disengaged families, wanted. Children from authoritarian families often feel variations of the theme, "If only I did better . . . ." In these circumstances, the child attempts to win or earn the parent's attention or approval. Children do this by adopting the coping strategies already described, such as striving to achieve, being perfect, taking care of the parent, and so forth. Children from disengaged families, on the other hand, are more likely to think, "I don't really matter," and cope by withdrawing or becoming invisible and never needing anything. In either case, the parent's responsiveness to the child is dependent on the child's efforts to elicit the caretaker's attention, and a secure attachment fails to develop. If the parent's responsiveness to the child is dependent on the child's efforts in one of these ways, the child is responsible for eliciting the caretaker's attention and affection and does not develop a secure attachment. This routinely occurs for children in authoritarian and in disengaged families, where children must find ways to cope with the anxiety generated by their insecure ties.

Let's explore this subtle but important concept more closely. To establish more secure bonds, *children attempt to control or manipulate their parents' feeling for them.* They do this by learning to employ certain interpersonal coping strategies—for example, by being compliant and pleasing, by compulsively striving for achievement, by perfectionistically trying to be good or not getting in the way of the parent, and so forth. As we will see in Chapter 7, this attempt to win attention and affection often becomes a *pervasive coping style* that clients' continue to use with others throughout their lives. These coping strategies come at a personal price and take a toll, however, and become central to the symptoms and problems that clients subsequently present in treatment. Furthermore, because of the schemas that have developed from these childhood experiences, these clients also are going to believe that they similarly have to elicit, or be responsible for, the therapist's interest in them—as they have with others in their lives.

Second, whereas secure children do not feel they have the power to break the parent–child tie, insecurely attached children often are told that it is their "bad" behavior that leads to their parents' rejection, contempt, or withdrawal. As we have seen, the child often is led to believe that they deserve and are responsible for disrupting the parent's loving commitment to them. For example, children in highly authoritarian and disengaged families learn that their anger, their tears, or even their questions can provoke their parent to cut off emotionally from them and disrupt the parent–child tie with threats such as these:

- Stop those tears right now or I'll give you something to really cry about!
- Wipe that angry look off your face. You'll do what I say, and like it, or else!
- Never ask me why. Just do what I say, when I say it.
- Now you've really done it! This time I've just had it with you—for good!
- Stop bothering me with all those questions. Leave me alone!

Even as adults, the sons and daughters of such parents who use severe love withdrawal techniques at times may be subjected to the same threats. If they do not do what parenting figures demand, they will (seemingly) destroy the relationship:

- If you marry him, we won't come to the wedding or visit you anymore.
- If you get a divorce, we are going to disown you and take you out of the will.
- If you do that, no one in this family will ever speak to you again.

The important distinction here is that authoritarian parents and others who use severe love withdrawal techniques to control their children are not just setting limits on unacceptable behavior; they are threatening to cut off fundamental relational ties. As attachment-seeking children try to cope with the intense anxiety this arouses, compliance becomes a generalized trait, pervasive personality constriction and inhibition occur, and obsessive/compulsive symptoms and other control issues—such as eating disorders, often develop. In contrast, some children ("avoidant" attachment style) distance themselves in relationships and, throughout their lives, learn to cope by closing down their feelings and never being vulnerable or needing anyone. They develop a pseudo-independence that gives the erroneous impression that they are in perfect control of their lives. These adults may become workaholics or select careers that do not involve emotional contact with others. In sharp contrast to this avoidant attachment style that dismisses legitimate emotional needs, we have seen that securely attached individuals can ask others for help. Based on their attachment history, they hold the expectation that certain others can help them when they have a problem—such as during a marital or health crisis. In contrast, avoidant individuals do not expect therapists or others to be able to recognize or respond to their distress. They do not want to experience or reveal an emotional need, and feel "weak" or ashamed of feeling vulnerable.

Permissive Parenting, Overinvolvement, and Insecure Attachment

Whereas authoritarian and disengaged/dismissive parents tend to use love withdrawal to create an insecure attachment with their children, permissive parents often become emotionally entangled or overinvolved with their children, at times smothering them with unwanted love and attention. As we have seen, the interpersonal strategy of authoritarian and disengaged/dismissive parents is to distance the child from them, especially when they, themselves, are feeling anxious or distressed. The permissive parent's strategy, conversely, is to draw the child closer to the parent, but often with the purpose of making the parent (not the child) feel better. Instead of instilling a sense of shame in their children, guilt is often the legacy of children raised in permissive homes.

Permissive parents overtly show affection to their children yet find it difficult to assert appropriate parental control over a child's behavior. In many ways, these parents act more as friends than as parents to their children. Thus, a client may describe her family of origin as a group of siblings living under the same roof, signaling the absence of clear intergenerational boundaries. When parental guidance or help is required, such parents may appear inept, even incompetent. To remain attached to such parents, children in these families often take on an adult role. During emergencies, for example, these parents may "fall apart," requiring the child to come to their assistance. Children in these families become responsible for the emotional (and sometimes physical) well-being of one or more parents. In families with more serious problems, a child may be given the role of "rescuer" in a family where one parent is physically abusing the other parent. Children may receive love and attention in the form of gratitude; however, the love is conditioned on the child's ability to aid the parent. It is a love that ultimately smothers a child and prevents her from developing a secure attachment or a viable identity.

Such clients may recall a parent for whom they were excessively important, and still feel guilt for not adequately meeting this parent's needs. To illustrate, Will grew up watching his alcoholic father get drunk, pick a fight with his mother, and sometimes end up hitting her. Unable to sleep without nightmares, Will was an anxious child—worried about his mother's safety. But Will was also a big kid, and, as puberty came on, he also became a strong kid. Now he no longer listened anxiously through his bedroom door to hear if his parents might be wrangling. Instead, he learned the best way to protect his mom was to just walk out and provoke a fight with his dad before he could get started with her. Continuing this role as the strong rescuer, Will eventually become a police officer. Regularly promoted, he earned the respect of his most seasoned colleagues for his willingness to risk his own safety and step into the line of fire to save a child or protect a bystander. Over the years, physical injuries accumulated from this work, of course, but the emotional toll was even greater. As an urban police officer, he regularly saw children die in accidents and youth shot on the streets. Will couldn't forgive himself when he saw this kind of tragedy; he felt guilty, as if somehow he should have done more or been more to prevent it. By age 40, Will was "burned out" and, almost unable to work, came to treatment for drinking too much himself.

A Diversity of Parenting and Attachment Styles

In their initial caseloads, most new therapists will work with some clients from authoritarian and dismissive backgrounds. For therapists who have had better developmental experiences, it may be hard to appreciate the emotional severity of highly authoritarian parenting, rejection, severe love withdrawal and other shame-inducing child-rearing practices. These therapists often wonder how the ostensibly normal and, in many other ways, decent parents of these clients can be so rejecting on occasion and can have caused such profound insecurities or even self-hatred in their clients. Other graduate student therapists may have difficulty with these client dynamics because they evoke the therapists' painful feelings about their own backgrounds and how they were raised. Thus this material can be challenging to work with because it evokes such strong countertransference reactions in many therapists. To help therapists understand and respond more effectively, let's illustrate what a range of secure and insecure attachment configurations look like.

While growing up, Molly benefited from secure attachment ties without fear of love withdrawal. For example, she recalled an incident when her mother was angry about something Molly had done when she was 7 years old. Her mother made strong eye contact with her, reached out and touched her on the shoulder, and said in a calm but firm voice, "I love you, but I don't like it when you act this way. Go to your room and stay there until you decide you are ready to come back out and get along." Testing the limits, Molly protested, "But you can't send me to my room. You have to want to be with me because you're my mom!" Her mother went on to clarify that it was Molly's behavior toward her sister that she did not like, and Molly recalled that she began to understand that concept.

Her mother was angry, and she got her point across that she did not like what Molly was doing, but Molly also felt secure in her mother's love. Her behavior wasn't acceptable, but she was—Molly was not coping with "conditions of worth." Looking back, Molly thought that her sense of secure attachment in this conflict was maintained primarily by the nonverbal messages that accompanied her mother's restrictions and explanations. Because of these secure relational ties, Molly was able to internalize her mother's loving feelings for her and to develop "object constancy," as discussed in Chapter 1. As an adult, Molly now is able to establish friendships in which she feels respected and affirmed and to establish a marriage in which aspects of this same loving affect are present.

Often, children will be secure in their emotional ties with one parent but struggle with a lack of constancy with the other parent. For example, Ellen recalled that she felt secure with her mother—even when her mother was mad or disapproving of something she had done. Her father was highly inconsistent, however. At times, he was affectionate and responsive, but at other times he could lose his temper and be overtly rejecting. Ellen recalled happy memories of her father patiently painting ladybugs on her roller skates, as well as painful memories of her father blowing up at her and yelling angrily, "Get out of here. I hate being around you!"

Ellen's secure base with her mother allowed her to cope with this intensely ambivalent relationship with her father. With the consistent or reliable emotional support provided by her mother, she could learn to anticipate her father's moods and stay away from him when necessary. Her feelings about herself were not based on the unstable fluctuations of her father's moods. Now, as an adult, Ellen is an especially perceptive and aware person; in this way, personality strengths often develop from such conflict-driven demands for coping. Ellen's brother, however, was not so fortunate. Ellen's mother seemed to enjoy raising girls more than boys, and her brother did not receive the same secure base with his mother that Ellen enjoyed. As a result, he was left to bob about unconnected at the mercy of his father's stormy emotional seas. Ellen describes her brother as feeling "very bad about himself" and being depressed a lot. Now in his late 20s, he has been unable to make commitments to relationships or a career and cannot find a life for himself. Ellen says she worries about him a lot.

Let's look more closely at insecure attachments. With certain clients, therapists may wish to explore directly about the security of attachment ties:

- How did your parents usually respond to you when they were angry or disciplining you?
- What did each parent tend to say and do when they were upset with you, and how did that usually leave you feeling?
- Would you seek help from your parent and talk to them when you had a problem? How did you expect them to respond when you were distressed or needed help?

In response to such queries, many clients will describe disciplinary techniques involving love withdrawal that threatened their ties to their parents. In addition, many learned not to expect help and, after a while, didn't even think of seeking assistance from their caregivers as an option or possibility. At the heart of the attachment story, *they were not secure in the expectation that their caregiver would be their ally when they were distressed and try to help them with their problem*—even if that meant only to hear their concern and be with them in it. Furthermore, these insecure clients often had to fulfill familial roles of being "bad," "too demanding," "too sensitive," and so on. Figure 6.2 presents some typical comments directed at clients by their authoritarian, dismissive or otherwise rejecting parents, along with the clients' statements about the impact these messages had on them. As painful as they are to hear, new therapists will find that they commonly hear comments like these from their clients. Again, it is important for therapists to anticipate that, oftentimes, *their clients will be able to remember and talk about such interactions, yet the painful feelings accompanying them are blocked off or unavailable.*

FIGURE 6.2 | PARENTAL COMMENTS AND IMPACTS ON CHILD

Parental Comment or Action	Impact on Child
"Stop it, or I'll send you to live with your father!"	The client, whose father was an alcoholic, said this threat made her feel that her mother didn't really want her or care about her. As a child, the client was "good" all the time and stayed away from her mother to avoid this threat; because of her *vigilance*, it was not often voiced.
"How could you do this to me? Can't you think of anybody but yourself? What's wrong with you?"	The client said she always felt guilty and tried to "earn love" by figuring out what her parent wanted and trying to provide it.
"I've had it with you! Just you wait until your dad gets home!"	After this comment, the client's mother would withdraw and not speak to her for the rest of the day. The client said it made her feel "very alone."
"Look what you did! Get away from me! I don't want anything to do with you!"	The client said he had always felt that he was a "terrible person."
"You did it wrong again; you always do it wrong. You're ruining my life!"	The client said she "hated" herself.
A father would wordlessly slap his son's face for "eating wrong."	The client said, "I don't remember having any feelings about it."

Highly authoritarian and dismissive parents shame and intimidate their children, break off emotional contact with them, and—in words and, especially, through tone—may communicate contempt for them. Children who suffer such developmental experiences—especially internalized shame from parental contempt—typically struggle with anxiety and depression throughout their adult lives. They often report feeling guilt, loneliness, and low self-esteem when they enter treatment, but without understanding why, and describe their parents in idealized or problem-free terms. As introduced earlier, these clients have developed interpersonal strategies to cope with this, such as always being good, taking care of the parent, being quiet and "going away inside," and so forth. As we will explore in the next chapter, *the therapist provides a treatment focus by highlighting these faulty interpersonal coping strategies.* Although these coping strategies once were necessary and adaptive, the therapist helps clients explore how they are no longer necessary or effective in many current relationships, and are disrupting relationships with others:

> THERAPIST: As you describe this argument with your wife, Bob, it sounds like she is complaining that "you always have to be right." I've heard this theme in problems you've described with others—think there's something to it?

In addition, the therapist is focusing on how these problematic coping strategies may be being reenacted in the therapeutic relationship along the process dimension:

> THERAPIST: You know, Bob, it feels almost like we're having an argument right now—like one of us has to be right and the other wrong. What do you see going on between us here? Any ideas?

These interpersonal defenses—always having to be right or on top— originally helped protect the client from experiencing the painful feelings that resulted from hurtful interactions, like those described earlier, that occurred repeatedly for them. If therapists begin to ask about these interpersonal coping strategies when they see them occurring with others or with the therapist, this accompanying pain will soon emerge. When therapists respond with affirmation and compassion, however, clients have the safety and support they need to reexperience these feelings. They will benefit greatly from the therapist's validating response and begin integrating these feelings that previously had been sequestered away. Only then can these clients stop protecting their caregivers at the expense of their own symptoms, denying their own feelings, and maintaining family myths of happiness and togetherness that only serve to confuse and disempower them.

Clinical Implications for Working with Disrupted Ties

Parenting is probably the most challenging task in life. The influential family therapist Salvadore Minuchin says forgivingly that parenting has always been more or less impossible. Almost all parents are trying to do the best they can for their children. Even many of the highly authoritarian parents we have been discussing, who indeed caused significant lifelong problems for their children, are not cruel or ill-intended people in most cases. In child rearing, as in other aspects of personality, people are uneven in their development. Most of these parents do other things well for their children, live by certain moral standards, believe they are trying to do what is best for their children much of the time, and usually are treating their children as well as or better than they themselves were treated. Although children certainly are hurt by such child-rearing practices, almost universally they still love their parents and seek their approval.

Therapists will not do well when they fail to appreciate adult clients' willingness to give their parents "another chance" and their lifelong efforts to improve or repair these flawed but all-important relationships. Therapists want to help clients realistically assess whether caregivers have changed over the years and to what extent they are capable of responding better now than before. That is, some caregivers have gotten better and can respond more constructively now than when clients were young, whereas others only continue to respond in the same problematic ways. The therapist's goal is to foster realistic expectations in their clients and help these current relationships between parents and adult offspring become as good as they can be. It is important that therapists encourage clients to express in treatment the full range of their positive and negative feelings toward their caregivers. Therapists who "foreclose" on this process by quickly echoing and remaining stuck on just the negative aspects of the caregivers, or just the positive characteristics, will rob their clients of the important experience of integrating both the beneficial and the problematic aspects of their caregivers. Over the long term, clients who cannot resolve their ambivalence and integrate the good and the problematic aspects of their relationships with their caregivers will also have difficulty accepting the good and the imperfect parts of themselves—and of their own children.

In sum, the therapist's role is not to foster splitting defenses by bashing parents who have been hurtful and making them bad, by encouraging clients to reject them or break off contact, or by subtly inducing clients to replace the parents with idealizations of the therapist. Nor, on the other hand, is the aim to deny or in any way to minimize the real impact of the hurtful interactions that have occurred. Instead, the therapist

1.   helps clients come to terms more realistically with the good news and the bad news in their family of origin,
2.   change their own responses to problematic others in current relationships, and
3.   facilitate clients' current attempts to establish new relationships that do not repeat the problematic relational patterns that have come before.

Therapists provide a corrective emotional experience when they can sustain a working alliance and remain emotionally available to their clients. Doing so provides many of the clients discussed in this chapter with the secure base that they missed developmentally. *Such consistent emotional availability, week in and week out over the course of treatment, usually has more effect on client change than do more dramatic but isolated incidents of compelling insight, important self-disclosure, or other significant therapeutic interventions.* As we will see, however, providing this consistent presence in the face of the client's maladaptive relational patterns is often difficult. The inevitable mistakes, misunderstandings, and reenactments that routinely occur will "rupture" the therapeutic alliance at times. The issue is not whether such ruptures will occur, but the therapist's willingness to acknowledge the conflict in the therapeutic relationship and engage the client in working together to restore the alliance (Safran & Muran, 1998). Let's examine this point more closely.

The clients whom we have been discussing share the experience that caregivers withdrew from them in one way or another or were not emotionally available to them at important times of need. This developmental deficit—and the maladaptive relational patterns that result from it—can readily yet subtly be reenacted in the therapeutic relationship. For example, some therapists may have trouble being emotionally available to clients' pain and vulnerability; they may not be able to remain present with certain affects, such as the client's raw shame or intense sadness. The client's individuation may make other therapists uncomfortable because success or independence was not supported in their own development or because the client's success evokes the lack of fulfillment in their own work or marriage. Some therapists may also feel sad or anxious about losing the caretaking role that they were scripted to assume in their family of origin. Most important, perhaps, many therapists disengage or "give up" on clients when the client doesn't change or somehow disappoints the therapist—routinely saying that "the client wasn't ready to change yet." Therapists often feel they are failing when the client fails or cannot change, and therapists commonly respond ineffectively to this countertransference reaction about their own performance by withdrawing from the client.

When any of these countertransference reactions occur, as they commonly do, clients are again left alone and unconnected in their experience. Clients' pathogenic beliefs are confirmed rather than resolved by this problematic interpersonal process: feeling sad or mad, or being "stuck" or successful, causes others to be hurt or angry, to disengage, and so forth. When such reenactments occur, without being addressed and rectified, they generate further insecurity for clients and impede change. Rather than being resolved, old schemas and problematic relational expectations are confirmed by what is being played out in the therapeutic relationship.

In addition to their own countertransference tendencies, certain features in the client also make it difficult for therapists to remain consistently available. For example, many clients will report that they like and trust the therapist but they may also believe that, "if the therapist really knew them," she would not respect or care about them. In line with their faulty schemas and problematic coping strategies, these clients believe they have deceived the therapist or manipulated the therapist's positive feeling for them. This is why it is necessary for the therapist to help clients adopt an internal focus and to draw out the full range of clients' feelings toward the therapist—including the client's perceptions of the therapist's reactions to the client. For example:

THERAPIST: What do you think I am feeling toward you as you tell me this?

Unfortunately, clients often hold the pathogenic belief that their despised, weak, or otherwise unacceptable emotions constitute their real selves. This far-reaching problem is resolved when the therapist sees the vulnerable, dependent, shameful, or other parts of themselves, which the clients believe are the damning proof or confirming evidence of their basic unacceptability, and still remains affirming and emotionally connected. *The clients' encounter with compassion and understanding, rather than the emotional withdrawal, rejection, and so forth, that they have learned to expect, is one of the most powerful relearning experiences that therapists can provide.* Thinking about conditions of worth, Carl Rogers (1951) emphasized long ago that clients do not change until they accept themselves and that they begin to accept themselves when they feel such acceptance from the therapist. Thus, by providing a safe holding environment that contains these conflicted feelings, as clients are experiencing them, the therapist allows clients to resolve them.

This pivotal experience not only permits the integration of those warded-off feelings but also opens up a variety of therapeutic opportunities. For example, in the next minute or two, clients often disclose important new material or make significant links to other related problems they have been working on. Or, clients may risk trying out previously threatening new behaviors with the therapist; for example, they may be more assertive or risk bringing up problems between them that had been too threatening to broach. In the week that follows, clients may act in these stronger ways with significant others in their lives. Clients may also express more self-acceptance, resolve some ambivalence, or make an important life decision they had been unable to act on. Furthermore, therapists can often successfully reframe a problem or help clients see a situation in a different way—a new perspective that did not seem realistic or relevant previously. Similarly, because they feel safer now, clients may be able to hear the therapist challenge a pathogenic belief about themselves or identify a problematic relational pattern. In short, clients often can make use of a wide variety of therapeutic and educational interventions at this point that they could not utilize before.

Finally, *some clients are adept at eliciting the same problematic responses from the therapist that they have received in the past.* As we have already emphasized, a basic tenet of the interpersonal process approach is that most clients' problems will be temporarily reenacted with the therapist along the process dimension and need to be identified and reworked. It is easy to be empathic, warm, and genuine with clients who are usually cooperative, friendly, and respectful. However, it is far more challenging for therapists to remain committed to a working alliance with difficult clients because their interpersonal coping strategies—for example, being hostile, dominating, passive, and so forth—eventually alienate, intimidate, or frustrate the therapist. With these clients, it is especially important for therapists to keep process notes and formulate working hypotheses about the clients' maladaptive relational patterns; specifically, *to hypothesize what the clients have tended to elicit from others and what they are likely to evoke in the therapist.* When therapists prepare such conceptual formulations, they increase their chances of providing a corrective experience, rather than responding automatically to the client and merely reenacting the problematic scenario with which the client is familiar (see Appendixes A and B). The following case study illustrates how the therapist was able to resist reenacting the client's maladaptive relational scenario and, instead, provide a corrective relational experience by remaining emotionally connected to him during a challenging disagreement.

John, an 11-year-old client, was being treated in a strict, physically punishing manner by his authoritarian stepfather. His mother reluctantly accepted her husband's harsh corporal punishment and tried to ignore his derision of the boy; she often walked away and left the room when the stepfather was angrily deriding him. In treatment, John defied his female therapist, responded contemptuously toward her, and repeatedly tried to push her away. Near the end of an especially frustrating session, in which he had repeatedly pushed every limit, John was unwilling to speak. Trying to find some way to salvage something of a working alliance with him, the therapist agreed that they would not have to talk together if he did not want to and that he could just throw the ball as he wished. Trying to join him in his chosen activity, she asked whether she could silently play catch with him. He agreed but, of course, began throwing the ball too hard. This process continued throughout the session: John tested, and the therapist set limits by saying, "You have to throw the ball below my shoulders," while still working hard to try and find some way to connect with him.

Near the end of the hour, the therapist invoked their standing rule and asked him to help her pick up the room. John refused; the therapist insisted. Exasperated, she took his arm lightly and directed him toward the mess in the room, whereupon he recoiled. At this moment, the same conflicted emotions that John struggled with at home were evoked toward the therapist and, in his mind, were about to be reenacted in the session. The therapist had been sorely pushed by John for weeks; but even though she was frustrated with him, she was still trying to find another way to relate. She did not withdraw in resignation as his mother did, although she felt like it. She also did not want to physically dominate him as he felt she was beginning to do, and thereby reenact a milder version of what his stepfather did. Instead, she met his eyes and slowly said, "You're a good boy, John, I still like you. I want us to work this out together." Even though she was upset, the therapist could still communicate that she felt for John and was remaining emotionally connected to him. This was a turning point in treatment: John was not able to reenact with the therapist the same painful scenario of domination by stepfather, withdrawal by mother, and angry but helpless isolation for John that had been played out so many times at home. John grudgingly helped the therapist pick up the room; at their next session, however, John was more respectful of the therapist and began to disclose more.

In closing, clients will present with an extraordinary diversity of developmental and familial experiences. Some clients' problems will be unrelated to the issues discussed here. However, faulty child-rearing practices, insecure attachments, and love withdrawal techniques will be heard pervasively throughout the experiences clients report in therapy, and they will inform therapists about the presenting symptoms and problems that clients are struggling with in treatment.

Relating the Three Dimensions of Family Life

Let's now examine some relationships among the three basic dimensions of family life: the nature of the parental coalition, the separateness–relatedness dialectic, and child-rearing practices. Authoritative parents have the emotional flexibility and range of parenting skills necessary to meet children's needs for both closeness and individuation. The experiential range of authoritarian and permissive parents is more limited—specifically, to opposing halves of the separateness–relatedness continuum. Thus, the range of authoritarian parents does not include nurturance, communication, and mutuality. These parents do not meet young children's attachment needs and older children's age-appropriate needs for warmth and understanding, although their emphasis on mature, responsible behavior often allows children to achieve. In contrast, some permissive parents communicate more effectively and meet children's emotional needs better, but they do not foster their children's successful individuation by expecting responsible behavior from them. Moreover, permissive parents are more likely to embroil children in cross-generational parent–child alliances, which further stifles their independence. Of these three parenting styles, permissive parents are least likely to establish a primary marital coalition, most likely to embroil children in cross-generational alliances, and most likely to parentify children. Children of disengaged parents are overtly rejected or passively ignored. These children are at risk for acting out, developing conduct-disordered problems, and becoming involved in the social welfare and judicial system.

The family dynamics presented in this chapter offer a parallel to the relationship between the therapist and the client. Therapists, like authoritative parents, are seeking to develop their own capacity to respond at both ends of the separateness relatedness continuum. On the one hand, effective therapists are those who have a broad experiential range that allows them to be empathic and emotionally responsive to clients. On the other hand, therapists also need to be able to set limits with clients, question their actions or challenge them, and tolerate clients' anger or disapproval at times. More specifically, therapists need to be able to respond to the emotional deprivation of clients who have been reared by highly authoritarian or dismissive parents. At the same time, they also want to be able to set limits with the provocative, demanding, or testing behavior of clients who have been reared by permissive or inconsistent parents. Thus, the personal challenge for all therapists is to examine their own experiential range, which has been shaped in their families of origin. Effective therapists are willing to explore the interpersonal spheres that are anxiety arousing for them, and work to broaden their own interpersonal range so that they can respond to the wide range of problems that clients present. To achieve this flexibility, therapists are encouraged to do their own family-of-origin work.

Such work is an integral part of training for those who specialize in family therapy. In particular, Bowen (1978) has written extensively about how clinical trainees can systematically study, and sometimes change, maladaptive relational patterns in their own families of origin. *By achieving greater understanding of their own family backgrounds, and perhaps greater empathy for all family members, therapists can make a significant contribution to their own clinical skills.* Readers are encouraged to read and embark on the great adventure of doing family-of-origin work (Boszormenyi-Nagy & Spark, 1973; Kerr & Bowen, 1988; Minuchin & Nichols, 1998). In particular, therapists can prepare a family genogram—a three-generational map of family rules, roles, myths, and structural family relationships (McGoldrick, Gerson, & Shellenberger, 1999). By learning about their families of origin in this way, therapists learn much about their own countertransference propensities, become more compassionate toward their clients' struggles to change, and better appreciate the profound influence of familial experience on adult life.

Closing

In this chapter, we have examined three basic dimensions of family functioning: structural family relations and the parental coalition, the separateness–relatedness dialectic, and child-rearing practices. These three aspects of family life will help therapists better understand many of the problems that clients present. For some trainees, it is disruptive to read this material because it violates familial rules and cultural prescriptions to question or examine what transpired in their families of origin. Also, many readers are parents or grandparents themselves. As they learn more about family functioning, and parenting practices in particular, guilt may be evoked as they recognize the limits of their own parenting skills and the problematic consequences that have resulted for their own children. Such readers may need to relinquish unrealistic expectations of themselves and forgive themselves for not knowing more about child rearing than they did at the time.

Suggestions for Further Reading

1. Clients make more progress in treatment when therapists can help them come to terms with the good news and the bad news in formative relationships with caretakers. However, it is especially challenging for both therapists and clients to do this when abuse has occurred. To help with this complex issue, readers are encouraged to read "Ambivalent Feelings: Case Study of Sheila" in Chapter 6 of the Student Workbook.

2. *Becoming Attached*, Robert Karen's (1998) highly readable overview of attachment, is an illuminating book that will help therapists better understand many of their clients' symptoms and problems.

3. Basic information on child development and child psychopathology will help therapists understand their adult clients' personalities and problems. One outstanding text is *Psychopathology from Infancy through Adolescence: A Developmental Approach* by Wenar and Kerig (2000); see especially Chapters 4–8.

4. Parentification is an essential concept to understand because so many clients and therapists have been scripted into this familial role. Further information about this important concept can be found in Chapter 9, "Helping Children Cope with Divorce," in Teyber (2001).

5. Structural family relations provide an illuminating road map for understanding family functioning. A classic work in this area is Salvadore Minuchin's (1974) *Families and Family Therapy*; see especially Chapters 3 and 5.

6. Monica McGoldrick et al.'s (1999) textbook *Genograms: Assessment and Intervention* provides a clear "how-to" approach for those interested in doing their own family-of-origin work. This interesting book provides illustrative genograms of multigenerational processes in notable families, such as Sigmund Freud, Bill Clinton, and Woody Allen. New therapists can learn much about themselves, and the countertransference propensities they will bring to their clinical work, from doing this important, personal exploration with genograms of their own families of origin.

CHAPTER SEVEN

Inflexible Interpersonal Coping Strategies

Conceptual Overview

This chapter provides an interpersonal model for helping therapists further conceptualize their clients. This model is based on clients' *core conflict* and their interpersonal strategies for coping with it. *Core conflicts are the central problems or key issues that pervade the client's life.* The core conflict is the hub of the wheel that links together the different problems and concerns that clients present, and provides a focus for treatment.

The core conflict (also called the *generic conflict*) has arisen out of the client's repetitive interactions with significant others, especially parents or primary caregivers. Following the developmental issues presented in Chapter 6, this core conflict often arises from the combined effects of insecure attachments, problematic childrearing practices, and faulty structural family relationships. When these types of problematic interactions with caregivers were repetitive or consistent, they give rise to pathogenic beliefs about oneself, faulty expectations of others, and a narrow or skewed view of what will usually occur in close relationships. To cope with the painful feelings, low self-esteem, and interpersonal threats that accompany these developmental experiences, clients often develop a repetitive or fixed interpersonal style. For example, the client is *always* pleasing others or trying to be perfect, *repeatedly* taking charge or being in control, *routinely* taking care of others and meeting their needs, *consistently* remaining invisible or avoiding all conflict, and so forth. The key here is that this interpersonal coping strategy is not just used sometimes or flexibly employed in situations when it is actually needed. It is pervasively and rigidly overused—even in situations where it is not necessary or adaptive.

Defenses such as these interpersonal coping strategies are necessary for all people at times—they help us cope in difficult situations. Clients will feel understood and affirmed when their counselors can grasp how one of these interpersonal coping styles really was a necessary and adaptive way to cope at one point in time. That is, recalling the concept of "honoring the client's resistance," therapists can recognize that these rigid or inflexible interpersonal coping styles once were an adaptive strength feature in the client's personality— they helped the client cope more effectively with reality-based problems. The problem for most clients, however, is that *these coping strategies are no longer necessary or adaptive in most current relationships.* In fact, rather than continuing to help, they now contribute to many of the symptoms and problems that clients present in treatment. Thus, the purpose of this chapter is to clarify this as an overriding treatment goal. On the one hand, the counselor aims to help clients discriminate or sort through current relationships where they still need to respond in this pleasing or avoidant way (for example, with an authoritarian, dominating boss). And, on the other hand, discern other relationships (such as with a new friend) where the old coping style (continuing to take care of everything or always be in control) is no longer needed and only serves to create new problems.

To reach this treatment goal, the therapist's intention is to

1. identify the formative patterns and painful feelings that make up the client's core conflict,
2. highlight the interpersonal coping strategy the client has used to cope with these developmental challenges,
3. clarify how the client's interpersonal coping strategy is being expressed in current interactions with the therapist and others, and
4. begin changing this pattern in the relationship with the therapist and others in the client's life.

As new therapists become more successful at formulating the client's core conflict and interpersonal adaptation to it, this conceptualization will help counselors recognize themes and find organizing patterns in the complex material clients present. As noted in Chapter 2, client conceptualization is an ongoing process that begins with the initial client contact. General hypotheses about clients' problems are formulated early in treatment, and these tentative working hypotheses are further refined or discarded as the therapist learns more about each client. However, as the client's conflicted emotions emerge, the key concerns that are central to the client's problems become clear. Having focused the client inward, the therapist can use the client's emotions that begin to emerge as guideposts to clarify this central conflict, and focus on how it is being played out with the therapist and others.

Interpersonal Model for Conceptualizing Clients

In this section, we explore an interpersonal model for conceptualizing the client's core conflict—how it originally developed, the client's interpersonal adaptation for coping with it, and how it is being expressed in current symptoms and problems. This model for conceptualizing the client's problems includes five components:

- The client's unmet developmental needs
- The original environmental block that engendered the problems
- The client's intrapsychic defenses against her own core conflict
- The client's interpersonal strategy to rise above or overcome the core conflict
- An interpersonal resolution of the core conflict

Adapted from Horney's (1970) interpersonal theory, these five components provide a useful model for conceptualizing client problems (see Figure 7.1). This model is an integrated system composed of several, complex psychological processes. After discussing each component of this model, we will apply it to treatment by illustrating the model with three case studies. Let's begin by examining how enduring problems result when one or more of a child's basic developmental needs go unmet. Be patient—this is a lot of reading to digest before it is illustrated with case studies.

Blocked Developmental Needs

This conceptual model begins with children's basic needs for secure ties. Part 1 in Figure 7.1 refers to young children's emotional needs of their parents and, in particular, to children's need for a secure attachment. If parenting figures are consistently responsive to the children's bids for attention and affection, children will be able to freely express this need. Of course, as children develop, they benefit from learning to accept limitations on parents' ability to respond and to tolerate delays in parental response. If parents are consistently punitive or unresponsive, however, anxiety will soon become associated with the child's needs. Children are then forced to fashion a solution; the *compromise solution* they construct often alienates them from themselves and their authentic experience. Although many have written about this, none have captured it more succinctly than Carl Rogers (1951) did years ago in writing about *congruence*.

Consider preschool children whose need for understanding and affection goes unmet from disengaged, authoritarian or even permissive parents. When such children approach their caregiver for comfort when distressed, their need is often rebuffed:

> DISENGAGED: What do you want now? You always need something—get out of here!

> AUTHORITARIAN: What's wrong with you? Stop that crying right now, or I'll give you something to cry about!

FIGURE 7.1 | ADAPTATION OF HORNEY'S INTERPERSONAL
MODEL

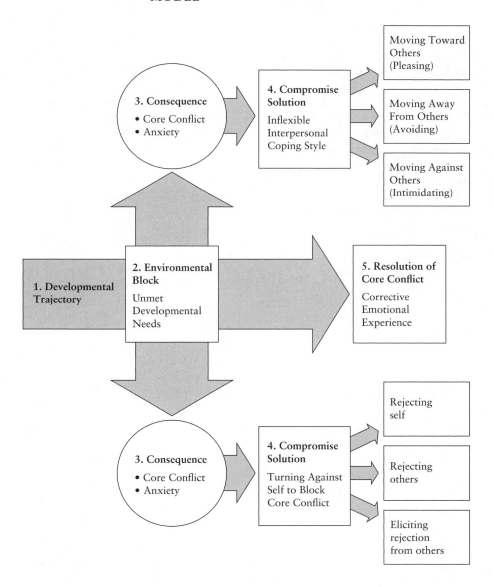

PERMISSIVE: Just go ahead and do whatever you need, honey. Can't you see I'm busy?

If these types of interchanges occur repeatedly, children will soon learn to anticipate rejection and feel anxious whenever emotional needs for comfort or soothing are aroused. Children are biologically organized to continue experiencing attachment needs, but their direct expression is blocked by the parents' unresponsiveness (Part 2 of Figure 7.1). The attachment-seeking child is then compelled to find some way to cope with this painful set of circumstances.

Although clients will present with a wide range of problems that result from the varied developmental challenges presented in Chapter 6 (referred to as an environmental block in Part 2 of Figure 7.1), enduring psychological problems often begin when basic childhood needs for secure attachments are not met. Most core conflicts also involve a failure to provide children with clear communication and consistent emotional access, on the one hand, and a predictable environment with consistent limits and discipline, on the other. Children vary in their developmental needs, experience different kinds of family environments and developmental problems, and develop their own individualized coping strategies. Most children experience anxiety when their emotional needs are blocked and, later as adults, also feel threatened when similar needs or feelings are activated in current relationships (for example, the client has a problem or emotional need and has to ask others for help). To cope with this lack of safety, many clients begin to construct a "compromise solution" designed to manage the continuing anxiety evoked by these unmet needs and the unwanted responses clients expect to receive from others now.

Parts 3 and 4 of our model illustrate how both sides of this compromise solution works. On the one hand, Part 3 illustrates how clients turn against themselves to block or minimize the feeling or need that has become anxiety arousing (for example, "I hate myself. I have such a big mouth—I'm too much for everybody"). Simultaneously, in Part 4, we will also see how clients implement different coping strategies to "rise above" the need and indirectly gratify it (for example, "If I just get a little thinner, then he'll never leave me"). Part 5 will refer to the corrective emotional experience that therapists provide to help clients resolve the core conflict and their compromise solutions to it.

We will now explore both sides of the client's compromise solution and learn about the complex psychological maneuvers that individuals adopt to cope with these core conflicts. In particular, we will see how their inflexible coping styles are replayed in the current relationship with the therapist, just as they are with others in their everyday lives.

Compromise Solutions

In Chapter 2, we proposed three domains for conceptualizing clients' problems: maladaptive relational patterns, faulty beliefs about self and unrealistic expectations of others, and core conflicted feelings. All three domains are intertwined, but the environmental block in Part 2 of Figure 7.1 principally evokes the core conflicted affect. As we will see, the faulty beliefs that arise from these core conflicts predominate in Part 3 and the interpersonal strategies used to cope with them are highlighted in Part 4.

Clients employ two psychological mechanisms in order to defend against the anxiety associated with expressing, or even experiencing, their blocked needs. First, we will examine three ways in which clients defend against their core conflicts by responding to themselves in the same way that others originally responded to them. These can be thought of as *cognitive schemas* that clients develop—that is, beliefs about themselves and others that, once adopted, regulate their behavior. Second, we will examine three ways in which clients adopt a general *interpersonal coping style* to try to rise above their personal or emotional needs. In treatment, this style will often quickly become apparent in the way the client interacts with the therapist (for example, pleasing, controlling, or withdrawing). Taken together, these two mechanisms of *blocking* and *rising above*, which represent attempts to minimize anxiety and provide a modicum of self-esteem, are clients' compromise solutions to their core conflict. Let's explore how this complex defensive system works.

Blocking the Unmet Need   With the limited skills and means available to them, insecurely attached children try urgently to get their caregivers to respond to their emotional needs but are unable to elicit the help they need. For example, this may reflect a general lack of parental responsiveness, parents respond on their own timetable rather than when the child initiates contact, punitive response, inconsistent response, and so forth. To gain some active mastery over their helplessness and to ward off the anxiety aroused, *such children begin to block their own need in the same way that the environment originally blocked it*. For example, children may deny it as insignificant and regard themselves as unimportant (for example, "I'm stupid") or may reject it and view themselves contemptuously as demanding ("needy"). Routinely, counselors will hear clients say the same hurtful responses toward themselves that others originally expressed toward them. In turning against themselves, children are able to ward off the unwanted anxiety that would be engendered if they felt that their developmental needs were reasonable and legitimate but still viewed as unacceptable to those they need and love most—their caregivers. As well as preserving attachment ties, this adaptation provides a sense of control and allows them to continue to view the world and their parents as reasonable and just. In this way, a certain kind of internal security or safety comes from viewing themselves as the problem, rather than the caregiver. This then becomes their schema for viewing themselves and others; by acting accordingly, they elicit responses consistent with the schema. This adaptation has been referred to as identification or modeling; some clinicians speak colloquially of "using the devil's tools to fight the devil." Thus, children begin to do to themselves what was originally done to them in three ways.

First, clients block their own need internally and respond to themselves in the same hurtful ways that others have responded to them. That is, when the children of nonresponsive parents feel a need for reassurance or affection, they will block their need by feeling critical or contemptuous of themselves and their own need. As adults, their own emotional needs are still unacceptable, and they will often say to themselves the same critical things they heard from others years ago. Often they adopt the exact words and tone—but without recognizing the source of these internalized templates or "tapes" as their own attachment figures who once said these words to them in this same tone of voice. Furthermore, *the judgmental, punitive, or rejecting affect that clients feel toward themselves as they replay these internalized tapes is the same affect that parenting figures originally expressed toward them years before* (for example, "What's *wrong* with you!").

Second, clients block their anxiety-arousing need by reenacting in current relationships the same scenario that was originally experienced in earlier, formative relationships. Thus, on an *interpersonal* level, clients will say and do to others what was originally done to them. For example, the child of strictly authoritarian or dismissive parents may grow up to be a parent who similarly disparages the same emotional needs in his or her own children—especially the child who most resembles the parent in gender, birth order, temperament, and so forth. Especially in the arena of parenting, clients who have not integrated or come to terms with their own conflicted feelings are prone to reenact the same maladaptive relational patterns that they experienced in their own childhoods. As parents, many clients feel helplessly dismayed as they watch themselves respond to their children in the same hurtful ways that their parents responded to them. Despite sincere pledges that "I will never do to my children what my parents did to me," variations of the old themes occur with clocklike regularity. Here again, in the interpersonal sphere, adult offspring gain some control over their own dilemma by repeating what was originally done to them rather than knowing and experiencing what they once felt.

Third, clients block the experience and expression of their own conflicted need by *eliciting* the same unsatisfying response from others in current relationships that they received in their childhood. For example, unless the children with problematic developmental experiences have other reparative relationships (with friends, teachers, and mentors) to help them resolve these concerns, they are likely, as adults, to select a marital partner who cannot respond well to their emotional needs. In the event that their spouse is capable of being affectionate or intimate, they may not be able to accept this emotional responsiveness. Why? The possibility of having a response to their old, unmet needs—even a well-intended, benevolent response—will often arouse shame or anxiety. That is, the spouse's affection is likely to arouse the pain of the original deprivation. Following their schemas, such clients also will anticipate that their spouse will reject their needs as others have done in the past—even though the spouse has never responded in this hurtful way. Operating on schemas or internal working models that fit past relationships better than current ones, such clients may simultaneously elicit and reject nurturance from their spouse. This will certainly frustrate and confuse the spouse, especially if these mixed messages activate the spouse's own problematic relational templates. Unresolvable marital conflict that results in emotional gridlock occurs when the spouses' schemas dovetail and each presents the other with a version of their old relational pattern. In such cases, the likely outcome is enduring conflict or divorce unless marital counseling helps the partners recognize this pattern and change their mutual interaction.

To sum up, clients defend against their unmet needs or unacceptable feelings by

- internally responding to themselves in the same rejecting or dismissive ways that attachment figures did,

- modeling the rejecting parent and responding to others in the same hurtful way, and
- repeatedly becoming involved with others who provide the same hurtful response to them that they originally received as children.

Although one of these three modes may be predominant for a particular client, many clients will employ all three mechanisms. In all three maneuvers, however, clients turn a passive experience that originally happened to them into an active experience over which they now have some control. Therapists can focus on how clients take up the same hurtful affect that was originally expressed toward them in childhood and actively turn it against themselves, or others, to block their own unmet need. This turning against the self (often expressed by clients as variations of "I can't") is only half of the conflict, however. Paradoxically, while defending against the anxiety of reexperiencing unmet needs or unacceptable feelings, clients simultaneously try to rise above it and have it met indirectly. Next, Part 4 of Figure 7.1 shows the other side of clients' compromise solution—clients' attempts to rise above the need and indirectly gratify it.

Rising above the Unmet Need   The most important component of this model is the defensive interpersonal styles that clients adopt to cope with their developmental problems. Children's significant unmet needs and developmental problems do not just dissipate or go away as they grow up. These disruptive needs and feelings may be denied or sequestered away but they continue to be evoked in adult relationships. Because these needs and feelings remain too anxiety arousing or unacceptable to be expressed or dealt with directly in current relationships, clients try to cope or "rise above" them by adopting various interpersonal coping styles. Horney (1970), Beck, Freeman, and Davis (2003), Millon (1981), Benjamin (2003), and others have used related terms to describe the interpersonal coping styles that clients employ. To illustrate these interpersonal coping strategies, we will focus on Horney's three lifestyle adaptations: moving toward, moving away, and moving against others.

Inflexible Coping Styles: Moving Toward, Moving Away, and Moving Against Clients adopt a fixed interpersonal style both to reduce the anxiety associated with having unacceptable needs and feelings revealed and to find some sense of value or self-worth. For example, in their families of origin, some clients learned to cope with problematic familial interactions by *moving toward* or pleasing people. These clients learned they could earn some needed approval, and diminish the threat of further rejection, criticism, and so forth, by excessively complying with their parents and being unfailingly good (always helpful and nice, and even perfectly well behaved). These children are not learning to be appropriately well behaved; in order to ward off anxiety and maintain self-esteem, they are giving up too much of themselves and their own identity. In this pervasively pleasing stance, clients have lost their own voice and the ability to "have their own mind." Generalizing this coping strategy they have adopted with caregivers, they carry over to others this pattern of relating in which they always try to please or accommodate others and avoid disagreeing, expressing their own interests or preferences, or asserting their own limits and boundaries. Thus, these clients have learned to defend against anxiety and win some approval by consistently moving toward others in a pleasing, servile way.

Other clients have learned that aggressiveness and resistance to parental wishes, if pursued long enough, will ward off pain or insecurity. They may have needed permissive caretakers to act more strongly and take charge of the family effectively, which would have provided the safety they needed to feel secure without having to try and take control of everything themselves. These expansive and dominating individuals cope by *moving against* others. They want to be in complete control of themselves and their emotions at all times and, even more problematic, they dominate or try to exert too much control over others. They approach relationships competitively with the orienting attitude that they must win what they deserve, and they often assert themselves aggressively for this purpose. Their mind set is that they should prevail in every situation or conflict. Thus, these clients adopt a repetitive interpersonal style of moving against others to protect themselves from the unwanted feelings evoked by needing someone, being in situations where they are not in charge or are unsure of what to do, or being close and intimate with someone.

In Horney's third interpersonal style, clients have learned that the best way to reduce the interpersonal threats they are growing up with and create some safety for themselves is by *moving away* from others through physical avoidance, emotional withdrawal, and self-sufficiency.

Because life is stressful for everyone and all people need defenses, healthy individuals also employ these three interpersonal styles at times. When they do not become exaggerated or used pervasively, there are adaptive strength features in them. However, they are not an identity or way of living for well-functioning people; rather, they are more flexible strategies used selectively to cope with a difficult person or problematic situation. With most clients who enter treatment, however, one style has become predominant or characteristic. It is overused, inflexibly, even in situations where this uniform mode is not adaptive or is not helping the situation any longer. The same coping style is also used unnecessarily—in safer situations where, realistically, this defensive coping strategy was never actually needed. As we will discuss later, *the therapist's role is to help clients learn to discriminate or assess more realistically when and with whom they need to cope in one of these ways and when they don't.*

To review, we saw that clients turn against themselves in different ways to cope with problematic developmental experiences within their environment. In these turning-against-self defenses, clients respond to themselves and others in the same problematic ways that important others originally responded to them. We then discussed how clients try to cope with their developmental problems by adopting a fixed or rigid interpersonal coping strategy. These coping strategies included the interpersonal styles of moving toward, moving against, or moving away from others. Although there are personality strengths in these coping styles when they are used flexibly, they become maladaptive for clients and create problems because they are employed inflexibly in all types of situations.

With both sides of this compromise solution in mind, we are now ready to look more closely at how clients use their interpersonal coping style to try to rise above their core conflicts. As we will see, however, even though clients derive self-esteem and an identity from their interpersonal coping style, it remains a defensive adaptation that avoids, rather than addresses, their basic conflicts and, in the long run, exacerbates rather than resolves their problems.

Being Special to Rise above Core Conflicts   When the developmental problems were pervasive or severe, clients often adopt one of the three interpersonal styles: moving toward, moving against, or moving away. This inflexible coping style is a habitual way of relating to others that is also reflected in vocational, marital, and other defining life choices. There are many exceptions, of course, but individuals who move toward may find that careers in nursing or counseling are a good match for their caretaking skills and needs. Individuals who move against may find that careers in law, medicine, or money management are congruent with their interpersonal style of taking charge and needing to be in control. Individuals who move away may feel comfortable as researchers, artists, technical experts who work alone, or individuals in a meditative lifestyle. In their self-concept, many clients will privately frame this defensive coping style as a special attribute or virtue and use it to feel special and rise above their core conflict. That is, clients do not experience their interpersonal strategy as a defensive coping system to avoid anxiety and low self-esteem but as a virtuous way of relating to others that may make them feel special. However, this unrealistic sense of being special is compensatory and actually reflects their low self-esteem or shame-based sense of self. *Their self-worth is brittle and vulnerable—it relies too much on their ability to rise above their core conflicts by pleasing others, by achieving and succeeding, or by withdrawing and feeling cynically superior.* In these ways, clients frame their defensive interpersonal coping style as a superior quality or virtue. More contemporary therapists write about the narcissistic element in many different types of clients who are defending against a shame-based sense of self (Stolorow, Atwood, & Brandchaft, 1994; Kohut, 1977). They describe what clients do to "restore" their sense of self-worth or power when they (too readily) feel shamed and overreact to feeling rejected or diminished by others or that they have failed in some way. Sensitive issues to work with, certainly, yet relieving for clients when counselors can take the risk and find respectful ways to broach these far-reaching issues with clients.

Thus, even though moving-toward clients are compliant and submissive, they do not experience themselves as servile. Instead, they often feel special—more caring than others who are not as compassionate. They accomplish this distortion by perceiving themselves as selflessly loving; always altruistic and sensitive to the needs of others or committed to ideals of peace or justice. In this way, the defensive interpersonal style becomes a way to cope with a lack of self-esteem—a virtuous quality that wins needed approval from others and wards off anxiety and shame.

Similarly, moving-against clients do not see themselves as angry, competitive, self-centered, or demanding, although they frequently behave in these ways. Instead, they see themselves as heroes or heroines—strong leaders entitled to direct others, who often are regarded contemptuously as weak, dependent, or incompetent.

Finally, moving-away clients prize their aloofness, perhaps as evidence that they are more refined or feel more deeply than others. They may be cynical and at some level feel better than the ordinary masses, whose involvement in everyday happenings is seen as mundane and small-minded.

Thus, to lessen their anxiety, clients inflexibly employ an interpersonal coping style that allows them to feel special; gain a sense of identity and self-esteem; and rise above their unmet needs, unacceptable feelings, and pathogenic beliefs. Their ability to turn a defensive coping style into a virtue or talent that testifies to their specialness is what Horney terms a *neurotic pride system*. For these clients, however, their self-esteem is brittle and vulnerable. It is based on an unrealistic sense of being special and rising above rather than learning to value themselves and their own authentic feelings.

We now have the two sides of clients' core conflict. First, clients block their own needs and authentic feelings by responding to themselves in the same critical, dismissive or rejecting ways that adult caregivers originally responded to them. Clients come to feel toward themselves the same affect that the caregiver originally communicated toward them (for example, disappointment, resentment, or indifference). This affect will usually be evidenced as low self-esteem when clients enter counseling, and clients are often seeking treatment to be rid of this intrapunitive feeling. Usually, it is relatively easy for therapists to recognize how clients are turning against themselves and to respond in more affirming ways.

Second, clients rise above their core conflicts and find ways to reframe their defensive strategy as evidence that they are special. This aspect of their compromise solution is usually more covert and is harder for therapists to address because the clients' neurotic pride is typically bound up with shame. Let's examine some other reasons why clients are reluctant to let go of their attempts to rise above their conflict.

Clients' interpersonal coping style has indeed granted some relief from the anxiety of unmet needs and unacceptable feelings by providing a sense of control over them. It has also allowed clients to earn some attention from others and respect for themselves. In this way, *clients' interpersonal coping style has become the primary source of their self-esteem and their principal avenue to succeeding in life*. Furthermore, clients have often developed impressive abilities in the service of their coping strategies. For example, many moving-toward clients have not only been able to make others like them, but have also become highly skilled at caring for others. Likewise, moving-against clients have often been able to achieve a great deal of success and power, and may develop significant technical or other abilities along the way. Similarly, although moving-away clients have often been able to become safely aloof, they may also have developed a rich inner life or been able to develop creative or artistic abilities.

One of the most important interventions therapists can make is to *help clients recognize the significant emotional price they pay for their interpersonal coping style.* Therapists do this by highlighting, as concretely as possible, how their interpersonal adaptation is creating stress and limiting their lives now:

> THERAPIST: You've just earned another award at work, where they have nicknamed you the "star," and you do so much for everybody at home. I'm so impressed by how capable you are and how much you are always doing for everyone. But I'm also concerned about how tired you look and how often you are sick. You often use the expression that you're "running on empty." I'm hoping that we can begin to listen to that feeling together here in counseling—I think it's telling both of us something important.

In this way, therapists want to affirm the genuine strengths that clients have developed through these coping strategies but simultaneously be compassionate toward the toll they take. Later in this chapter we will see how clients can retain the skills and strengths they have developed in these compensatory modes, but without utilizing them in the same inflexible or unidimensional ways as in the past. As before, therapists need to help clients honor their interpersonal coping styles and appreciate rather than pathologize them. They were absolutely necessary at one time and represented the best possible adaptation they could have made at that point in their development. Although they are still adaptive in some problematic relationships, they are overused now and are no longer necessary or helpful in many current situations. An important treatment goal for many clients is to learn that more flexible or varied interpersonal coping styles will serve them better.

To sum up, by turning their defensive interpersonal adaptation into a sense of being special, clients often gain significant secondary benefits and genuine personality strengths. It's no wonder, then, that some clients initially may be reluctant to explore this adaptation with the therapist. It has become the basis of their identity, the primary source of their self-esteem, and the only way they have had of defending against their seemingly unresolvable conflicts. *Once therapists fully grasp all that this defense has provided for clients, and help clients to appreciate both what it has accomplished and what it has cost, both therapists and clients will be less judgmental or impatient with the clients' symptoms.* Both will have greater empathy for the clients' struggles to change and will be able to work together collaboratively to develop more flexible coping strategies that will be more adaptive. Understandably, clients will not want to risk giving up what they have without having something better to replace it. As we will see, however, clients never really succeed in rising above their core conflicts through these compensatory adaptations. By acknowledging conflicts and addressing them more directly, clients can develop a wider experiential range that will be an important part of resolving many problems.

Shoulds for the Self, Entitlement from Others   Clients both gain and lose something by their compromise solutions. As a result of feeling special and virtuous, clients gain what Horney terms *false pride* and a certain type of *entitlement* from others. For example, self-effacing, moving-toward clients hold unspoken expectations of unwavering loyalty and approval from others, and they often demand constant reassurances. Expansive, moving-against clients claim leadership and control, and they expect deference from others. And, although it may look as if detached, moving-away clients place no demands on others, they silently hold expectations that others never criticize them or make demands upon them. Of course, the world doesn't go along very well with these neurotic claims. They create ongoing conflict and leave clients exposed to continuing frustration, as others frequently do not accept such unrealistic demands and unspoken expectations.

Accompanying clients' unrealistic demands on others is a harsh and uncompromising set of "shoulds" that they place on themselves; these faulty beliefs reflect their impaired self-esteem and diminished sense of self. For example, moving-toward clients suffer under the self-imposed demands that they should be the perfect lover, teacher, spouse, and so forth. They must always be caring and responsive to the needs of others and never feel angry, compete, or advocate on their own behalf. Moving-against clients demand of themselves that they should be able to overcome all difficulties and dominate. They must win and be on top, control their feelings at all times, and overcome personal insecurities and even physical illness simply by an act of will. Moving-away clients pay for their specialness by believing that they should be able to work tirelessly and always be productive. They demand that they should be able to endure anything without becoming upset and that they should never need help from anyone. As life and aging goes, such individuals will not be able to meet at all times these rigid demands they place on themselves. For example, inevitable failures occur:

- *Moving toward*: They cannot win someone's love.
- *Moving against*: They do not succeed in an expansive new venture.
- *Moving away*: They cannot remain aloof but are evaluated critically or forced to compete in some way.

When such inevitable failures occur, clients often feel "defeated" and often suffer great shame. Intense anxiety, depression, and other symptoms result when their coping strategies fail, and some will enter treatment in crisis.

Clearly, the "tyranny of the shoulds" is an exacting price to pay for the client's sense of specialness. One of the most helpful interventions counselors can offer is to be able to articulate, both accurately and empathically, how this coping strategy is central to the clients' basic sense of self—the basis of whatever self-worth they may possess, and nothing less than the organizing theme in their life. At the same time, therapists need to help clients *appreciate* these interpersonal coping strategies. They do in fact contain real strengths, and therapists can help clients recognize how once they really were necessary and adaptive.

Overused and overgeneralized, however, they are no longer necessary or adaptive in many current situations. Therapists help by clarifying the unnecessary price in symptoms and stress that the client is paying for this coping strategy that no longer serves them well.

Resolving the Core Conflict

Part 5 in Figure 7.1 illustrates how the client's core conflict (key issue or primary problem) can be resolved. Ultimately, clients' compromise solutions to their core conflicts will not succeed. The client can block the developmental problem (turn against self) and try to rise above the core conflict by moving toward, away, or against. However, these compensatory maneuvers will not resolve the client's problems—no matter how successful the client becomes through these interpersonal coping strategies. As they repeatedly employ these interpersonal styles—even in situations where they are not needed or are not effective—they will engender problems in current relationships, and create personal stress that often contributes to health crises. Let's examine three reasons why the client's compromise solution will not succeed and then explore what can be done to help the client instead.

First, the "tyranny of the shoulds" places unrelenting demands on the client. These can be heard, for example, when clients enter counseling exclaiming, "I can't do all of this anymore! I can't be a perfect parent, a great spouse, and the most successful professional in my office. I can't be everything to everybody anymore!" An individual's quality of life is diminished by these unrealistic expectations, they are depleting and fatiguing—even if these unrelenting demands they place on themselves (and on others) do not provoke a crisis that brings them to treatment.

Second, clients' sense of entitlement and being special will usually fail as well. For example, self-effacing clients will not be able to win the love or approval of everyone they need. Expansive clients will not always succeed or be able to make others defer to their demands, and detached clients will encounter criticism from others or situational pressures that force them to compete. Inevitably, clients' attempts to become special and rise above their conflict will lead to conflict at work and at home. Typically, clients enter counseling when situational life stressors, developmental transitions in adulthood, or aging/illness have caused their interpersonal coping strategies to fail.

Therapists help clients change when they grasp the pain that clients suffer when their attempts to be special and rise above their conflict fail; Kohut (1977) writes eloquently of the *narcissistic wound* that is so painful to clients' core sense of self. Even though the demands that clients place on themselves and others are unrealistic, it still can be excruciating when their interpersonal coping strategy fails. Tragically, some individuals will even attempt suicide in response to such a failure, and many others will contemplate it.

Why are some clients driven to such an extreme response? In a word, shame. The interpersonal coping style that many clients have adopted is the basis of their identity and their primary source of self-worth, but it is a brittle substitute for the genuine self-esteem they lack. From these clients' point of view, they *are* their ability to please, to achieve, or to remain superior and aloof. Such clients overreact to seemingly insignificant events (for example, gaining 3 pounds over the holidays or losing their patience with a trying 2-year-old). This overreaction occurs because their entire sense of self is threatened when their interpersonal coping strategy fails and their shame-based sense of self is revealed by their mistakes—which they frame as "failures."

When such students get an A minus rather than the A they demand of themselves, they may feel like a complete failure. Too often, the therapist's initial response to this disappointment is to challenge the client's overreaction to the grade and provide a more realistic appraisal of the situation. This well-intended, reality-based response misses the deeper meaning for this client. Such a reality check is certainly useful but, initially, a more productive response may be a bid for an empathic connection. That is, wondering aloud, the counselor may reflect that there seems to be no room for error for this person—it has to be perfect or, seemingly, everything is lost and the consequences seem so dire. In this way, the therapist is trying to empathically reflect how all-important this "failure" or disappointment seems to be from the client's point of view. Delving further, with a compassionate and curious attitude, the therapist invites the client to join and wonder aloud as well about how this grade seems to indicate something very important about the client and who they really are. Collaboratively exploring in this way, the counselor and client often find that this failure experience is viewed as "proof" or confirming evidence that something bad or unwanted about them is true. For example:

CLIENT: When I got passed over for the promotion, it proved to everyone that I couldn't make it.

OR

CLIENT: When I got mad, I knew none of them would ever like me again.

Working collaboratively in this way, the therapist and client are trying to hone in and clarify, as precisely as they can, what this disappointing grade seems to say or mean about the client. For example, together they may find that it proves to the client that he is just an "imposter" who really is inadequate and can never do it right; or is letting someone down whose approval he "must" earn, and so forth. This pathogenic belief is true, in the client's mind, despite the client's Herculean efforts to rise above or disprove it by achieving A's. In other words, the therapist's intention is to discern more specifically what this grade really means or says to this client about herself. By exploring and *clarifying together* how completely this client's sense of self depends on his ability to achieve or please (which is a different interpersonal process than interpreting or reframing), it reveals the anxious insecurity and emotional deprivation that originally led to this coping strategy. Thus, therapists use this understanding to share their compassion both for the pain of the original problem and the burden of trying to rise above it that the client has suffered with for so long. By being able to help the client articulate this dilemma and holding a compassionate or respectful stance toward it, therapists behaviorally demonstrate that they grasp the real meaning of this prototypical experience that occurs over and over again for the client. This kind of understanding usually diminishes presenting symptoms of anxiety and depression. *It provides safety for clients—a secure base from which they can begin to question this coping strategy, appreciate how and why they learned it, and begin to change this pattern of relating—first with the therapist and soon with others in their lives.*

In sum, by appreciating the broader life context of this "failure experience" when their rising above defense fails—an impact that has always seemed so irrational to others as well as to clients themselves—therapists understand clients in a way that most clients have not experienced before. This compassionate understanding is a corrective emotional experience that helps clients free themselves from the tyranny of the shoulds and the unrealistic demands they place on themselves, and helps them be able to come to terms more realistically with who they are and how they choose to live.

Why is it such a crisis for many clients when their interpersonal coping style fails? It triggers the unwanted feelings and faulty beliefs that were originally engendered by problematic familial interactions (that is, "I really am worthless" or "It's my fault they didn't want me"). One way that the therapist helps clients expand these narrow and repetitive coping styles, and discriminate how to use them more *flexibly* in situations where they actually are necessary or adaptive, is by helping clients understand the family roles and interactions that originally led to these interpersonal defenses of moving toward, moving against, or moving away. Usually, it is more effective to respond to these developmental problems *after* therapists have responded to the current crisis brought on by the clients' pathogenic beliefs that—if they are unable to successfully please, achieve, or remain aloof—they are not love-worthy, cannot protect themselves, are too needy or demanding for others, and so on.

Third, the most important reason the client's compromise solution cannot succeed is that the client can never rise above the original conflict. No matter how saintlike or loving, how successful or accomplished, or how self-contained and independent, the client is still defending against the original conflict rather than resolving it. These interpersonal maneuvers are defensive coping strategies. They are attempts to *avoid* the problem by blocking it, or to manage the problem by trying to rise above it, rather than address it directly. This is clearly illustrated when the client's increasing success in life is not met by good feelings of satisfaction or accomplishment, but by increasing depression, cynicism, or unrelenting pressures to achieve and accomplish more.

Let's review this complex sequence. When the child's basic developmental needs are not met by caregivers, the individual will adopt a "compromise solution" that temporarily relieves the anxiety of having basic emotional needs unmet but does not resolve the reoccurring problem. On the one hand, clients try to block their own needs, just as they were originally blocked by significant others. Clients turn against themselves and take on the same self-punitive feelings and attitudes that attachment figures once held toward them. These clients lose their self-efficacy to feelings of helplessness and hopelessness.

On the other hand, clients try to rise above their core conflict by becoming special, turning their defensive lifestyle adaptation into a virtue, and obtaining some compensatory self-esteem. Clients will be highly invested in trying to overcome and rise above their core conflict; typically, to have to face it will feel shameful, frightening, or hopeless. Thus, Part 4 of Horney's model (Figure 7.1) indicates clients' exaggerated sense of being special and defiant attempts to rise above the conflict. For example, this client might vow to himself, "I will never need anything from anybody ever again." This rigid, defensive stance reflects an act of will to overcome these developmental problems and rise above them, rather than trying to resolve them by addressing them directly.

To resolve their core conflicts, the counselor works to help clients recognize and relinquish their compromise solutions. Attempts to block the original need and to rise above it are fragile solutions that cannot be sustained; failure experiences that precipitate crises and symptoms are inevitable. Thus, instead of avoiding their real problems in these two ways, the therapists' goal is to help clients approach them. Therapists help by allowing clients to experience their core conflicts in a more supportive relationship that does not meet the original need but does give clients the understanding and validation they need to live through the feelings that were so unwanted. The interpersonal solution provided by the therapist's more effective response also allows clients to integrate other reactive feelings, helps clients disconfirm pathogenic beliefs about themselves, and alters clients' faulty expectations about what may occur in relationships with others. Now, as adults, clients are able to understand and resolve the original conflict by reexperiencing and reevaluating with the therapist what was once so hurtful or problematic (for example, a client might say, "I'm really not being selfish or doing something wrong if I become more independent and successful, even though this does make my parent feel sad and withdraw sometimes"). In this way, what was once so unacceptable or threatening can now find compassion and understanding. Clients can integrate these difficult feelings and question these pathogenic beliefs and faulty expectations, rather than continuing to defend against them or be ruled by them. When clients allow the therapist to help contain the conflicted feelings, old relational patterns are broken and new, more flexible schemas become possible. Therapists' compassion and validation comprise a new response that disconfirms these faulty beliefs and, in turn, allows clients to be more self-accepting and forgiving of themselves. Thus, clients can now integrate the previously unacceptable parts of themselves and can better accept both the good qualities and the limitations in other relationships that have been important but conflicted.

As clients receive a more effective response from the therapist than they have received from others in the past, they will not need to block their core conflict or attempt to rise above it anymore. The therapist's more accepting and affirming response to the client's original predicament alters the maladaptive relational pattern and alleviates the anxiety that was originally evoked. As this occurs with the therapist, clients are able to begin generalizing this experiential relearning with the therapist to others in their lives. With the therapist's help, that is, clients can begin responding in more adaptive ways that improve their relationships with others and expand the narrow coping style that they origi-nally adopted to cope with their dilemma. In the chapters ahead, we will ex-amine more specifically (1) how the therapeutic relationship can be utilized to help clients resolve their core conflict and especially (2) how this emotional re-learning with the therapist can be generalized to other relationships in clients' lives. An extended case study of a moving-toward client, and two case sum-maries of the other coping styles, will help therapists use this conceptual model to guide their interventions.

Case Study of Peter: Moving Toward Others

Developmental History and Precipitating Crisis

Peter was an insecurely attached child, and his emotional needs went unmet in his authoritarian home. Further stressors accumulated when his parents di-vorced when he was 6 years old. Although this family change could have been an opportunity to improve parenting and family relationships, it was not. Es-pecially problematic, his parents could not work out a cooperative parenting relationship following the breakup. His father did not take an active parenting role and, by 2 years after the divorce, Peter saw his father infrequently and when they were together, their relationship was superficial.

Peter's mother was overburdened by the demands of raising three children on her own, working full-time, and trying to make some kind of personal life for herself. She was involved with several men in the years following the divorce but was never able to establish an enduring love relationship. Frustrated by the many demands and few pleasures in her life, Peter often became the target of her resentment. Impatient and irritable toward him, she criticized Peter when things went wrong in her life and often felt resentful of his needs. Although she tried her best to be fair to the children and give them a good home, she could not be very responsive to Peter or affirming of him.

After his father left, Peter quickly learned that taking care of his mother was the best way to ward off her disapproval and to win whatever affection he could. Following the anxious/ambivalent attachment style, he was "preoccu-pied" with monitoring his mother's moods and concerns. By the age of 10, Peter had already adopted a pervasive interpersonal style of moving toward people. His teachers described him as an especially responsible and well-behaved boy who was a pleasure to have in class. Let's turn the clock forward 15 years and see how these developmental challenges are being expressed in Peter's early adulthood.

At the age of 25, Peter is a graduate student in psychology. Becoming a therapist felt like a perfect career choice to Peter. He prided himself on his sen-sitivity and concern for others and took pleasure in being able to help those in need. And now that he was carrying his own client caseload, it was great to find that he enjoyed being a therapist as much as he thought he would. Best of all, Peter was finding that he was often able to help his new clients. At least, all of his clients seemed to like him, and, unlike with some of his classmates, they kept returning to their counseling sessions each week. As his second-semester practicum got underway, Peter felt that he was on his way.

Later that semester, though, Peter had a setback. After presenting a video-taped recording of one of his therapy sessions in group supervision, Peter received some unexpected feedback. The practicum instructor told Peter that he was being "too nice" to his clients and that he seemed to need his clients' approval too much. The instructor went on to say that Peter seemed to be afraid to challenge his clients, reassured them too often, and tended to avoid conflicts in the therapist–client relationship that needed to be addressed.

Peter was stunned. Although he had some awareness of his aversion to conflict, he did not truly understand what the instructor was talking about, and he felt hurt and confused by the criticism. It was important to Peter that his instructor like him and approve of his clinical work. Peter tried carefully to explain that the instructor did not understand the close relationship that Peter was developing with his clients or recognize all of the important issues that his clients had been revealing. The instructor responded that this was probably true, but Peter was missing the point, and he repeated that Peter needed to think about his reluctance to address interpersonal conflict or say things that his clients might not want to hear. To make matters worse, two students in the practicum group chimed in and agreed with the instructor's comments. With that, Peter's anxiety became so high that he could no longer defend or explain himself, let alone try to understand or learn from their comments. Peter stopped arguing with them, looked down, and quietly nodded agreement throughout the rest of the supervision session.

Throughout the next few days, Peter felt just sick inside. He was so dismayed by the criticism that he couldn't think about anything else. For a while, he thought that he should drop out of the practicum group, but then he decided that, if he only tried hard enough, he could make the instructor see that his criticism was unfounded. Peter kept searching for a way to discount the feedback and stop the anxiety that was churning inside.

A week later, Peter found out that his girlfriend was having an affair with another student in the program. Although he tried to be understanding at first, he felt shocked and betrayed. He alternated between withdrawing and announcing that their relationship was over and desperately trying to win her back. Peter felt shattered. He became so anxious that he was unable to eat or sleep, let alone study. It felt as if a motor were racing inside of him—accelerating out of control. Peter began hyperventilating, experiencing heart palpitations, and having anxiety attacks. By the end of the week, he was unable to drive his car.

To make matters worse, Peter tried to keep all of this turmoil to himself. He thought he should remain calm and together, and he was afraid that his supervisors would not want him to see clients if he was "so messed up" that he was having anxiety attacks himself. But despite his attempts to cover up his distress, his individual supervisor soon asked him what was wrong.

Although he never could have allowed himself to ask for it, Peter desperately wanted his individual supervisor's support, and he was greatly relieved to receive it. Peter explained how his practicum group was becoming one of the worst failure experiences of his life and how hard it was to accept what his girl-friend had done. The supervisor was supportive but also found a tactful way to say that the practicum instructor's comments fit with some of his own obser-vations. Because he knew that his individual supervisor liked him, Peter was able to consider the feedback this time. The supervisor suggested that these were important issues for Peter to work with but that they could be dealt with better in his own therapy rather than in supervision. Peter agreed and began seeing his own therapist.

It helped that Peter's supervisor thought he could be a fine therapist, in spite of his anxiety attacks. However, Peter's anxiety remained paralyzing as he began his own treatment. Fortunately, Peter was assigned to a skilled and experienced therapist who soon conceptualized the predisposing vulnerability, interpersonal coping strategy, and situational stressors that precipitated his crisis. Peter's cop-ing strategy of moving toward people had worked well for him up to this point. However, both of the crises that Peter had just experienced ran headlong into the heart of his core conflict and his interpersonal style for coping with it. As his abil-ity to rise above his core conflict by pleasing others failed, the anxiety associated with his profoundly insecure childhood broke through, and anxiety attacks re-sulted. Let's try to understand more fully why Peter developed these symptoms.

Precipitating Crisis, Maladaptive Relational Patterns and Symptom Development

Many people would have coped with the two stressful events that Peter experi-enced without developing such significant symptoms. As we have seen, however, we understand more when we apply the concept of client response specificity. We will use Peter to illustrate how symptoms develop when situational stressors reenact maladaptive relational patterns, confirm pathogenic beliefs, or cause in-terpersonal coping strategies to fail.

The first stressor for Peter was his practicum instructor's critical feedback. Criticism from a respected authority figure would be unsettling for most people, but generally they could cope with it. For someone like Peter, however, such dis-approval carries far more weight. Because of Peter's strong need for approval, his history of receiving excessive criticism, and his lifestyle adaptation of trying to rise above these developmental conflicts by being helpful and nice in order to win approval, the practicum instructor's criticism felt huge.

Second, the practicum instructor specifically challenged Peter's interper-sonal coping style of moving toward others. This confrontation not only aroused the anxiety of receiving criticism and feeling rejected, but also undermined Peter's primary means of warding off these unwanted feelings (pleasing others and accommodating to their needs). Thus, *the instructor's feedback triggered Peter's core conflict and, simultaneously, weakened his defenses against it.* In other words, Peter's developmental experiences made him acutely sensitive to criticism. He adopted the attitude toward himself communicated by his early caretakers—that he did not matter enough to be committed to (from his father) and was not good enough to be approved of and responded to (from his mother).

In the most important relationship in his life, he learned to expect that his mother often would be frustrated, critical, and disappointed with him. His attempt to defend against the anxiety evoked by this set of circumstances was to try to be perfect—to please others and meet their needs, so that this self-critical schema could be refuted by their appreciation. This coping style was, of course, a formula for failure: It was impossible for Peter to be liked and appreciated at all times by everyone he met. There was no such thing as constructive criticism for Peter—he construed all disapproval as rejection, and he felt very uncomfortable whenever he could not win everyone's approval. When teaching his first undergraduate course, for example, Peter seemed only to notice the two or three students in the class who seemed disgruntled. He did not meaningfully register that the large majority of students enjoyed the class and thought he was doing a good job.

If not for the subsequent crisis with his girlfriend, Peter probably could have recovered from the first setback without developing symptoms. Most likely, he would have reconstituted his interpersonal coping style and been somewhat successful in winning the approval he needed from others in his life—different instructors, friends, and perhaps even clients. As we will see, however, the subsequent stressor with his girlfriend also struck at the same core conflict. At that point, his moving-toward coping style toppled, and the intense anxiety associated with his insecure attachment history was evoked. More specifically, this was "shame-anxiety"—anxiety over the threat of having his unlovable and unworthy self exposed. This anxiety over having his shame-worthy self revealed, both to himself and to others, became too intense to be blocked; it broke through in anxiety attacks and symptom formation.

A partner's infidelity will be highly stressful for almost everyone. Here again, though, this particular stressor held far greater significance for Peter when viewed within the personal context of his developmental history and subjective worldview. Peter had to cope with far more than just the loss of trust with his girlfriend; he also suffered a blow to his identity and basic sense of self-worth. In Horney's neurotic pride system, Peter's coping style of moving toward entitled him to be special, so that others would love and prize him inordinately. Although Peter was not very aware of his own unrealistic claims, he expected his girlfriend to idealize him as the most sensitive and loving man there could be. Clutching to this defensive sense of being special, Peter couldn't believe that his girlfriend could actually be interested in someone else. Peter's neurotic pride system was painfully shattered when he found that he was not the commanding center of his girlfriend's life that he felt entitled to be.

Multiple stressors occurred for Peter in a short period of time. In and of themselves, these events would precipitate a crisis for many people and lead them to therapy. However, if these situational stressors do not tap into preexisting schemas, repeat maladaptive relational patterns, and confirm pathogenic beliefs, they will not usually provoke such strong symptoms. The client will often be able to recover in a relatively short time with crisis intervention or short-term supportive therapy. In contrast, when stressful life events tap squarely into a client's core conflict, a client such as Peter has to cope with far more than just the demands of the current situational stressors.

Thus, when Peter's girlfriend became sexually involved with another man, he had to cope with much more than just the pain of this betrayal. He had to cope with the even bigger, developmental problems that were triggered: his strained attachment history; his interpersonal coping strategy of pleasing to ward off the criticism and rejection that he expected; and with the wound to his self-esteem when his sense of being special was shattered. Clients such as Peter do not have a secure sense of self-esteem to fall back on in times of crisis. As unrealistic as it is, the sense of being special and the secondary gains that clients can earn with their interpersonal coping style is the only self-worth that some clients have been able to garner. It is understandable, then, that Peter was over-whelmed by anxiety attacks. Both of the situational stressors intensified Peter's original conflicts, and they both took away his coping strategy for defending against them. In this light, *we see that clients' presenting symptoms are not ir-rational.* Per client response specificity and subjective worldview, they make sense once they are understood in the broader context of the client's develop-mental history and the cognitive schemas, pathogenic beliefs, and inflexible coping styles that result from this history.

Course of Treatment

Fortunately, Peter's therapist was both knowledgeable and kind. He was gen-uinely empathic to the pain that these situational crises brought on for Peter but also grasped what made Peter so vulnerable to them. The therapist recognized that the current crisis provided an opportunity to resolve the more important developmental problems that left Peter prone to reexperience crises such as these whenever others rejected or disapproved of him. In the months that fol-lowed, Peter was able to resolve the precipitating crisis. More significantly, he made progress in coming to terms with the impact of his developmental history and the problems that followed from it. As a result, Peter was able to expand his unidimensional moving-toward coping style and adopt a more flexible in-terpersonal repertoire. Here and in the next chapter, we will review the course of therapeutic events that allowed Peter to make these far-reaching changes.

Peter entered therapy in crisis, and the therapist was able to respond effec-tively to his distress. From the first session, Peter felt understood and cared about by the therapist. As his accurate empathy led to a strong working al-liance, Peter's anxiety attacks stopped. The anxiety and distress that had been so disruptive for him steadily subsided as the therapist continued to provide an effective holding environment. However, as Peter began to feel more secure and function better in his everyday life, he began to be resistant to the feelings he had been sharing with the therapist and wondered about ending treatment.

The therapist observed that, as Peter felt understood and affirmed by him, Peter's unmet needs for acceptance and support were assuaged and the attendant anxiety diminished. As a result, Peter felt safer, his symptoms abated, and he began to function better. At the same time, however, he also began to reconsti-tute his interpersonal coping style of moving toward others. Rather than termi-nating therapy at that point—which might need to occur in crisis intervention or time-limited treatment—the therapist was able to help Peter go further and address the developmental conflicts that made him so vulnerable to situations where he faced disapproval or the threat of being left. The primary way that the therapist did so in the beginning was to help Peter identify his habitual coping style of pleasing and recognize how automatically and frequently it came into play. Thus, the therapist *named it*, so they could talk about it and begin to explore it together—compassionately and without shame. This opened the door that changed everything. From these conversations, for example, Peter learned to focus inward to identify what he was feeling, and what was occurring with others, at the moment just before he felt compelled to employ this coping pattern and please. (Cognitive behaviorists use such "self-monitoring techniques" to increase self-awareness of the situations, thoughts, and emotions that trigger faulty coping strategies.) Usually, at such moments, Peter found that he was expecting others to be disapproving of him or to go away and cut off from him in some way.

Additionally, as he listened to Peter's narratives, the therapist would highlight when instances of Peter's key relational themes were occurring. That is, the therapist would punctuate the interaction when Peter was (1) pleasing others, (2) avoiding anger and other interpersonal conflicts, and (3) expecting others to be critical or rejecting. More specifically, the therapist used process comments to point out these three repetitive relational themes whenever he saw them occurring in their interaction together or with others. For example, when the therapist thought that Peter was trying to please him, too, he made this overt and inquired about this possibility:

> THERAPIST: When we're talking, Peter, sometimes I find myself wondering if you might be thinking too much about what I or others might want to hear. What do you think—is there anything to that?

The therapist also highlighted whenever he thought he heard one of these three themes in Peter's relationships with others. For example:

> THERAPIST: As I listen to you describe this interaction with them, it sounds like you are trying very hard to keep them from being angry at you. I feel like I've heard this before. Does it seem to you like this comes up a lot?

This feedback provided direction for their work together—a treatment focus—and served as the basis for developing mutually held treatment goals. For example, the therapist encouraged Peter to risk not pleasing him and focus instead on just saying and doing what he wanted in their relationship. Not having to be vigilant about what the therapist liked or wanted gave Peter the opportunity to "live inside his own skin" and "have his own mind" rather than be thinking about what the other person wanted to hear. This was a corrective emotional experience for Peter. He could enjoy this relationship where he could be cared about even when he was not trying to please the therapist or act in ways that, at some level, were designed to keep the therapist engaged with him and ensure that he would not leave him or be critical.

The therapist continued to focus Peter inward. In particular, they explored the feelings that were evoked for Peter when, instead of employing his usual coping style of pleasing others, he risked new, more assertive responses that he had rehearsed with the therapist. Gradually, by focusing on what he was experiencing just before he employed his usual coping patterns, Peter began to clarify the emotional deprivation he had suffered as a child but had always been too ashamed to reveal to others or even acknowledge very fully to himself. Relatively quickly, the therapist's accepting presence allowed Peter to overcome the shame of having his pain revealed. Now, for the first time in his life, Peter could be fully seen by someone with all his strengths and problems and continue to be respected. Peter began to understand how he had coped with his childhood predicament by taking care of and pleasing his mother, and then others as well. As this significant progress occurred, however, Peter still continued to struggle with feelings of anxiety, shame, and sadness.

These contradictory feelings continued to wax and wane for Peter over the next few months. Peter was increasingly recognizing how unwanted and alone he had felt as a child and how ashamed of himself he had always been. (Poignantly, he recalled that the animal he felt most similar to during elementary school was a worm.) Accompanying these feelings, Peter recognized and began to challenge the false belief that he was not loved because he was somehow unworthy—that he did not matter enough to be important to others. As Peter continued to feel held by the therapist's understanding, his long-withheld feelings of sadness over this deprivation and invalidation could be expressed for the first time. Peter moved through the original environmental block as he received a far-reaching, corrective emotional experience. Repeatedly, the therapist's validating responses disconfirmed Peter's old relational expectations and were deeply comforting, offering him the affirmation for which he had always longed. Securing his own identity, this validation also provided him with a sense of being seen or known for the first time and accepted for who he really was.

Peter began to feel sadness and compassion for himself and mourn his childhood losses—rather than continue to blame himself for what occurred. In response to this primary affect of sadness, two reactive feelings followed closely behind. Peter found himself feeling furious: at his father for walking away from him, and at his mother for making him feel responsible for trying to earn her love and trying so hard to be "good" to ward off her disdain. As soon as this anger emerged, however, Peter became exceedingly anxious—afraid of being left on his own by the therapist, his girlfriend, his parents, and anyone else for being angry with them and "protesting" how they treated him. Each of these three feelings in Peter's affective constellation—the shame surrounding his unmet emotional needs, his anger over being rejected and dismissed so readily, and the anxiety of being left or emotionally cut off if he protested—was repeatedly aroused in treatment. With the therapist's help, Peter was gradually able to integrate and resolve these feelings. This occurred by being able to fully experience or know them himself, to risk sharing or revealing them to the therapist, and discovering that he could live through or contain them with the therapist's support. This was not a smooth or simple process, however. As the primary feeling of sadness or deprivation emerged, Peter's own shame and internal blocking defenses were activated. Let's examine how the blocking side of Peter's compromise solution was expressed in therapy.

Peter's situational crises toppled his rising-above defense. He could no longer be special and win the approval he needed from either his practicum instructor or his girlfriend. As his interpersonal coping strategy failed, it exposed his shamefully unacceptable attachment needs (which Peter derided as "weak") and evoked intense anxiety. However, the other side of Peter's compromise solution was still operating: To defend against the painful feelings being evoked, Peter turned the same type of block that was originally imposed by the environment against himself.

In treatment, and throughout his life, Peter had tried resisting each feeling in his affective constellation. He did this by blocking both the subjective experience ("they don't want me") and the interpersonal expression (pleasing others to ward off disapproval) of his core conflict (excessive criticism and rejection). For example, as the sadness of his childhood deprivation emerged, Peter thought that the therapist would be critical and rejecting. He was afraid that the therapist would want him to "stop crying, grow up, and act like a man." This *transference reaction* again served to block his sadness until the therapist was able to draw it out more fully and clarify this misperception by talking it through with Peter. More important, Peter was able to experience the therapist's continuing respect for him in the face of his, seemingly, unacceptable emotional needs. Peter learned that in this relationship he did not have to be perfect or take care of the therapist to be cared about. Although his schemas led him to expect that the therapist would be irritated by his emotional needs and resent them, as his mother had done, Peter experienced a reparative new response of respect and empathic understanding.

Next, as Peter's anger toward his girlfriend, and then, both of his parents, emerged, Peter felt guilty. He felt that he should be understanding of them, as his coping style demanded. He also felt ashamed of his inability to sustain this forgiving response. This reflected his moving-toward adaptation and served to block the expression of his profound need for more secure relational ties. The therapist was skillful in balancing Peter's protest or anger while still preserving what was good in their relationship. For example, he affirmed the validity of Peter's anger but also acknowledged that Peter still loved his mother and had received many good things from her. With support for both sides of his ambivalence, Peter gradually felt safe enough to risk experiencing both the longing and the anger that had always been present but too anxiety arousing to acknowledge.

Why was it so threatening for Peter to experience his hurt and anger or share it with the therapist? While growing up in his family, Peter had learned that relationships were not resilient. His emotional needs in general, and any angry or sad feelings in particular, were simply unacceptable. He knew that, because of his father's weak commitment to him and his mother's readiness for such strong love withdrawal, he could easily lose the little emotional support he had. Aware that many different feelings could rupture his tenuous attachments, Peter kept what little he had by agreeing with them and believing that his needs were unimportant. In particular, his anger aroused strong separation anxieties.

Realistically, he feared that he would be emotionally cut off if he expressed even disappointment, let alone the outrage he felt. Peter then tried to block these threatening feelings by telling himself that he should be calm and accepting—especially if he were going to be a therapist and help other people. The therapist knew that Peter was on his way when Peter said he had told his girlfriend that he no longer wanted to try and work things out or get back together—it was time for him to move on and explore new relationships that would be a better match.

To sum up, the therapist provided a skillful balancing of the good news and the bad news in Peter's development. He was able to acknowledge the reality that Peter loved his mother and still felt loyal to her despite all that occurred, and that some of the qualities he valued most about himself were qualities derived from her. However, these reality-based strengths were accompanied by the painful realities of Peter's emotional deprivation and invalidation. By exploring with Peter and then affirming what was legitimately good in his relationship with his mother, rather than simply blaming or rejecting her, the therapist helped Peter preserve important aspects of this tie. This gave Peter the security he needed to risk feeling his anger at both of his parents for what really was wrong in their relationship (to "protest," in Bowlby's attachment terms). Being able to protest with the therapist what had been so unfair and hurtful was not to reject his parents, hold a grudge, feel sorry for himself, or make his parents bad people. Instead, it was to change the self-narrative of his life story and know that these things actually did happen, really were painful, and that he didn't cause or deserve them. Rather than deny the reality of his circumstances any longer, being able to have the validity of his own feelings and perceptions for the first time led Peter to a stronger stance that he was able to generalize to other relationships. Peter became more confident and assertive, developed firmer boundaries and more clarity about who he was, and began to disconfirm the pathogenic belief that he would be criticized or rejected if he wasn't taking care of others. Peter came to understand that his mother's rejection was unwarranted, had absolutely nothing to do with him, and was just the projection of her own hopelessness about ever being loved. Significantly, Peter was able to feel compassion toward her and appreciate the predicament she was in coping with her own long-term depression, yet without denying the reality of the painful consequences this engendered for him.

As a result of these changes, Peter felt less compelled to make others like him, less threatened by interpersonal conflict, and more self-contained. Finally, Peter was also able to keep the best parts of his moving-toward style: He remained a genuinely caring and responsive person, however this was no longer the unidimensional response pattern that it had been in the past. Clearly, Peter had grown through this crisis and become a more resilient person. Therapy helped immunize him against his predisposing vulnerability to rejection, criticism, and loss. By experiencing a reparative response from the therapist and finding that some relationships can be different from those he had experienced and expected, Peter expanded his schemas of what can occur in relationships. In this way, the therapist helped Peter to stop carrying emotional baggage from his past into his present relationships and from reenacting the same maladaptive relational patterns and inflexible coping strategies. Peter will certainly have further crises in his life, especially when circumstances again tap into his "old wound," but the changes he has made will help him respond to future problems in a more empowered, present-centered manner.

Two Case Summaries

In the case study of Peter, we have seen how the interpersonal model for conceptualizing clients is used with a moving-toward client. Next are two case summaries that highlight salient features in the treatment of a moving-against client and a moving-away client.

Carlos: Moving Against Others

Carlos had been riding high, on his way to becoming a real estate "king," when his business failed. Buoyed by his initial success in a booming real estate market, Carlos had been woefully overextended when interest rates rose unexpectedly. His business went bankrupt, his new home and sports car were repossessed, and party-friends deserted him as creditors and the IRS aggressively pursued him. When his dreams of wealth and power were dashed, Carlos became seriously depressed. Reluctantly, he entered therapy with an older Hispanic therapist who was widely respected in his community for helping migrant families and children. Even though Carlos felt that people should solve their own problems and that therapists were in this business just for the money, he decided to try it with the resolve that he was not going to let the therapist, or anyone else, take over and "tell him what to do."

After 3 months in treatment, however, his depression had in fact improved. Carlos could see how helpful the therapist had been and had recently told a friend that he didn't know where he'd be today if not for the therapist. On this particular day, however, Carlos was ready to quit. He was angry that he was still coming to therapy—he did not want to need the therapist, or anyone else, anymore. Carlos often felt frustrated and resentful like this. On this day, however, he was fed up with the therapist's fees in particular. As Carlos often did with others, he began the session by taunting the therapist:

> CARLOS: Who do you think you are to charge so much? Do you really think you're worth all that?

The therapist acknowledged the feeling in Carlos's provocation:

> THERAPIST: You're angry about my fees. Let's talk about them.

> CARLOS: (*in a provocative and insulting tone*) What are you doing—trying to prove that you can gouge as much money as the psychiatrist next door? Did your mother want you to be a "real" doctor and make a lot of money or what?

Fortunately, the therapist was able to remain *nondefensive* and stay empathic to Carlos during this critical incident. The therapist did not retreat (flight) or counterattack (fight), which is what Carlos's provocations usually elicited from others. Nor did the therapist act on his own automatic response tendency—which was to try to explain and convince Carlos of his good intentions. Instead, the therapist tried to maintain a neutral stance, keep from overreacting, and offer an empathic bid as a way to change this maladaptive relational pattern that Carlos was trying to reenact again:

> THERAPIST: Yes, you're angry with me. I do charge you for my time, and it's hard for you. And there are still problems in your life that haven't changed yet.

The therapist's accepting response only frustrated Carlos further, and he tried harder to embroil the therapist in conflict with him:

> CARLOS: I hate your understanding; I hate coming here; I don't want to do this anymore! What do you do—be nice to people when they're falling apart so they become dependent on you? Then they can't leave you and you can keep taking all of their money. There's something wrong with you—you're sick!
>
> THERAPIST: It seems right now that I am trying to make you dependent on me, so that I can take advantage of you financially?
>
> CARLOS: (*loudly*) Yes, you idiot, I want to leave and get out of here, but I can't. I'll just get depressed again if I leave. I need you and you know it, and you use that to take the little money I have left. I'm trapped here. You've got me and I hate it. I hate you!

The intensity of Carlos's reactions told the therapist that, in addition to whatever reality-based issues needed to be addressed regarding fees, Carlos's core conflict was being activated right now. On the basis of other hypotheses that he had already formulated, the therapist reasoned that, for Carlos, having a need of someone meant being used or exploited by them. Trying to articulate what Carlos was experiencing in their relationship at that moment, the therapist responded:

> THERAPIST: It's as if you don't experience the decision to remain in therapy as your own choice, Carlos.
>
> CARLOS: What do you mean?
>
> THERAPIST: It seems like you feel trapped by your own need to be helped or understood. (*pauses*). I'm wondering if you have learned that to let yourself need somebody puts them in control of you and leaves you subject to their own selfish needs.
>
> CARLOS: Bingo! That's how the world is and that's what people do. It's about time you figured it out!
>
> THERAPIST: So, if that's the only way relationships have gone for you, your only choice is to resist by leaving the help you still want or to stay in treatment with me but feel exploited. What a tough bind you've had to live in.
>
> CARLOS: (*becomes quiet and nods in agreement*)
>
> THERAPIST: You have experienced something like this with others, Carlos, but I want you and me to have a different kind of relationship. I want to respond to your need for help, not to use your trust and vulnerability for my own gain. You've been hurt in this way too many times before.

Unable to provoke the therapist and elicit the fight-or-flight response that he had usually succeeded in getting, Carlos calmed down and was gradually able to accept what the therapist was saying. Carlos's transference reaction was that the therapist had manipulated Carlos's dependency in order to meet his own financial need. This faulty schema, that selectively filtered and biased his perceptions, slotted every relationship into the same unwanted but familiar pattern. Although this pattern certainly distorted and disrupted Carlos's current relationships, developmentally, it accurately encapsulated his life story. Carlos had grown up in an orphanage—often feeling unwanted and alone. Sadly, when he was finally placed with foster parents during second grade, they took advantage of his need to finally belong to someone.

By the time he was 9 years old, Carlos had become an exploited laborer on his foster parents' farm. To earn their approval and affection, he worked sunup to sundown on the days he wasn't in school. At age 18 and ready to leave home, Carlos had long realized that his foster parents were taking advantage of him and were not to be trusted. At some point during his early adolescence, Carlos's anxious efforts to please his foster parents had evolved into an angry defiance toward them and others. He had vowed to himself that he was never going to need anybody again. Attempting to disconfirm the shame of having felt used, Carlos resolved that he was going to be rich and powerful—strong enough that no one would ever be able to take advantage of him again. Carlos's feelings of exploitation and betrayal were further fueled by his racial and ethnic history. A Latino, Carlos was the child of undocumented migrant workers. According to the story he had been told, his babysitter had been instructed to leave him at the orphanage if his parents were to be suddenly deported. Carlos was angry at their exploitation by farmers but also angry with his parents for never returning to claim him. These factors all contributed to Carlos's determination to never need anyone again. He was going to be rich so that no one would ever use or betray him; he would simply buy whatever he needed.

The dramatic confrontation with his therapist proved to be a corrective emotional experience for Carlos. Although his fear of being used was aroused in relation to the therapist, as it was whenever Carlos began to get close to anyone, his coping strategy of hostility and intimidation did not succeed in pushing the therapist away. The therapist was able to tolerate Carlos's antagonism and remain concerned about him. The therapist validated the betrayal in Carlos's familial and adoptive experiences, and how these original injustices were compounded by seeing the impact of racial exploitation in his community today.

As the therapist repeatedly provided a different response to his moving-against style than Carlos expected, the next feeling in his affective constellation emerged: His primary feelings of hurt and betrayal became the predominant issue in treatment. Just as the therapist was able to tolerate Carlos's provocative anger, he was also able to provide a holding environment to contain these painful feelings of vulnerability and loss that were so unacceptable to Carlos. Before long, they, too, became less threatening. Carlos's shame about being helped, feeling sad, or concern that he was being used continued to be activated in the therapeutic relationship. In this way, Carlos experienced the same fears and concerns over and over again with the therapist. Each time these issues came up between them in some way, however, Carlos received a genuinely empathic and understanding response from the therapist that was different from what he expected and usually received from others.

In addition, it was helpful to Carlos when the therapist validated his perception that he had indeed been taken advantage of as a child, that people of his culture were often devalued and exploited, and that elements of devaluation and exploitation were present in some of his current relationships. However, the therapist patiently but repeatedly questioned whether others in his current life were *always* trying to exploit him. Meaningful changes started to take place in his life as he began to find that there were some others in his life who were trustworthy and interested in sharing a real friendship.

Honoring his interpersonal coping strategy, the therapist also observed that, as a child, Carlos had made the best adaptation that he could to his unsolvable predicament. The therapist pointed out that there was real strength in Carlos's attempt to rise above his circumstances by adopting an aggressive and defiant interpersonal style. This attempt to cope by rising above his abandonment and exploitation was no longer necessary or effective in many situations. However, it had indeed once served a necessary purpose—to protect him from the vulnerability he experienced as a child and was determined never to experience again. Eventually, Carlos understood what the therapist meant when he suggested that it took just as much courage for Carlos to face these heartbreaking feelings now, as an adult, as it originally did to defend against them as a child by compulsively striving to achieve power and control.

With the therapist's support, Carlos was gradually able to reexperience his old feelings of urgent need and simultaneous outrage at being used by his foster parents and, later, at being left by his biological parents. As he gradually let go of his combative and competitive stance with the therapist, and let himself undergo those painful feelings of his childhood that had always seemed disgusting and weak, Carlos began to change. First, his provocative, argumentative style gave way to a friendlier camaraderie with the therapist. Carlos began to disclose more easily and felt less competitive with the therapist. In particular, he became less concerned about needing the therapist or being taken advantage of by him.

These changes with the therapist began to carry over to other relationships as well. With the therapist's encouragement, Carlos began to establish male friendships for the first time; his need to dominate and always be right had precluded such friendships in the past. Carlos's relationships with women had always been short-lived, superficial sexual contacts, which often were manipulative. During counseling, Carlos realized that he felt less threatened in these casual, short-term relationships. If he became more committed in a love relationship, he again felt vulnerable to being left and being used. As Carlos stopped defending against his shame over being left and exploited as a child and began to feel compassion instead for his own childhood predicament, his pattern of using women changed. At the time treatment ended, however, he was still unable to sustain an intimate love relationship on more egalitarian terms.

Other evidence of change came from Carlos's business associates, who told him that he seemed "less driven" than in the past. His attempts to rise above his problems through achieving wealth and power were modulated and gave way to an ambitious work ethic. When an established banker in town offered Carlos a good job as a loan officer and appraiser in his real estate division, Carlos felt ready to terminate therapy. All of his problems were certainly not solved, but Carlos now felt capable of successfully moving on in life on his own. It was important to Carlos to be reassured that he could return to therapy if he needed to in the future, and that doing this would not signify failure but rather strength in recognizing when he needed help. Assured that the therapist would welcome seeing him again in the future if he chose, Carlos terminated and did not recontact the therapist.

Maggie: Moving Away from Others

Six months after her adolescent daughter had been raped, Maggie was still in crisis—as if it all happened yesterday. Nightmares stole her sleep; migraine headaches persecuted her days. Feeling so helpless and inadequate to comfort her daughter, and enraged at the casual indifference of the lawyers and police, Maggie was afraid her life was spinning out of control. At work, her supervisor's evaluation said she was "irritable, sullen, and difficult for coworkers to interact with" and suggested that she seek counseling. Although she had always "hated" to ask for help with anything, Maggie contacted a therapist when she realized her job was in jeopardy.

The therapist was responsive to Maggie's feelings of rage, guilt, and helplessness evoked by her daughter's tragedy. To her surprise, Maggie felt understood by the therapist. Time and again, the therapist "got it"—understood what something really meant to her—and Maggie began looking forward to their meetings. She gradually came to trust the therapist and increasingly invested herself in their relationship.

During one session, Maggie reported a dream from the previous night. In the dream, Maggie was alone in a vast desert night. No other people existed in this great, silent space. The desert night was black; no light shone from stars or moon. As she walked across the endless sand, a cool, dry wind began to move lightly across her face. Maggie laid down on the sand, closed her eyes, and silently slipped away into the darkness.

Maggie relayed that variations of this dream had recurred throughout her life. Because this was a recurrent dream, the therapist knew that it held much meaning and, in ways she couldn't fully understand yet, probably encapsulated her core conflict. Hoping that the dream could also provide an avenue for joining her in her central feeling of aloneness, the therapist tried to bridge Maggie's moving-away orientation:

THERAPIST: Can you close your eyes and find the dream again?

MAGGIE: Yes. (*settles back and closes her eyes*)

THERAPIST: Describe what you see to me.

MAGGIE: I'm walking, but I'm tired of it; there isn't any place to go anyway. It's quiet and dark, empty. I'm alone. I can feel the breeze. Now I'm lying down on the sand. I close my eyes, and just slip away somehow, like going to sleep forever. It's right.

THERAPIST: I don't want you to be there alone. Will you let me join you?

MAGGIE: (*pauses, then speaks slowly*) Well, yeah, if you want.

THERAPIST: Keep your eyes closed, and hold that same, familiar image. But you're not alone this time. I am walking toward you, and I reach my hand out toward yours. Will you take it?

MAGGIE: Mm-hmm.

THERAPIST: Good. Now I'm in the dream with you—you're not alone there anymore. We're holding hands and walking together through the desert night. Can you see that?

MAGGIE: Yes, we're walking together. (*opens her eyes and looks kindly at the therapist*) I'm not alone. It's good to have you with me.

This joining experience was a turning point in therapy. Maggie had grown up in a silent void, much like the setting in her dream. She had never known her father and her mother was often "away." By 10 years of age, Maggie was regularly spending most of the weekend alone, fixing her own meals, and putting herself to bed while her mother was "out." The dream reflected both the emptiness of her childhood and her lifestyle adaptation of moving away from others. However, the crisis with her daughter overwhelmed her coping strategy to withdraw and be self-sufficient and aloof. In letting the therapist join her in her dark, empty space, Maggie took the enormous personal risk of accepting the human contact she longed for yet had long since resolved to hold away. Of course, this single, corrective emotional experience did not resolve her core conflict or change her interpersonal style for coping with it. However, in big and small ways, similar incidents of sharing continued to occur with the therapist.

Maggie's sense of aloneness was heightened by her consciousness of being "different" because she was bicultural, as her therapist, herself a bicultural woman, was well aware. The therapist's support and ability to articulate all the factors, familial and cultural, that contributed to Maggie's sense of isolation helped her begin to change in three ways.

First, Maggie's conflicted affective constellation emerged. The profound *sadness* resulting from her childhood neglect began to come to her. To ensure that the lack of response that Maggie had suffered as a child was not reenacted in their relationship, the therapist was careful to let Maggie know that her feelings were being heard this time. The therapist was skillful in communicating to Maggie that she was not in a dark, silent void anymore but rather in a caring and responsive relationship. Maggie became more comfortable sharing her loneliness with the therapist—and began to question the accompanying beliefs that she "didn't really matter very much" and that others "wouldn't be very interested in helping" if she had a problem and needed them. Feeling safe with the therapist allowed her to experience her deprivation and face it rather than dismiss it as she had always done before. She was able to register or know more realistically how painful this had actually been for her. This deep sharing and resonance from the therapist was relieving for her, but it evoked another threatening feeling—the intense *anger* she felt toward her mother for, in effect, walking away from her. The therapist was affirming of her anger and validated her protest (for example, Maggie would say, "I didn't deserve this—a mother shouldn't leave her daughter home alone to go chasing after the next stupid boyfriend!"). Feeling such anger toward her mother evoked the third feeling in her affective constellation—*anxiety*. Painful separation anxieties were aroused by this protest, as Maggie feared that being angry with her mother would only further push her away and leave her back where she started—feeling even more alone. The therapist affirmed each feeling in her affective constellation and, as she did, Maggie progressively became more comfortable with each feeling. In turn, Maggie also became more forthright and confident in her interactions with others.

Second, as the therapist continued to understand and support Maggie through each feeling in her affective constellation, Maggie began to make important behavioral changes in how she responded to the therapist. Although Maggie had developed a strong working alliance with the therapist, she was still deeply reluctant to accept help from anyone and held a part of herself back. Interestingly, the therapist's bicultural identity had given her an *ascribed credibility*, based on perceived similarities of race and gender, which helped Maggie remain in treatment initially when she was so unsure of this process. Now, however, the therapist's *achieved credibility*—her skillfulness and the effectiveness of her responses—took center stage. As their working alliance continued, Maggie incrementally relinquished her coping style of moving away and gradually allowed the therapist to help her. Maggie became less reserved toward the counselor and, instead, more expressive and responsive than she had been with others. She talked more freely about herself and found herself being curious about the therapist's own personal life. The therapist saw this as positive for this particular client and responded willingly at times. Maggie's nightmares and headaches already had been alleviated, but now a sense of humor and graciousness were emerging as well.

Third, as Maggie talked about feeling "fuller" inside, the therapist observed that changes were occurring in other relationships as well. At work, her supervisor said he was pleased to observe that Maggie was "less irritable and sullen than before"—it was easier for others in the office to talk and work with her. When tensions surfaced with others, Maggie's initial reaction still was to withdraw and "go away inside." However, Maggie became better able to approach problems with others and talk through the interpersonal conflicts that inevitably come up in a busy workplace. For Maggie, it was especially helpful for her to describe the conflict she was having with a coworker or customer and have the therapist suggest new or different ways for her to respond. Together, they would role-play different responses that would keep Maggie involved in the situation, rather than to withdraw or remain aloof as she had always done in the past. Maggie loved to take the part of the difficult customer or coworker and have the therapist role play what Maggie could say or do in that situation. Maggie found this behavioral training invaluable. With success, she regularly found herself using almost the same words and phrases with others that the therapist had modeled.

Especially meaningful for the therapist to observe, Maggie became more accessible—more emotionally present—with her two adolescent children. For the first time, she talked more about her own interests and personal history with them. The nearly grown children welcomed this sharing and, in turn, began to say more about themselves and what was going on in their lives as well. Maggie was feeling closer to her children and more capable of helping her daughter with the aftershocks of her assault.

In addition, Maggie began to change how she responded to her boyfriend. This had been a superficial relationship, like most others in her life had been. Even though they had known each other for more than 2 years, Maggie now invited more personal sharing and emotional closeness in their relationship and asked for more commitment from him. Specifically, she asked him to talk more about himself and to spend more time with her, which he was able to do. It was exciting but anxiety arousing to take each of these steps forward. Her old coping style of moving away was activated each time she took a step toward her boyfriend or others, especially when others did not respond in completely positive ways. Repetitiously, the therapist celebrated her successes in each of these arenas, and patiently helped her work through the setbacks and recover from the disappointments that came along as well. As she continued to make progress and participate more fully in life, however, Maggie tentatively suggested that it might be time to end counseling.

Although things had been going well, Maggie became depressed as they started talking about ending. Even though she had brought up the topic of termination, she missed their next session and arrived late to another. As before, the therapist continued to focus Maggie inward, and her profound feelings of loneliness emerged again. The therapist suggested that they put off setting a termination date for a while and work further with this sadness. With more vividness and detail than before, Maggie recalled her childhood depression. Painful recollections returned—for example, the memory of being 8 years old, sitting alone on the living room couch, and listening to the clock tick the empty afternoon away. Maggie sobbed as she recalled her aching wish for someone to come home to, which was only reenacted in her marital choice. Following her early maladaptive schemas, Maggie had married a salesman whose job took him away from home for extended periods—and he had been preoccupied and emotionally unresponsive to her when he was home as well. Thus, in her first significant love relationship, Maggie chose someone who repeated, rather than resolved, her history of aloneness. However, touching the pain of her childhood neglect so directly, and sharing so fully how it was echoed in the disappointment of her marriage, relieved the depression that the suggestion of terminating had precipitated. Before long, Maggie again felt ready to end and, this time, successfully terminated.

Closing

In this chapter, we have studied an interpersonal model for conceptualizing clients. In an extended case study of Peter, a moving-toward client, we saw how this model can be applied to treatment. We also looked at two case summaries that illustrate its application to moving-against (Carlos) and moving-away (Maggie) clients.

Therapists in training need further guidelines for conceptualizing clients and help in using this formulation to shape treatment plans that guide where they are going in treatment and what they are trying to accomplish with each client. To do this, further information about clients' current interpersonal functioning will be provided in Chapter 8 to help therapists conceptualize their clients. At this point, however, readers are prepared to utilize the guidelines for keeping process notes provided in Appendix A and for writing case conceptualizations in Appendix B. These guidelines have two overarching goals. First, they will help therapists formulate treatment plans by clarifying the maladaptive relational and cognitive patterns that provide a focus for treatment. Second, they will help therapists recognize the process dimension and intervene more effectively by linking the current interaction with the counselor to the problems that clients are having with others.

The therapeutic relationship is the most important tool that therapists can use to help clients resolve problems and change. In Chapters 9 and 10, we will discuss further how the therapist can provide a corrective emotional experience and generalize or apply this experience of change with the therapist to relationships outside therapy.

Suggestions for Further Reading

1. In the current managed-care milieu, it has become essential to write effective case formulations with clear treatment goals and programmatic intervention strategies. In Chapter 7 of the Student Workbook, an extensive case formulation of the character Hap Loman, from Arthur Miller's play *Death of a Salesman*, is provided that illustrates the format for writing case formulations provided in Appendix B.

2. Useful guidelines for conceptualizing clients are found in Chapters 5 and 7 of *Psychotherapy in a New Key: A Guide to Time-Limited Dynamic Psychotherapy* by Strupp and Binder (1984). Readers are also encouraged to examine Wiger's (1999) practical book for all aspects of psychotherapy documentation: The

Psychotherapy Documentation Primer. An especially informative article on case formulation and treatment plans is by Persons, Curtis, and Silberschatz (1991) entitled "Psychodynamic and Cognitive-Behavioral Formulations of a Single Case."

3. *Coping with Conflict: Supervising Counselors and Psychotherapists* (Mueller & Kell, 1972) is a classic in the field, and unsurpassed in helping trainees use their own reactions and the process dimension to conceptualize client dynamics and effect change.

4. Karen Horney's work is essential reading for interpersonally oriented therapists; it has been widely incorporated by psychodynamic and cognitive-behavioral

theorists alike. Readers may be especially interested in her *Neurosis and Human Growth* (1970) or *Our Inner Conflicts* (1966). *Cognitive Therapy of the Personality Disorders* by Beck, Freeman, and Davis (2003) effectively links Horney's interpersonal coping styles to cognitively based interventions for clients with difficult-to-treat personality disorders.

5. Readers may also wish to explore the extensive literature on the "interpersonal circle" (Kiesler, 1996). This rich clinical and empirical literature helps therapists conceptualize interactional patterns and has been widely incorporated by contemporary interpersonal theorists, especially Lorna Benjamin (2003).

CHAPTER EIGHT

Interpersonal Themes and Patterns

Conceptual Overview

In this chapter, we are going to learn more about interpersonal patterns and themes that cause problems in clients' lives. We will explore maladaptive relational patterns that provide useful information for conceptualizing clients and understanding what is occurring in the therapeutic relationship. In particular, we will learn how therapists can use their own personal reactions toward clients as the best source of data for understanding clients' problems with others. That is, clients often make the therapist feel like responding in the same problematic way toward them that they elicit from others in their lives (for example, leading others to feel bored, controlled, helpless, and so forth). In this way, *we will look more closely at how the problematic patterns that are disrupting clients' relationships with others are brought into the therapeutic relationship and played out with the therapist along the process dimension.* With this awareness of the interpersonal process, we will see how therapists can utilize clients' current interpersonal functioning to formulate treatment plans and guide their interventions. Throughout, we will consider different ways that therapists can intervene and change these problematic reenactments by making process comments, providing interpersonal feedback, and other immediacy interventions.

How Clients Reenact Their Problems with Others in the Therapeutic Relationship

The symptoms and problems that clients try to resolve in treatment originated from, and are now being expressed in, close personal relationships. Readily, the same relational patterns that are causing problems with others (such as acting excessively controlling or competitive, responding passively or compliantly, being distant or withdrawing, and so forth) emerge in the therapeutic relationship. As we have seen, clients do not merely talk with therapists about their problems with others. In a far more experiential way, they recreate with the therapist the same maladaptive relational patterns that they play out with others. The challenge for the therapist, then, is to find a way to alter or change this problematic pattern in their real-life relationship. Providing this experience of change with the therapist will greatly facilitate clients' ability to change these same patterns with others. Thus, throughout the course of treatment, the therapists' overriding intention is to

1. identify these maladaptive relational patterns that keep recurring with others,
2. provide a new and better response that does not repeat the familiar scenario in their interaction, and
3. generalize this experience of change in the therapeutic relationship to the clients' interactions with others in their everyday lives.

Therapists practicing in varying psychodynamic, interpersonal, and cognitive/schema-oriented approaches work with three closely related but distinct ways in which clients bring their conflicts with others into the therapeutic relationship: eliciting maneuvers, testing behavior, and transference reactions.

Clients often employ *eliciting maneuvers* (Sullivan, 1968) to avoid anxiety and cope with their problems. At other times, clients will employ a second interpersonal strategy, *testing behavior* (Weiss, 1993), to approach and try to resolve their conflicts. Testing behavior reflects a healthy, growth-oriented attempt to resolve problems. Clients behave in certain ways to assess whether the therapist is going to respond in the familiar but problematic ways they expect or in the more helpful ways they actually need. Working with these complex and sometimes contradictory maneuvers can be one of the most interesting aspects of clinical work. The third (and better-known) way that clients bring their conflicts into the therapeutic relationship is through *transference reactions*—the clients' systematic misperceptions or cognitive distortions of the therapist. As we will see, much confusion arises from two different definitions of the term *transference*. We will clarify important distinctions between the traditional psychoanalytic usage of this term and contemporary interpersonal/psychodynamic definitions that are more useful.

In sum, these three interrelated concepts reflect the most fundamental aspects of interpersonal psychotherapy. In all three concepts, the basic assumption is that the client re-creates the problematic early relationship patterns with the therapist either as a way to *confirm or reject* that the therapist will act in the same way (for example, indifferent, judgmental, demanding) as the early caregiver (Gelso & Hayes, 1998). Let's take a closer look at these very interesting interpersonal processes.

Eliciting Maneuvers

Many cognitive and interpersonally oriented therapists have described how clients develop fixed interpersonal styles to avoid anxiety and defend against the unwanted responses they expect from others (see, for example, Beck, Freeman, and Davis, 2003; Benjamin, 2003). Routinely, clients are inaccurate in their perceptions of others and restricted in their range of emotions. As we saw in the last chapter, they often respond to others inflexibly, using fixed coping styles of moving toward, moving away, and moving against. In part, clients systematically employ these interpersonal styles to (1) elicit desired responses from others that will avoid conflict and ward off anxiety and (2) preclude threatening or unwanted responses from others that will activate or trigger their core conflicts. An *eliciting maneuver* is an interpersonal strategy that wards off anxiety and brings about certain desired, safe responses. To illustrate how eliciting maneuvers serve to defend against problems, we will return to the case study of Peter, the moving-toward client in Chapter 7.

Peter's moving-toward style tended to elicit approval and kindness from others. As a way of life, Peter was helpful; he sympathized, agreed, and cooperated with people. For example, Peter tried to understand his girlfriend's infidelity and to win his instructor's approval. When Peter began to see a therapist, he continued to employ the same interpersonal style that he had used with others throughout his life. As many moving-toward clients do with their therapists, he tried to elicit his therapist's approval by being a good client who was quickly getting better—at least until the therapist began using process comments to question this compliance. For example, "Peter, sometimes I wonder if you are trying too hard to figure out what I want you to do. Any thoughts about that possibility?"

In contrast, Peter was rarely forthright, appropriately assertive or angry, or skeptical with anyone in his life. These responses, which a well-functioning person needs to employ at times, were not part of Peter's interpersonal repertoire. For example, he did not communicate how angry he was with his girlfriend or even recognize how angry he was with both of his parents. Similarly, he did not set limits with his practicum instructor or classmates regarding how much critical feedback he could incorporate at one time. Thus, *his pleasing interpersonal coping style was designed to discourage angry or critical responses from others that would arouse intense anxiety for him.*

If Peter's therapist had merely responded automatically, without first considering what Peter's interpersonal style tended to elicit from others, Peter would not have changed much in therapy. That is, if the therapist had merely supported Peter and met his need for approval, Peter would have only reenacted in therapy the same rising-above defense that he had used throughout his life. Shoring up these defensive coping styles is often the goal in some time-limited or supportive therapies and helps many clients in crisis regain their equilibrium. However, the therapist wanted to do more than this with Peter. Having the opportunity to seek more enduring changes, and not merely symptom relief, the therapist used process comments to make Peter's moving-toward style overt and a focus for discussion. Thus, instead of automatically responding to what Peter elicited and providing only approval, support, and reassurance, the therapist focused on this eliciting maneuver as a part of Peter's problems that needed to be addressed in treatment. How did this intervention occur?

Over the course of several months, the therapist repeatedly pointed out, in a supportive and noncritical way, instances where Peter's coping style of moving toward others might be operating with him and with others in Peter's life. Working together in a collaborative alliance, Peter and his therapist considered both how this was serving to protect him from his problems on the one hand, and how Peter suffered as a result of this coping style on the other. Once they began to identify and look for this behavior pattern, Peter began observing himself using this moving-toward style with others in his life and recognized how it elicited from others the approval and support he had missed as a child. In addition, he began to understand how this style discouraged the critical or rejecting responses from others that were so intensely anxiety-arousing for him. As Peter made progress in treatment, he gradually became less intent on winning others' approval and was able to expand his interpersonal repertoire. He felt more resilient and began to be appropriately assertive and more forthright with others for the first time in his life.

The eliciting maneuvers used by moving-toward clients like Peter are familiar and may not be especially difficult for new therapists to work with. In contrast, moving-against clients employ eliciting maneuvers that are often challenging for beginning therapists. In the first few minutes of the first session, these clients often do something to take command of the counseling relationship. The moving-against client seems to quickly find some way to intimidate or disempower therapists, making them feel insecure or one-down in the therapeutic relationship. For example, clients may insist on sitting in the therapist's chair, questioning the therapist's adequacy, or criticizing what the therapist has just said or done:

> CLIENT: So, if you're just a trainee, what makes you think you know enough to be able to help me?
>
> THERAPIST: Well, I've had some experience before I entered the program, and I have a supervisor to help me.
>
> CLIENT: Uh huh. Does your supervisor always tell you what to say, or do they let you newbies say what you think sometimes?

This provocative presentation elicits anxiety in many new therapists—as it often succeeds in doing with others in the client's life. Many therapists will respond by withdrawing emotionally; others will respond in kind and become competitive; a few will counter with their own hostility. Treatment will not progress when moving-against clients succeed in eliciting one of these defensive responses from the therapist. By reenacting this maladaptive relational pattern with the therapist, these clients have successfully neutralized the therapist's ability to help them (or to hurt them). In this way, they have defended against their anxieties associated with having problems, asking for help, relinquishing some control, and so forth. Thus, clients' eliciting maneuvers protect them from their core conflicts, but at the price of change.

Unfortunately, therapists tend to respond "reflexively" or automatically to the client's eliciting maneuvers with their own countertransference propensities—that is, with their own characteristic tendencies toward flight or fight. What can therapists do instead? First, they can try to retain legitimate control of themselves internally by trying to understand what is occurring right now in their relationship with the client. They can begin formulating working hypotheses about

- the impact this moving-against behavior might have on others in the client's life (such as making others feel like arguing with or withdrawing from them) and
- what conflicts or anxiety arousing situations the client might be avoiding by these maneuvers (feeling "weak" or ashamed, for example, of needing help or having a problem).

Second, on an interpersonal level, *therapists can try to engage the moving-against client in some way in which the maladaptive relational patterns of domination, intimidation, competition and so forth, are not enacted.* There are many possibilities:

- One possibility is for the therapist simply to withstand the client's criticism, without avoiding the challenge or becoming defensive.

  THERAPIST: Your previous therapist was an experienced psychiatrist, and you're not sure that a younger social worker such as myself can help you. Let's talk together about that—tell me more about your concern.

- Another alternative is for the therapist to make a process comment and engage the client in talking together about the current interaction.

  THERAPIST: You're speaking to me in a harsh, loud voice right now, and you sound angry. What do you see going on between us here?

- With some clients, the therapist may wish to use a self-involving statement or provide interpersonal feedback to begin exploring the effects of their interpersonal style on others and highlighting the relational patterns that ensue.

  THERAPIST: Let me take a risk and talk about the impact you have on me sometimes. Right now, I'm feeling criticized as you correct me again, as I've felt other times. How do others in your life usually respond to you when you are critical like this—what usually happens next?

Some of these clients will be unaware of the unwanted impact their criticalness has on others, and this interpersonal feedback can provide a new opportunity to begin exploring this collaboratively. Others will say that they hit first before someone can get them. To emphasize, however, the purpose here is not to win the battle for control and get one up with the client. On a process level, this interaction would only continue, or recapitulate, the same maladaptive relational pattern that recurs with others. Rather, this approach is intended to help the therapist find some way to engage with the client other than by automatically responding with counterhostility, control, or withdrawal to the eliciting maneuvers—as others usually do. It will also help clients develop increased awareness of when they employ their coping style, its impact on others, and the personal and interpersonal price they pay for this coping style. As clients hear interpersonal feedback from the therapist and become more aware of the impact they are having on others, it often begins to introduce more flexibility and choice about responding this way.

- Still another possibility is for the therapist to explore how the client's eliciting maneuvers affect the client.

   THERAPIST: You have insisted on sitting in my chair, even though I asked you not to. Now that you are there, how does it feel?

Routinely, moving-against clients become aware of feeling alone, empty, or anxious at these moments, and the hollowness of their victory provides a new shared point of departure for the therapist and the client. Thus, as these four different types of responses illustrate, there is no simple formula for the "right" way for the therapist to respond, of course. Per client response specificity, one type of response will work well with one client and not at all with another. Therapists need to be flexible, assess how the client responds to each intervention, and modify their responses to find what works best for a particular client. Thus, to respond effectively to these challenging clients, the therapists' intentions are

- to attend to their own subjective reactions and recognize what the client is eliciting in them (such as feelings of incompetence or competitiveness);
- to formulate working hypotheses about how others would typically respond and begin to identify the maladaptive relational patterns that are occurring with others and causing problems in the client's life (for example, others tend to withdraw or argue);
- to try to find another way to respond (for example, via process comments, self-involving statements, or interpersonal feedback) that does not go down the old familiar path and reenact the same problematic scenario that the client usually elicits from others (Therapist: "If your secretary were watching us through a one-way mirror right now, would she say that you are talking to me in the same contemptuous tone of voice that causes so much trouble for you at work?").

Although moving-against clients may intimidate new therapists, this effect usually does not last very long. With a little more experience and a supportive supervisor, new therapists learn not to charge so readily at a waving red flag. With more exposure to moving-against clients, therapists learn to appreciate that the intensity and rigidity of clients' eliciting maneuvers, no matter how alienating initially, are commensurate with clients' degree of anxiety and conflict. As therapists better understand the purpose of these distancing maneuvers, they may begin to find some compassion for the predicament such clients are living out, which is usually the best way to step out of the old relational pattern being elicited. By flexibly trying out different approaches, therapists can usually find ways to engage these clients that do not reenact either the intimidated/withdrawal or competitive/hostile patterns that they usually succeed in eliciting from others.

As we have seen, the confrontational eliciting maneuvers of moving-against clients—such as Carlos in Chapter 7—place considerable demands on beginning therapists. However, the meeker, moving-away clients—such as Maggie in Chapter 7—often pose a greater challenge. Most people who select counseling careers have strong needs for close personal relationships. When moving-away clients continue to maintain emotional distance, many therapists are discouraged and ultimately give up—disengaging as many others in the clients' lives have done before. When this occurs, the clients' eliciting behaviors have successfully defended them against a relationship with the therapist and thereby allowed them to avoid the anxiety or threats evoked by risking more involvement in relationships. Thus, the clients' conflicts will not be activated, but the therapist's only real vehicle for resolving these problems is closed as well. Rather than responding automatically to these clients' aloofness by giving up or turning off ("Maybe this client doesn't really want to change or be in therapy with me?"), the therapist may use process comments to try to find other avenues for establishing a relationship:

THERAPIST: You're very quiet, Mary. I sometimes feel as if you are hardly in the room with me. What's it like for you to come here and try to bring me in on your life?

CLIENT: I'm not sure what you mean. Am I doing something wrong?

THERAPIST: No, not at all. I guess I'm saying that sometimes I feel like I'm not reaching you or getting close to what really matters in your life, even though I want to. Has anyone else ever said something like this to you?

CLIENT: Well, kind of. My boyfriend finally left because he said he never really felt very important to me, that he gave up trying to "pursue" me.

THERAPIST: This is a very sensitive thing for both of us to be talking about, but I think this is important. It sounds like other people feel held away from you in some way—or that you can "go away" so readily. I'm trying to understand how this goes in your life. Can you help me with this?

CLIENT: I think I withdraw from people too easily. I know I'm too sensitive—my feelings get hurt pretty easily.

THERAPIST: OK, your feelings get hurt too easily, and you "go away." I'm glad you're choosing not to withdraw right now but having the courage to stay with me and be so present. This feels different to me than how we usually are. I feel more connected with you. What's it like for you?

CLIENT: I like it, too. I just get afraid with other people, and I know I go away too easily. I think I've always just done that without really thinking about it.

THERAPIST: Well, right now you are doing it differently with me, and that makes me think you'll be able to do this better with others, too. Tell me more about how you are going to get hurt, by me or by other people, if you stay engaged or connected like this. What's going to go wrong?

To sum up, the therapist can attend to what clients systematically elicit in others and hypothesize how this interpersonal strategy is likely to effect others and cause problems in their relationships. In doing this, the therapist attends to the clients' developmental history, current life situation, and to the cultural context in which coping styles and eliciting maneuvers are embedded. Importantly, *therapists are also attending to their own feelings that are elicited by clients.* This self-awareness is one of the therapist's most important sources of information about clients and what their interpersonal style tends to elicit from others. To be able to learn from one's own reactions to the client, the therapist aims to be *nondefensive* enough to be able to observe what clients tend to elicit or "pull" in them (such as feeling attacked, ignored or pushed away). Attending to their own personal reactions to the client in this way will often provide important information about the clients' conflicts, coping strategies for defending against them, and how the relational themes being played out in the therapeutic relationship may be effecting other relationships.

We have already seen that the therapist can generate working hypotheses to understand why clients characteristically elicit certain responses from others, such as reassurance (moving-toward clients like Peter), competition (moving-against clients like Carlos), and disinterest (moving-away clients like Maggie). Restating this challenging task, the therapist's intention is to keep from responding "reflexively" or automatically with the types of response clients usually elicit and, if possible, to try to respond in a new and better way that does not reenact the clients' maladaptive relational patterns. Over time, the therapist can begin to help clients identify their eliciting mechanisms, recognize the situations or types of interactions where they tend to employ these measures, and understand how these maneuvers have served to create safety in the past. It does take courage for therapists to do this, however, especially as they begin their clinical training. On the one hand, therapists are trying to tolerate their own unwanted feelings (irritation, impatience, boredom, uncertainty) long enough to sort through whether they have more to do with their own personal feelings (as we will explore later when we discuss countertransference). Or, on the other hand, do our own feelings inform us about the clients' eliciting maneuvers and help us understand what this client tends to evoke in other relationships as well? The key element here is for therapists to remain nondefensive toward their own reactions so that they can reflect on and learn from them. Let's explore this important issue further and see how the client's eliciting maneuvers can tap in to the therapist's own personal issues or countertransference reactions.

As we are just beginning to see, things quickly become far more complicated when the clients' eliciting behaviors tap into the therapist's own issues or personal situation. With this new development, we enter the arena of countertransference—it's not just about what this client tends to elicit in others, but how the therapist's own life and personality have been brought into the equation as well. For example, suppose that the client acts helpless and escalates his distress, and the therapist tries to help but fails at each attempt. Ideally, the therapist would remain involved and keep exploring other ways to respond—for example, by making the process comment that whatever the therapist tries seems to disappoint the client or fails to help. However, what if this therapist grew up with critical caregivers and, no matter what she did, she could never quite do it right or please them. If the therapist's own personal concerns regarding performance demands or criticism are activated in this way, the therapist may lose her effectiveness because the client's eliciting maneuvers have tapped into her own countertransference issues. For example, the therapist who grew up being criticized and is now overreactive to clients' disapproval may feel inadequate and withdraw or become impatient with the client and become critical toward the client in turn. *When clients successfully elicit the same type of response from the therapist that they usually elicit from others, the therapist and client are reenacting rather than resolving the problem.* In different terms, they are recapitulating maladaptive relational patterns rather than providing a corrective emotional experience. This is a complex and challenging issue for beginning therapists to work with, but let's keep at it. As you learn more, it readily becomes a powerful way to understand what's going on in the therapeutic relationship and what therapists can do to help clients change.

Scenarios such as this, in which the clients' eliciting maneuvers tap into the therapist's countertransference propensities, are commonplace—for beginning and experienced therapists alike. When this type of reenactment occurs, clients often switch to the other side of their "ambivalence" and try to reengage the therapist, perhaps by reassuring the therapist about how helpful the sessions are. At some level, clients do not really want to succeed in immobilizing the therapist or ending the relationship by repeating the same problematic patterns, and they will often try to get the disengaged or discouraged therapist involved in the relationship again. Premature and unnecessary terminations routinely occur when the therapist (and perhaps the supervisor) fails to hear clients' renewed bids for relationship because the discouraged, frustrated or disengaged therapist has given up on the relationship.

> THERAPIST: This client is hopeless; he doesn't want to change. He just wants to tell everybody else what they are doing wrong. No wonder his third secretary just quit and his teenaged son doesn't want anything to do with him!

Especially with more troubled or conflicted clients, therapists will need to keep both of these disengaging and subsequent engaging maneuvers in mind.

In closing, it is key for therapists to learn how to distinguish between their client's eliciting maneuvers and their own personal or countertransference reactions toward the client. As we will continue to discuss, this distinction is referred to as *client-induced versus therapist-induced countertransference* (Springmann, 1986). How do therapists know whether the client's eliciting maneuvers (client-induced), their own personal issues (therapist-induced), or some combination of both are evoking their strong reactions? If this client makes others feel or react similarly (others in the client's life are also bored, impatient, or intimidated, for example, as the therapist is made to feel in the session), it is probably client induced. If not, this reaction probably has more to do with the therapist's own personal life (therapist-induced countertransference). At times, both are operating, and supervision is usually the best place to sort this through. We will be exploring these two different conceptions of countertransference further, but first need to learn more about the closely related concept of testing behavior.

Testing Behavior

We have seen how clients employ their inflexible coping style to systematically elicit responses from others that may seem self-defeating at first yet have served to effectively avoid anxiety in the past. These eliciting maneuvers—or security operations as Sullivan (1968) sometimes called them—are interpersonal defenses.

Although eliciting maneuvers are used defensively to avoid anxiety, at other times clients use different interpersonal strategies to approach and try to resolve problems. Although this point may sound contradictory, it reflects clients' ambivalent feelings about treatment—the hope that things could change and get better versus the anxiety that the same unwanted outcomes will only occur again. Clients often carry out the healthy, approach side of their ambivalence by means of *testing behavior* (Weiss, 1993; Weiss & Sampson, 1986). In direct or in more covert ways, clients systematically reenact the same interpersonal patterns and themes with the therapist that cause conflict in other relationships in their lives. Clients "test" to assess whether the therapist will respond in the familiar but problematic way that others often have in the past, or whether the therapist can respond in a different and better way that changes the old predictable scenario, expands schemas and expectations, and allows clients to make progress on their problems. There is nothing casual about this testing for clients; *they are vitally interested in assessing whether the therapist is going to reenact or resolve maladaptive relational patterns.* Many clients have histories of exploitation, denigration, and rejection; their trust in others has been betrayed. Their testing behaviors often reflect their intense wishes for healthier, safer relationships, as well as their fear of again being disappointed or hurt. Through testing, in other words, clients are trying to assess *safety/danger* in the therapeutic relationship.

As we will see, the change process is set in motion when therapists respond effectively and "pass" important tests by behaviorally disconfirming faulty beliefs and expectations. Clients feel safer when the therapist does not respond in the familiar but problematic way that others often have, and that they have come to expect from the therapist as well. Immediately following this corrective emotional experience, clients often make visible progress in treatment.

For example, in the next 2 or 3 minutes, the therapist may observe that clients act stronger or behaviorally improve in one of these ways:

- Expressing more confidence in themselves or in the therapist ("You know, I think I'm just going to go ahead and talk to my wife about this. I've gone around about it in my head for too long now. I'm going to bring this up with her end and put it on the table—it's not good for me to keep avoiding it like this.")
- Feeling better as symptoms such as anxiety or depression diminish ("I feel more hopeful right now. Maybe she won't ever stop being so bossy or demanding, but that's just her. I don't have to feel so ruled by what she says or how angry she gets.")
- Taking the risk of bringing up threatening new material with the therapist ("I've been ashamed to tell you about something that's happening: I've been having an affair for a long time. I know I should be talking to you about it, but I've been afraid to tell you. I guess I've been worried that you won't respect me anymore if you knew about this.")
- Acting stronger by being more honest or forthright with the therapist and discussing problems between them ("Therapy isn't working very well for me. You're so quiet—I don't know what you're thinking a lot of the time. I think I need more feedback or something from you.")

In these ways, the new or corrective response from the therapist creates safety for clients. Observing clients' reactions to their interventions, therapists will often see that this allows clients to make progress in the next few moments and change how they are responding to the therapist and to others in their lives. Later in this chapter, we'll return to this important concept of assessing the client's responses to the therapist's interventions.

For now, let's restate this important concept of testing behavior and explore it further. In a healthy attempt to address and resolve problems, clients often actively (but not consciously) elicit the same types of problems with the therapist that they are having with others in their lives. Routinely, clients are actively testing to see whether they can change these maladaptive relational patterns and disconfirm the pathogenic beliefs that stem from them. However, it is difficult for clients to find disconfirming evidence because repeated confirmation of the beliefs has made clients deaf to such evidence. Cognitive therapists have clarified the selective attention and biased filters that lead clients to keep slotting experiences into the same familiar categories (Beck, 1995). However, *the power of immediate, in-vivo relearning with the therapist allows clients to begin questioning, or sometimes even relinquish, deeply held but faulty beliefs about themselves and expectations of others.* Thus, the therapist can fail the client's test by responding in a way that reenacts maladaptive relational patterns and confirms pathogenic beliefs. Alternatively, the therapist can pass the test and provide a corrective emotional experience by behaviorally demonstrating that, at least sometimes, some relationships can be different.

To illustrate, suppose that a young adult client is being seen in a college counseling center for depression. Her therapist has formulated the working hypothesis that she is ruled by the pathogenic belief that, "If I have my own mind and do what I want with my life, my mother will be sad and withdraw. I deserve to feel bad because I have been selfish and hurt her." How can this hypothesis guide the therapist's interventions as he tries to pass the tests that she is apt to present? The therapist would likely fail this client's test, repeating the maladaptive relational pattern that the client lives out with others, if he expressed caution, worries, or doubts when this inhibited client said something about becoming more independent, acting on her own behalf, or pursuing her own interests and goals. For example:

CLIENT: I don't think I want a career in business after all. I'm not liking these business classes I've been taking very much. The other students are fine, but I'm not really like them. I think I might like teaching better—kids have always been fun for me.

THERAPIST: Well, maybe education would be a better major for you, but that job market is tight right now, and it hasn't ever paid very well.

CLIENT: (*sinking*) Yeah, I wasn't thinking about that. I guess I need to be more practical. Maybe teaching wouldn't work very well for me after all. . . .

In contrast to this reenactment, which exacerbated her presenting symptom of depression, the therapist would pass this test—and empower this cautious and compliant client—by expressing interest in the client's own interests and supporting the client's initiative whenever it emerged. For example:

CLIENT: I don't think I want a career in business after all. I'm not liking these business classes I've been taking very much. The other students are fine, but I'm not really like them. I think I might like teaching better—kids have always been fun for me.

THERAPIST: Kids are fun for you—that's great. I'd love to hear about your interest in teaching. Tell me more.

CLIENT: Well, I don't think my parents would like to see me be "just a teacher." I was a National Merit Scholar, you know. But I think I'd love having my own classroom, and I could have a more balanced life with summers off—you know, so I could have time to raise my own family.

THERAPIST: You sound excited about this, like it would give you the kind of lifestyle that you choose. You sound different right now, so alive as you talk about this possibility.

Testing behavior such as this pervades all counseling relationships from the first session to the last. However, testing is especially likely to occur in the initial sessions, in situations where strong feelings are being evoked, and when the current interaction between the therapist and client parallels relational problems that the client is having with others. Once therapists have the concept of testing, many of these tests will be easy to recognize and pass, allowing clients to progress in treatment and make changes in their lives.

Let's consider a similar type of testing. Suppose a rather passive and dependent woman says to her male therapist, "Where should we start today?" Even though he has the best of intentions, the therapist will fail this test and reenact a maladaptive relational pattern if he says, "Tell me about _____." By telling this client what to do, even though she has asked for his direction, the therapist has confirmed the client's pathogenic belief that she needs to take a subservient role with men or authority figures and be told what to do. The problematic belief that she ought not to assert herself or take the lead is confirmed when the therapist tells this client where to begin or what to talk about (recall that, per client response specificity, this same directive may be helpful to a different client). This client is likely to remain unassertive and dependent in her relationship with the therapist and, in turn, with others in her life. In contrast, the therapist is likely to pass this test and disconfirm this faulty belief if he repeatedly looks for ways to encourage her own initiative and actively support her when she takes it. Thus, when this client asks him what to talk about or where to begin, the therapist can provide an experience of change by responding as follows:

- Tell me about the issues or concerns that are most important for you right now.
- Where do you think is the best place for us to begin today?
- What do you think would be the best way for us to use our time?
- I'd like to talk about what you think would be most helpful.

Most clients will not accept these invitations so easily, of course. Oftentimes, they will not accept the therapist's first invitation but continue testing to see whether the therapist really means it when he says he is interested in her choices and what seems most important to her:

CLIENT: (*vaguely*) Oh, I don't know, I can't really think of anything. What do you suggest?

Trying again to change the old relational pattern in their current interaction and engage the client in a new, more egalitarian way, the therapist might pass the test by giving a second bid for the client to take the lead:

THERAPIST: OK, let's just wait a minute then. Take a moment for you to be with yourself, and see what comes up for you.

CLIENT: (*pause*) Well, OK, uh, I guess I haven't been sleeping very well lately. I keep having this bad dream.

THERAPIST: The same bad dream keeps coming back. That does sound important—tell me about it.

With each of these responses, the therapist's intention is to demonstrate that he genuinely wants to engage her in a more collaborative way where they can share control over the treatment process. Behaviorally, the therapist is saying that he does not share her belief that she needs to remain dependent and be told what to do. Repeated interactions of this sort—which consistently disconfirm her faulty belief that she isn't capable of acting independently or that others need her to depend on them—will be reparative. As they continue to occur in treatment, these corrective experiences will bolster the client's self-esteem and self-efficacy—experientially disconfirming her belief that she needs to act dependently and have the therapist and others lead, guide, and take responsibility. As she continues to find that it is safe to act stronger with the therapist, she can begin to try this new way of relating with others in her life as well. She and the therapist then will begin to sort through others in her life with whom it is safe to act more assertively and independently and others who undermine her confidence and do in fact want her to remain one down.

Usually, it is relatively easy for new therapists to identify this type of testing around dependency/compliance issues and respond effectively. In contrast, tests that involve issues of power and control are often far more challenging for new and experienced therapists alike. Clients with different developmental experiences—for example, those who could run over their caregiver or who wielded too much power and influence in family matters—often test the strength or resolve of the therapist. When these moving-against and other confrontational clients can successfully dominate, demean, or control the therapist, things quickly go downhill for therapist and client alike. Before they feel safe disclosing personal or vulnerable issues, or even beginning to engage in a working alliance, these clients need to assess safety/danger in the therapeutic relationship. For these clients, this means testing whether the therapist can handle them. For example, when clients are insulting, provocative, or push limits with the therapist, they often are testing whether they can manipulate or run over the therapist as they could their caregiver. Or, is the therapist able to tolerate their disapproval without becoming defensive, set limits on acting out with the therapist, and provide firm treatment boundaries—thereby offering the interpersonal safety they did not receive before? Let's consider a case example of this more challenging, provocative type of testing.

Graduate students in Carl's training program were supposed to videotape all therapy sessions. Early in treatment, however, Carl's client Lucy complained that the videotaping made her "self-conscious." She said it was already hard for her to talk about what was going on in her life and that the videotaping just made it "impossible" to talk about really important things. With each successive week, Lucy voiced her increasing exasperation with Carl and disappointment in treatment. She told Carl that he was not helping her and threatened to stop if he didn't turn it off.

Carl was torn and did not know what to do. He knew the clinic rules and that his supervisor would abide by them, yet he also felt he should be flexible enough to do what his client needed. Without consulting his supervisor, he reached a compromise with Lucy and agreed to turn the videotape off for the last 5 minutes of each session. Lucy was appreciative and did indeed disclose new material to Carl. Carl felt badly about doing this without his supervisor's permission, but he also thought it was the only way to help his client and keep her in treatment.

Almost immediately, however, Carl observed that treatment was not going well. Even though she brought up some relevant new issues, Lucy became scattered and started jumping from topic to topic. Carl tried to pin her down and focus on one thing at a time—such as her ever-changing feelings or scattered thoughts, but she slipped away each time. Lucy complained that she was "very upset" and needed more guidance from Carl, yet whatever he did failed to help. Realizing that things were getting worse rather than better, Carl decided to talk with his supervisor.

First, Carl and his supervisor spent some time honestly discussing what Carl's decision to turn off the tape meant for their relationship and their ability to work together collaboratively. With their relationship restored, Carl and his supervisor began exploring what had been going on between Lucy and Carl, and trying to understand what this interaction pattern might mean. To begin, they clarified how badly Carl had always felt about himself when others were critical of his performance. Carl relayed that he just "dreaded" to be in situations where, no matter what he did, he "couldn't do it right." Further clarifying their mutual interaction, the supervisor helped Carl recognize how this countertransference issue in his own life was activated by the critical and demanding stance that Lucy took with Carl—and with others in her life. Carl felt empowered by these new ways of thinking about what was going on in the client–therapist relationship and accepted his supervisor's firm limit that the videotape must remain on throughout every session. To his dismay, however, Lucy did indeed feel angry and betrayed when Carl reinstated this limit at the beginning of the next session. She accused Carl of being "untrustworthy" because he broke their agreement, told him that she didn't think he was going to become a very good therapist, and informed him that she could no longer talk with him about certain important topics. All of this was so difficult for Carl, but the biggest threat came when she told him that she would have to think about whether she wanted to keep working with him.

Carl felt "horrible" about her threats and accusations yet, with his supervisor's support, he tolerated his discomfort, tried to remain as nondefensive as he could, and stuck to the rules this time. Although she continued to blame and complain, Lucy did return. In fact, despite her protests, she actually seemed calmer than before and began to focus more productively in therapy. Over the next few weeks, Lucy succeeded in making some important changes in her life. She began to share with Carl how "sickening" it had felt for her to be able to "boss my father around." Following this disclosure, which Lucy had felt so ashamed about, she returned to the next session and said that she had apologized to her boyfriend for being "so bossy." Lucy told her boyfriend that she "didn't want to be this way anymore" and was going to change it.

It is adaptive for clients like Lucy to play out their maladaptive relational patterns with the therapist and test whether the therapist will reenact or resolve them. Clients cannot progress to more vulnerable material if the current therapeutic relationship is not safer than past relationships have been. However, there is very limited value in verbally reassuring clients. Moreover, most clients are not consciously aware of testing therapists in these ways, and usually it is not very helpful to point out testing behavior to them. Instead, as in all human relations, *it is what the therapist does that counts, rather than what the therapist says.* Therapists pass tests and provide a corrective emotional experience when they repeatedly respond in ways that behaviorally disconfirm pathogenic beliefs and maladaptive relational patterns. Fortunately, even though all therapists will fail tests at times, they will have many opportunities to recover—just as Carl did so effectively. It is especially important for new therapists to be reassured that they can readily recover from mistakes, failed tests, and reenactments. Therapists can recover by acknowledging what went wrong and talking it through with the client, or by responding more effectively the next time this issue or theme comes up—often later in the same session or in the next session. In particular, therapists can recover from these inevitable ruptures that regularly occur for even the most effective and experienced therapists if they

- stay closely attuned for overt communications and subtle messages that clients give about the therapist and their interaction together (Client: "You probably wouldn't approve of this, but . . .") and
- in each interaction sequence, attend closely to how the client responds to what the therapist has just said or done (Therapist: "I just asked you to clarify that for me, and you became quiet. I wonder what went on right there for us—maybe something didn't feel quite right?").

When the therapist passes a client's test by responding in the way the client needs, the client experiences a new degree of safety in the therapeutic relationship (such as providing limits for Lucy, who held too much power over adults while growing up and did not experience consistent boundaries). An exciting process to participate in, therapists will often observe clients respond to this new safety by beginning to remember more clearly, feel more deeply, and share more fully. In the next few sessions, therapists will also observe clients taking risks, as Lucy did, and trying out new and better ways of responding with others in their lives. Many clients will initiate these changes on their own:

CLIENT: Hey, guess what I did this week? I spoke up and told my roommate. . . .

With other clients, therapists can encourage them to try out the changes that have occurred between them in other relationships:

THERAPIST: You just did this differently with me. That makes me wonder if you couldn't speak up for yourself better and say more of what you want with your husband, too. What would it be like for you to try this with him, like you just did with me?

How can therapists accurately identify the specific interpersonal response that this particular client needs in order to pass a test and provide a reparative experience? To guide their responses, therapists can (1) consider client response specificity and (2) track the moment-to-moment interaction sequences and assess the client's positive or negative reactions to the therapist's responses or interventions. Let's consider both carefully.

Client Response Specificity   As discussed in Chapter 1, therapists need the *flexibility* to be able to respond differently to different clients. In counseling relationships, there are no cookbook formulas or "right" ways to intervene. One size does not fit all, that is—the same therapeutic response that passes a test for one client may reenact conflicts for another client with different schemas and expectations. For example, in response to pressure from her client, Beverly agreed to reduce her fee substantially. Almost immediately, her client began wearing more expensive jewelry, delighted in telling Beverly about shopping trips, and described how wonderful it was to fly first-class on her vacation. Realizing that the client was taking advantage of her, just as the client had done with her doting mother and indulgent husband, Beverly renegotiated the fee. Later in the session, the client brought up important new material about, first, feeling "pushy" and, later, "lonely" when she felt she had too much control over others. Working together, they entered a productive new phase of treatment where the client began to explore how this pattern of being demanding or dominating was being played out and disrupting other relationships in her life. Based on the client's willingness to begin looking at these unwanted parts of herself, Beverly concluded that, by being firm, she had passed the client's test and provided her with the corrective response she needed. This client was unaware of her own testing behavior with Beverly, however, and unable to articulate it in any way. Like most clients, the client did not feel safe enough to recognize or discuss this sensitive issue until she was *behaviorally convinced* that she could not manipulate Beverly as she could her mother, her husband, and her previous therapist (mere interpretations, reassurances, reframing, and other verbal responses usually are far less potent).

Per client response specificity, in contrast, the same response of lowering the fee can be helpful for a different client. Chris was coping with a diminished income following a recent divorce and was having trouble making ends meet. Knowing that she grew up with demanding, authoritarian parents who gave her little, the therapist realized that offering Chris a more flexible fee arrangement could be a significant overture. This client would never have asked the therapist for a reduced fee (or asked anyone for help with anything) and was initially unwilling to accept any of the alternative fee arrangements that the therapist suggested. When the therapist asked about her reluctance to even consider his offer, Chris began to cry. As their discussion soon revealed, the therapist's generous response disconfirmed her pathogenic belief that she "didn't matter" and evoked the longing she had felt for her parents to "see her" and reach out to help when she had a problem. Chris had grown up under the demanding expectations that she had to do everything perfectly and also without ever needing help of any kind. Clearly, changing the payment schedule with this client disconfirmed pathogenic beliefs about herself and faulty expectations about others. Within this session, treatment progressed into more intensive work about her emotional deprivation and resulting belief that she "did not matter much" to others. As these two examples illustrate, *the same response by the therapist will often have very different effects on different clients.*

Clients' cultural/ethnic backgrounds add complexity to response specificity. For example, self-disclosure of an issue that the therapist had great difficulty resolving in his own life could be viewed as unprofessional by some traditional Asian clients. However, the same disclosure with some African American clients may be welcome and convey that the therapist is human and less likely to be judgmental. In this way, racial, gender, religious, or other differences between therapist and client will affect how the client responds to the therapist's offer. Following our previous guidelines, however, the therapist's intention is to pay close attention to how the client responds to this intervention. For example, if the therapist decides that self-disclosure may be productive for this client, the therapist will closely observe how the client reacts to it—with a helpful new perspective or more involvement in the working alliance, for example, or with greater reserve and emotional distance from the therapist.

Client response specificity puts more demands on the therapist. It takes away the security of having a simple, rule-based approach to therapy. Instead, it requires the therapist to remain close to the unique and personal experience of each individual client. However, within this highly "idiographic" approach, guidelines are available to help therapists discern more accurately the specific needs of differing clients. In particular, the therapist can learn to assess the specific relational experiences that each individual client needs by identifying the maladaptive relational patterns that tend to recur in three arenas:

- current client–therapist interactions,
- relationships with important others in the client's current life, and
- the client's family of origin that originally engendered these schemas and patterns.

Furthermore, therapists learn how to pass tests and provide clients with the specific relational experiences they need by assessing how clients react to different interventions from the therapist.

Assessing Client Reactions   How do therapists know whether they have passed or failed the client's test or simply responded in a useful way to what the client just said or did? Are there guidelines to assess whether the therapists' response has been helpful or has reenacted some aspect of the client's conflict? If clinical trainees rely only on their supervisor's advice on how to intervene with their clients, they will remain dependent on guidance and insecure about their own clinical work. Similarly, new therapists will have a false sense of security, and risk becoming rigid and dogmatic, if they settle on one theoretical approach too readily. New therapists do not want to readily adopt any single approach to treatment. Too often, new therapists "foreclose" on this aspect of professional identity development in order to keep from feeling incompetent or to avoid the anxieties and ambiguities inherent in clinical training. Instead, trainees want to try out different ways of responding and explore competing theories in order to sort through for themselves what works best for them and for this particular client. Thus, even though supervisors and theories certainly are useful, beginning therapists develop more autonomy, and find their own identity as a therapist, by learning how to *assess client responses* and determine for themselves the effectiveness of their interventions.

Beginning therapists often suffer under the misconception that there is one "right" or correct way to intervene with clients—and feel worried or insecure that they haven't done it right. A better approach is to observe how the client reacts to the therapist's response. In this approach, the client is determining the best way to intervene—based on how well the client can use what the therapist just said or did to make progress in treatment. With each successive turn of the conversation, therapists are trying to decide what is the best way to respond to this client right now—and choosing to respond by challenging, affirming, reframing, exploring, making links, providing feedback, and so forth. As soon as they respond, however, the therapist's intention is to *watch closely* and see how the client seems to be responding to this intervention (with disinterest, agreement and further exploration, bringing forth new material, and so forth). As the therapist tracks the moment-to-moment interaction sequences—how the client is responding right now to what the therapist just said or did—clients inform the observing therapist of the responses they can utilize or benefit from. Clients do not usually verbalize this outcome and may not be conscious of it, but *they reliably guide the therapist by their behavioral response to how the therapist just intervened*. Let's look more closely at how this goes.

When the therapist responds effectively to a client's test, disconfirming pathogenic beliefs about self and faulty expectations of others, the client often feels safer and begins to act stronger with the therapist, behaving in new and more adaptive ways. This progress can be expressed in many different ways. For example, in the minutes after the therapist passes a test, clients may stop complaining about other people and begin talking more directly about their own thoughts and feelings. Or, they may look more honestly at their own contribution to the conflict, as Beverly's client just demonstrated. Perhaps the client will relate a vignette that is clearer or more meaningful than the material they have produced before. The interaction or dialogue with the therapist may acquire more immediacy or intensity and become less intellectualized, superficial, or social. Clients will often engage the therapist more directly or be more present, perhaps by expressing warm feelings toward the therapist or by taking the risk to bring up a problem they are having with the therapist. Oftentimes, the client's affect will emerge or intensify. In the next minute or two, the client's feelings about whatever has happened—happy, sad, or angry—may be experienced more fully or shared more openly with the therapist. Therapists want to take note when clients act stronger in any of these ways. Yes, it is a pleasure to see the client is improving, but the client is also telling therapists what they have just done, and what they can do in the future, to make a difference.

Therapists also know that they have responded effectively and passed tests when clients bring up relevant new material to explore. When clients initiate new topics in this way, they are often especially productive. In particular, clients may make useful connections on their own between their current behavior and past relationships—without interpretations or guiding the client back to historical or familial relationships. The insights that clients generate in these circumstances are enlivening and hold real meaning for the client, and they are behaviorally translated into new ways of thinking and acting in their everyday lives. In contrast, it is not usually productive for therapists to make historical interpretations or to lead clients back to try and make familial connections. Even when these historical links are accurate, clients do not usually find them to be relevant to their current problems—just abstract ideas or distant possibilities that do not hold real meaning or lead to behavior change (for example, a client might exclaim, "My mother! Why are you talking about my mother! I'm 32 years old—what the hell does my mother have to do with anything?"). As a rule of thumb, it is not usually helpful to make familial/historical interpretations to help clients see the roots of their problems (Therapist: "I think you're afraid to leave him and move out on your own now because you felt so alone as a girl—and you just don't want to feel that way again"). In contrast, meaningful behavioral changes often result when clients spontaneously make these developmental links on their own (the "aha" experience)—which is most likely to occur just after the therapist has passed a test or provided a corrective experience (Client: "I think I've been afraid of moving out and living on my own because I don't want to feel all alone again—like I did when I was a girl").

In sum, the therapist has responded effectively or passed the client's test when the client responds by acting stronger in one of these ways. To illustrate, when Carl turned the videotape on for the full session, Lucy complained and even threatened to terminate. But by maintaining these appropriate professional boundaries, Carl demonstrated that he was different from Lucy's father and was therefore safer. Carl could tolerate Lucy's disapproval and do what she actually needed rather than what she demanded. Although Lucy threatened and complained, she also began to work more productively with Carl and readily began changing this maladaptive pattern with others in her life.

If therapists observe how clients respond to their interventions and try to identify what they have just done to pass (or fail) the client's tests, they will be able to formulate treatment plans more effectively. These conceptualizations can then be used to shape subsequent interventions and better provide clients with the responses they need in order to change. For example, at other anxiety-arousing points in treatment, Lucy will again push the limits with Carl. If Carl has learned from their previous interaction, however, he will be better prepared to provide Lucy with the clear limits she needs.

What about the other side of the coin—how do therapists know when they have failed the client's test or unwittingly reenacted the client's conflict in some way? On having failed a test, the therapist will often observe that clients act weaker in some way—such as, in the next moment, becoming more anxious, dependent, confused, externalizing, distant, and so forth. The client may also backtrack from healthy new behavior, retreat from recent successes, or become less of a partner in the working alliance. If the therapist is closely assessing how the client is responding to what they have just said, they will observe that these reactions to failed tests occur almost immediately. They are usually evident within the next minute or two and may continue into subsequent sessions. By tracking their interpersonal process in this moment to moment way, therapists are much better able to recognize when their current interaction may have just failed the client's test. In particular, therapists should consider the possibility that they have unwittingly reenacted some aspect of the clients' conflict when

- clients begin talking about others rather than themselves;

- the therapeutic process becomes repetitive or intellectualized; or
- clients become compliant, lose their initiative, or cannot find meaningful material to discuss.

We have just seen that when therapists pass tests (for example, take pleasure in the client's success rather than feeling envious or threatened as others have been), clients usually acknowledge the therapist's achievement by acting stronger and making progress in treatment. In parallel, when a therapist repeatedly fails a specific test (for example, by brushing aside the potential validity of a client's criticism of the therapist, just as the client's parent always needed to be right), *the client may inform the therapist that something is awry by recalling other relationships in which the same conflict was being enacted.* That is, the client may begin to tell the therapist vignettes about other people the client has known, or perhaps characters in books or movies, who are enacting the same maladaptive relational pattern. *Therapists always want to consider the possibility that, whenever the client is talking about another relationship, the client may be using this as a metaphor, analogue, or encoded reference to describe what is going on between the therapist and the client as well* (Kahn, 1997). As a sustained intention throughout each session, therapists are listening for the relational themes that characterize the vignettes clients choose to relay. For example, if clients repetitively share stories in which trust is betrayed, control battles are being enacted, others are not responsible and need to be taken care of, and so forth, they are often expressing how these same patterns are currently being activated or played out with the therapist. Yalom (2003) describes this as listening with "rabbit ears"—always considering the possibility that whenever clients are talking about another person they may be making a covert reference to the therapist or what is going on between the therapist and client as well. When this analogue occurs, the therapist can use a process comment to inquire about this possibility, make this relational theme overt, and bring it back into the immediacy of the therapeutic relationship—where something can be done to address and resolve the misunderstanding or repair the rupture in the working alliance. For example:

THERAPIST: Both of these people you have been telling me about failed to hear something important that others were trying to tell them. I'm wondering if something like that could be going on between you and me right now?

CLIENT: I don't have a clue what you're talking about.

THERAPIST: I'm wondering about the possibility that, like with these other two people, you might have told me something earlier that was important to you, but I didn't hear it very well. Do any possibilities come to mind?

CLIENT: Well, not quite, but since you mention it, I guess sometimes I think you do try to make things seem better than they really are. Like you need me to be happier than I really am or want to make my problems less than what they really are.

THERAPIST: OK, it seems like I'm trying to make things look better than they really are. Like I don't want to see how bad it really is for you sometimes? I can sure see why you wouldn't like that, and I'm glad you're taking the risk to talk with me about this. Help me understand it better, though, I want to get it right—and change it. Are you saying I'm too optimistic, or maybe just looking at the world through rose-colored glasses? How would you say it?

CLIENT: (*laughing*) Well, I wouldn't say you put a yellow, happy-face sticker on my forehead when I come in the door, but sometimes you don't seem to want things to be as bad as they are.

THERAPIST: Was there a time today when I did that?

CLIENT: Uh huh. I told you I was "terrified" when I woke up and they were gone. And you said, "Yeah, I can see how anxious you felt." Maybe you prefer that I only feel anxious, but I didn't. I felt terrified—that's what I said and that's what I meant, but I don't think that's what you wanted me to feel.

THERAPIST: OK, I think I'm seeing better what you mean now. Let me try that again. You weren't anxious; it was much more than that—you were terrified. (*pause*) Am I capturing it accurately this time—not making it less serious than it really was?

CLIENT: Thank you, that does feel better. You know, my mother was always so uncomfortable whenever I was upset. I don't think she knew what to do when something was really wrong, so she just tried to make things OK even when they weren't—like when my dad died. . . .

To sum up, just as clients use eliciting behaviors to avoid and defend against their problems, they also use testing behavior to try to resolve conflicts in the therapeutic relationship. By tracking the therapeutic process in the ways suggested here, therapists can assess whether they are providing clients with the corrective responses they need. The therapist's consistent intention is to track their moment-to-moment, back-and-forth interactions and assess how the client seems to be responding to what the therapist has just said or done. Although this can seem overwhelming to new therapists in the beginning, it often becomes second nature with practice. As therapists learn to closely assess client reactions in this way, they will be able to choose more effectively how they want to respond next to what the client has just said.

Transference Reactions

The third way clients bring their problems with others into the therapeutic relationship is through *transference reactions*. Transference is one of the most important but misunderstood concepts in therapy. Problematically, the field has used the same term to convey different meanings and, over the years, two competing definitions have evolved. Traditionally, transference has referred to the thoughts, feelings, and perceptions that the client has toward the therapist. This psychoanalytic model of transference is based on the notion of the therapist as a blank screen. Because the therapist is neutral, whatever emotional reactions the client has toward the therapist are *distortions* that have been unrealistically or inaccurately transferred to the therapist from historical relationships. Especially during vulnerable or affect-laden moments, most clients do indeed have predispositions to misperceive the therapist and others along old familiar lines.

These persistent distortions of the other (as wonderful, withholding, demanding, judgmental, and so forth) reflect the client's schemas or relational templates. They repeatedly disrupt their close relationships and are almost certain to be evoked toward the therapist at times. When these types of traditional transference reactions occur, they systematically distort clients' perceptions of, and emotional responses to, the therapist.

In contrast, the contemporary definition of transference is broader and more useful. It refers to *all* of the feelings, perceptions, and reactions that the client has toward the therapist—both realistic and distorted. These reactions, which may be positive or negative, range on a continuum from fully accurate and reality-based perceptions of the therapist to distorted responses that are based on past relationships and have little to do with the reality of the therapist and how the therapist has actually responded. Thus, they do not necessarily represent pathology. Although transference distortions in the traditional sense certainly occur, the contemporary definition is emphasizing that many of the client's reactions toward the therapist are reality based and, most important, *almost all hold at least a kernel of truth*. For example, the client's affection for the therapist may be a genuine response to the very real caring the client feels in this relationship. Similarly, the client's anger or competitiveness may be based on the reality that the therapist sometimes has been late for sessions, emotionally unavailable, unable to grasp what is most important to the client, invested in the client making certain decisions or taking certain actions, and so forth. Therapists make a significant error when they too readily attribute the client's positive or negative feelings toward them to traditional transference distortions. Before doing so, we want to be nondefensive enough to be able to consider the potential validity or, more often, partial accuracy of those reactions. Maybe the client does have a valid complaint and something we have said or done has *contributed* to the client's misperception or misunderstanding. With this more contemporary approach to transference, there is much more receptiveness to the mutual interaction or reciprocal influence between therapist and client. That is, transference is not just a one-way street that only says something about the client but, more often, reflects something about what is going on between the client and therapist—something that both share responsibility for and participate together in co-constructing.

As we will see, therapists can use transference reactions both to conceptualize their clients better and to intervene and help resolve problems within the therapeutic relationship. More specifically, contemporary short-term dynamic therapies have encouraged an important change in how therapists work with transference reactions. In decades past, transference reactions were used as a springboard to explore historical relationships and the genesis of conflicts with parents and early caregivers. Now they are used instead *to understand and change the current interaction between the therapist and the client* (Mills, Bauer, & Miars, 1989; Levenson, 1995). As noted earlier, therapists will often recognize important connections between current problems and historical relationships, yet few clients will find these links meaningful. It will be more productive to address and resolve these distortions first in the real-life relationship with the therapist. In this more contemporary approach, the therapist is working with transference in the immediacy of the current relationship, rather than through the more distant or intellectualized approach of trying to interpret or lead the client back to historical relationships. In this approach to transference, the goal is to

1. help the client explore and discuss potential misperceptions and other personal reactions toward the therapist;
2. work collaboratively to explore how this repetitive pattern (that is, frequently experiencing the therapist as impatient, when the therapist is not) disrupts the therapeutic relationship, and how the same theme may be contributing to misunderstandings and problems with others; and
3. work together to sort through and change this repetitive pattern or faulty expectation first with the therapist, and then with others in their lives.

In most cases, clients' reactions toward the therapist accurately capture *aspects* of the interaction with the therapist. As emphasized, it is important for the therapist to acknowledge the potential validity of the client's perceptions—without realizing it, we have often said or done something that contributes to the misunderstanding or rupture. Therapists want to enter into a genuine dialogue with the client to clarify what each was thinking, feeling, and intending during any misunderstanding or rupture of the working alliance. If therapists are willing to explore *nondefensively* what they may have done that activated the client's transference distortion or contributed to the misperception, the client sees that the therapist doesn't always have to be right and is willing to risk a genuine relationship that goes two ways. Clients feel respected and empowered in this egalitarian relationship, and they often are able to become more authentic with the therapist and, in turn, with others in their lives.

Thus, in the interpersonal process approach, the therapist affirms the validity of many client reactions and tries to resolve actual distortions in the current relationship, rather than focusing back on the historical genesis of conflicts. Ironically, just after the therapist and client have addressed and resolved a significant misperception or misunderstanding between them, clients will often go on to make meaningful connections between formative and current relationships on their own. Before going further with how therapists can intervene with transference to help clients change, let's become better acquainted with the everyday expressions of transference.

Transference in Everyday Life   Transference reactions are commonplace. To a greater or lesser degree, they occur for all people at times—including therapists, of course. We all systematically misperceive others as a result of overgeneralizing previous learning to the present situation—especially when we feel vulnerable or distressed. For example, imagine a male university professor who stands in front of a large lecture hall about to convene the first class meeting of Psychology 101. Before he has even begun to lecture, many students will be transferring expectations from significant male authority figures in their own lives onto the professor. Thus, in the first row of the lecture hall, 18-year-old Mary looks up to the professor admiringly. Just as she still somewhat idealizes her father, Mary respects the middle-aged professor, even though she actually knows little about him.

In contrast, Joe looks down on the professor disparagingly from high up in the last row of seats. Joe anticipates that the professor does not know as much as he thinks he does and that he will not be prepared very well for class. Joe expects that, just like his father who always had to be right and on top, this professor will probably try to "blow a lot of smoke" and get away with not knowing as much as he thinks he does in class. Joe has already decided that this know-it-all professor is not going to get away with much if Joe can help it.

Let's hope that the professor is aware that people tend to transfer expectations onto authority figures who hold power and onto those who have high public visibility (movie *stars*, sports *heroes*, and political *leaders*). If the professor is not sensitized to the pervasiveness of transference reactions, his self-confidence as a teacher will take a roller-coaster ride. One day, Mary will follow him out of class and tell him what an interesting lecture he gave; the next day, Joe will drop his books loudly when the professor starts to speak and demand that he defend half of what he says.

Transference reactions such as Mary's and Joe's are everyday occurrences. The more conflicted or troubled one is, however, the more one distorts the current reality by inaccurately displacing reactions from the past onto the present. For example, Joe and Mary have strong initial reactions to the professor that are based on little real-life experience with him. If Joe and Mary function fairly well psychologically, their cognitive schemas will be flexible and varied, and they will become more realistic in their assessment of the professor as they have more exposure to his teaching. In contrast, the more thoroughly Mary idealizes her father and the more pervasively Joe disparages his father, the more likely it is that they will not be able to perceive the realistic strengths and limitations of the professor. Instead, their schemas will be more fixed and self-fulfilling, and they will maintain their initial expectations despite the fact that they are inaccurate or "overdetermined."

Transference reactions are also more likely in emotion-laden situations. For example, transference reactions are especially pronounced when two people are falling in love. Many married couples recall their courtship as the happiest time in their lives. During this time, each member may systematically misperceive the partner as someone who can fulfill unmet developmental needs and make one feel love-worthy or capable when one has not felt this way before. This romantic period is usually short-lived, however. The transference projections soon break down—they both realize that the other person cannot fulfill their unmet needs, and their old schemas are reinstated. If the unmet needs and accompanying transference distortions are strong, the result is often dashed dreams or angry feelings of betrayal. In contrast, if the transference distortions of the partner have not been extreme, they can readily be shed, and a more realistic and enduring relationship can develop. Relinquishing these transference distortions of the new spouse is one of the major psychological tasks of courtship and early marriage.

In treatment, the therapist can systematically utilize the client's transference reactions to better understand the client's problems with others. One of the most effective ways to discern the client's schemas and faulty expectations is to check in regularly and ask about how the client might be perceiving or reacting to the therapist or what is going on between them right now. Routinely, aspects of the same patterns and themes that cause problems with others will be activated toward the therapist or played out in the therapeutic relationship. If the therapist tracks these transference reactions and successfully engages the client in a dialogue about them, they clarify faulty schemas to the client and help change the problematic relational patterns that are being reenacted with the therapist. As this important relearning occurs with the therapist, the client is empowered to begin changing this pattern with others.

Using Transference Reactions to Conceptualize Clients   Some therapists are uninterested in the client's transference reactions because they do not see them occurring in therapy. Therapists will not see the client's transference reactions, and the valuable information they provide about the client's schemas and expectations, unless they actively intervene to draw them out. How can the therapist highlight the client's transference reactions? Let's look at two ways of making the client's transference distortions overt so that the client can join with the therapist collaboratively to understand and change them.

First, *the therapist is looking for tactful but direct ways to inquire about the client's feelings and reactions toward the therapist.* As we saw in Chapter 3 when discussing resistance, it is important for both therapist and client to learn about the client's reactions to being a client and seeking help. Now, further along in treatment, it is also important to learn about the client's perceptions and reactions toward the therapist. This is especially important when

- conflicts or misunderstandings arise between the therapist and the client, and
- the client has just shared strong emotions or self-disclosed.

During these two types of sensitive interactions, the client is most likely to misperceive the therapist as responding in the same problematic or unwanted way that others have in the past. For example, if clients disclose something they feel embarrassed about or enter a strong feeling, they will often be convinced that the therapist feels burdened or depressed by their sad feelings. Or, they may inaccurately believe that the therapist doesn't respect them anymore now that they have disclosed this information about themselves. Clients believe the therapist is responding in this unwanted way because this is how important others have responded in the past—even though the therapist doesn't think or feel anything like this and has not communicated disapproval in any way. It is difficult for most graduate student therapists to grasp this in the beginning. Because the interaction or sharing seemed positive to them, new therapists find it very hard to believe that it could have been misperceived in some problematic way by the client. However, if the therapist doesn't check in with the client and ask how it has been for them to experience this significant feeling or disclose this important issue with the therapist, clients routinely miss the next session or return but keep the next session(s) superficial. Therapists need to clarify the potential transference distortion with the client by making overt that they are not thinking or feeling the same unwanted things that others have in the past:

THERAPIST: That was a lot to share—there's so much sadness there.

CLIENT: Yeah, I guess I was crying pretty hard.

THERAPIST: Yes, and I'm honored that you can share such important feelings with me. Can I check in with you for a minute about this—about how it feels right now to have shared so much and cried so hard with me?

CLIENT: Well, you're always very nice, but this has to be pretty tiring for you, too.

THERAPIST: Tiring? No, it's not tiring at all. Actually, it's enlivening for me—I feel close to you right now—that we've shared something that's meaningful for both of us.

CLIENT: Really? I don't wear you out with all this?

THERAPIST: No, I don't feel worn out or tired in any way. Actually, it's just the opposite; I like being able to respond to you like this.

CLIENT: Boy, this is really different for me, that it's okay for you if I feel sad or cry sometimes. I think my mom needed me to be happy all the time, so I acted that way, even though I wasn't. When I look at pictures of myself, other people often say I look pretty or have a nice smile, but I see a sad face. . . .

In this way, *the therapist wants to debrief with clients after sensitive interactions, and check whether the client is distorting or accurately perceiving the therapist.* However, at other, less significant times as well, therapists also want to check in with clients and explore their reactions toward the therapist. Routinely, the following type of questions will open some of the most important, and unexpected, information that clients share during treatment—revealing perceptions and reactions that are central to the client's problems but that would not become accessible if the therapist did not inquire in these ways:

• As you were driving to our session today, how did you feel about coming to see me?
• When you find yourself thinking about therapy; what kind of thoughts do you have about me and our work together?
• I'm wondering what you might be thinking is going on for me right now as we talk about this sensitive issue?
• I wanted to check in on you and me and see how you think we're doing together. What's going well in our work together, and what isn't?

Therapists can also explore what the client *projects* onto the therapist—what the client believes the therapist is thinking, expecting, or feeling toward the client at that moment:

• What do you think I am feeling toward you as you tell me that?
• How do you think I am going to respond if you do that?
• What do you think I expect you to do in that situation?

Oftentimes, it is uncomfortable for the beginning therapist to break long-held familial and cultural norms and invite clients to talk about "you and me" so directly. Doing so, however, often brings up key issues and concerns that are central to the client's problems and the therapeutic relationship, even though the therapist had no idea this was going on for the client. Although it may not seem this way in the beginning, therapists can learn to become more safely forthright with their clients than they have been able to be with others in their social life.

The distinction between everyday social interaction and clinical intervention is significant here. Because many therapists have a moving-toward interpersonal style, it is especially difficult for them to become more forthright in this way and risk the client's disapproval. By role-playing and rehearsing this interpersonal skill with classmates and supervisors, and by viewing themselves on videotape, new therapists can practice being more direct or straightforward with clients without crossing the line and becoming demanding or intrusive—which we don't want or need to do. If trainees practice this intervention with colleagues and take the risk to pose such questions, clients' transference reactions will be made overt, and the therapist will be taken right to the core of the client's concerns. The new issues that are revealed are often rich in meaning and routinely surprise the therapist as the client replies:

- I think you are feeling disappointed in me.
- I think you'll act nice, but really wish I would stop coming to see you.
- You're probably thinking that it's wrong of me to even be thinking about getting a divorce—that it's selfish for me to do that to my children.

As these schemas and misperceptions are made overt, therapists can clarify their actual responses to the client, which in itself often provides the client with a reparative experience. When transference reactions are revealed in this way, therapists have the opportunity to respond to the client with an immediacy and authenticity that is not often found in other relationships. For example, the therapist might clarify the transference distortions just listed by replying as follows:

- No, I'm not feeling disappointed in you at all. Actually, I'm very pleased with how well you have been doing. I'm wondering where that feeling of "disappointment" comes from though. Did I do something today that made you think I was disappointed in you, or does this come up in other relationships too?
- I want you to stop coming? Oh, no, that's not going on for me at all—I enjoy working with you very much. But I'm so glad you're telling me this—so we can work it through. Do you know when you started thinking this?
- No, I wasn't thinking that you are being "selfish." I was thinking about how hard you are trying to do the right thing—to find a balance between your needs and what you want to do for others. Tell me more about "selfish." Is that an important or familiar word for you?

To emphasize, most beginning therapists find it difficult to consider or explore the possibility that their client is having significant feelings and faulty perceptions toward them that are not based on the reality of how they have responded to the client. With experience, they often become more familiar with clients' distortions and may be better prepared to explore and clarify them.

In contrast, beginning therapists usually find it hard to accept that their clients have deeply held but unspoken beliefs that the therapist really doesn't like them, feels overwhelmed or bored by them, believes they are childish, demanding, or foolish, and so forth. Even though the therapist may never have responded to the client in any of these problematic ways, the discerning therapist still works under the assumption that such client reactions are likely to be present yet covert. If these distortions are not made overt and rectified, the potential for change is limited. For some trainees, this may be the critical point: The beginning therapist can accept intellectually that the client may hold such faulty expectations of others. However, it is often too threatening—evoking too much guilt over making a mistake or having done something wrong—to explore the possibility that the client may have such unwanted reactions toward the therapist.

To review, exploring clients' reactions toward the therapist reveals three important types of information. First, just as adopting an internal focus will reveal issues that would not have become accessible otherwise (Chapter 4), exploring clients' reactions toward the therapist will also uncover important new issues. As we have seen, the faulty perceptions and expectations that are revealed are often central to the clients' presenting problems and now can be addressed and resolved in the immediacy of the real-life relationship between the therapist and client. Several examples of this essential intervention have been provided here, and more extensive illustrations of how this change with the therapist can be generalized to others beyond the therapy setting will be given in the chapters that follow.

Second, exploring transference will also highlight how aspects of the same problems that clients are having with others—and that they originally learned in formative family relationships—are being reenacted with the therapist. Unless the clients' reactions toward the therapist are directly invited, these distortions and reenactments are not likely to be identified or resolved. Thus, transference reactions reveal clients' key concerns and, as we will explore further, also provide an opportunity to resolve them—first, in the here-and-now, real-life interaction with the therapist, and then with others.

Third, as paradoxical as it sounds at first, transference reactions ultimately provide evidence to clients that their feelings and perceptions really are trustworthy. We're back to our theme of self-efficacy. Because clients' expectations and emotional reactions may not accurately fit the current situation in which they are activated, they often do not make sense at first to clients or others. As a result, many clients experience their emotional responses toward spouses, children, employers, and others as "irrational," "stupid," or "crazy." Consequently, they do not trust the validity of their own experience and, as a result, their self-efficacy is undermined. This also occurs in the therapeutic relationship when clients have emotional reactions toward the therapist that seem inappropriate or exaggerated. However, clients' emotional reactions are not irrational. Though they may not fully fit the present circumstances, they do make sense when they can be understood within their original context. Real-life events have legitimately caused clients to feel and expect what they do. When therapists help clients trace feelings and expectations back to where they learned them, *they will prove to be an appropriate reaction to someone else in another time and place* (by asking, for example, "When is the first time you can remember feeling this way? Who were you with and what was happening?"). Once the therapist and client first have been able to clarify and resolve the transference distortion in their current interaction, this developmental link can often be made readily. In contrast, the developmental context will not come alive or be meaningful to the client if it is not first resolved in the current interaction with the therapist. In this way, transference reactions, like resistance, ultimately reveal that the client's feelings make sense.

To sum up, clients do transfer emotional reactions from past formative relationships onto the current relationship with the therapist. However, therapists begin by clarifying these misperceptions in the current, here-and-now relationship with the therapist. As they are sorted through in the therapeutic relationship, the client can use this real-life experience of change to begin changing these same patterns and faulty expectations with others:

CLIENT: He doesn't want to listen to me, maybe I am just too much—too needy and too demanding.

THERAPIST: Have you ever felt that way with me—that I find you "too needy and demanding"?

CLIENT: Well, sure, I think that a lot, but you can't say anything because you're a therapist. You know, you have to be nice—that's your job.

THERAPIST: Well, I don't think that being nice is my job—but giving you honest feedback is. If I thought you were being demanding or doing something else that was causing problems in your life, I'd tell you. It wouldn't be easy for me to say that, but I would take the risk and try to talk with you about it in a constructive way. So, no, I don't find you needy or demanding with me. I like it when you speak up for yourself—I think it's a strong and healthy part of you.

CLIENT: I'm glad you're saying that. I think I go along with other people too much, and don't say what I want, because I'm so afraid of seeming, you know, "demanding." I guess it's easier for me to say what I want in here with you than it is in the rest of my life.

THERAPIST: I'm glad it's easier with me—but what would you like to be able to say the next time you feel like your boyfriend isn't listening to you or taking you seriously?

CLIENT: I don't know—I'm just not good at this, but I guess I do have a right to speak up.

THERAPIST: Yes, you really do. Here's a suggestion: What if, next time, you just said to your boyfriend, "I don't think you're really listening to me right now"? That's a good way to see if he can meet your needs in this reasonable way or if this relationship can't give you what you want.

CLIENT: You know, I can say that to him. And you're right, if he can't really listen to me when I ask him to, I probably shouldn't be thinking about moving in with him. It's just not a good match.

THERAPIST: Yeah, if he can respond when you ask him for what you need, green light; if he doesn't hear you and can't change and listen when you ask for it, red light.

CLIENT: This makes sense. I'm going to start checking this out with him.

This type of transference clarification gives clients a real-life experience of change; they learn that their inaccurate expectations or problematic schemas do not fit in this relationship. As they have a real-life experience of change with the therapist, they find that at least some relationships, some times, can be another way. This experiential relearning, which comes from clarifying the misperception with the therapist, allows many clients to make two important steps. First, they often are able to make meaningful links to past or formative relationships that shed light on current patterns and problems. Second, clarifying the transference distortion and finding that the therapist is not responding in the familiar but unwanted way, clients are empowered to try out these new ways of responding with others that have succeeded with the therapist. Together, the therapist and client then begin the process of sorting through who responds well to these changes (for example, the client's husband) and who in their life doesn't (the client's father).

Optimum Interpersonal Balance

In this chapter, we are trying to better understand clients' current interpersonal functioning. In particular, we are exploring different ways that clients play out with the therapist the same relational patterns and themes that cause problems in other relationships. However, as we consider the client's eliciting maneuvers, testing behavior, and transference reactions, it draws us further into the closely related issue of countertransference. By striving to maintain an "optimal interpersonal distance" with clients, therapists can manage their own "client-induced" and "therapist-induced" countertransference reactions more effectively and be better prepared to provide clients with a corrective response that does not reenact the same, problematic relational scenarios. Let's return to the separateness–relatedness dialectic that was discussed in Chapter 6 and discuss two common types of countertransference—enmeshment and disengagement.

To develop a relationship that is more capable of producing change, therapists aim to establish and maintain an effective degree of involvement with their clients. On the one hand, if the therapist becomes overly invested in the client's choices or change, this will be an unwanted reenactment for many clients (enmeshment). Especially for clients who grew up with too much parental control or intrusiveness, or too little support for pursuing their own interests and goals, their fear and expectation that they will again be controlled by the therapist (and others in their life they get close to) will be confirmed. And, even for different clients who do not share this developmental background, it will not be safe for them to risk letting the therapist influence them or help them change. On the other hand, if the therapist is too distant or removed, a corrective emotional experience cannot occur, either (disengagement). Although this will be especially problematic for clients who had authoritarian, rejecting, preoccupied or otherwise distant parents, it will also diminish the potential for treatment gains for most clients because the relationship is too insignificant to effect change. That is, even if the therapist responds in new, healthier ways than clients have found in the past, *it doesn't have much impact on the client because the relationship does not hold much meaning or relevance*. Thus, throughout the course of treatment, the therapist's intention is to monitor the balance of separateness versus relatedness in the therapeutic relationship and maintain an optimum interpersonal balance.

At times, all therapists will have difficulty maintaining an optimum interpersonal balance with some clients—whether because of client-induced or therapist-induced countertransference. To manage these pulls, the therapist's aim is to be both a genuine participant in the relationship and an objective observer of it. Sullivan (1968) captures this double role in his enduring term *participant-observer*. However, combining real empathic engagement with an "observing ego" or objective viewpoint is no easy task for beginning therapists. As in the separateness–relatedness dialectic, it is a challenging paradox indeed to be close (provide empathic understanding) and separate (maintain clear boundaries) at the same time (Jordan, Kaplan, Miller, Striver, & Surrey, 1991).

Without question, this balancing act is one of the most maddening challenges facing most beginning therapists. Perhaps the best analogy is being a 16-year-old student-driver in a car with manual (stick shift) transmission. Simultaneously, the new driver is trying to press down on the clutch with her left foot, brake or accelerate with her right foot, steer the car with her left hand, and shift the gearshift with her right hand—all the while she is trying to look ahead down the road, see where she is going and attend to what is rapidly approaching, and listen to instructions from the driving instructor next to her! In just a short time, of course, the new driver will be able to do this almost automatically—and even change the radio station and carry on a meaningful conversation at the same time. Similarly, new therapists need to be patient with themselves. With practice and experience, it will become much easier to maintain this self-reflective capacity and simultaneously be both a participant in the therapeutic relationship and an observer of what is going on in it.

Let's look more closely at both sides of this particular form of countertransference—enmeshment and disengagement—and link it to our two differing types, client-induced and therapist-induced.

Enmeshment

At times, all therapists will become overidentified—enmeshed in family systems terms—with certain clients. For example, if this client is skillful at getting therapists and others to rescue, direct, or take responsibility for them, this is client-induced countertransference. A supportive supervisor can help the therapist recognize that the client is making the therapist feel responsible for them—just as they do with others in their life. It is relatively easy to begin talking with the client about this relational pattern and changing how it is potentially being reenacted in the therapeutic relationship. However, this becomes a more complicated problem if the client's attempts to make the therapist take responsibility for their decisions taps into the therapists' own personal issues. In this type of therapist-induced countertransference, the therapist loses his objectivity and neutrality with the client because of his own personal concerns. For example, the enmeshed therapist becomes invested in the client staying married as the therapist is feeling how hurt he had been by his own divorce, his parent's divorce, or other losses that he has had difficulty coming to terms with.

Thus, certain therapists, because of their own history or personal dynamics, will have a tendency to repeatedly become overinvolved with many of their clients. How can therapists recognize when this traditional or therapist-induced countertransference is occurring? Whenever therapists find that they are angry or critical of a client for not changing, they are probably overly invested in the client and trying to manage a personal issue through the client. Oftentimes, therapists are also enmeshed when they have dreams about a client, often think about a client outside the therapy session, feel depressed (rather than appropriately concerned) when a client is not changing, or become envious or elated over positive changes in the client's life. Usually, enmeshment is also occurring when therapists see the client as being "just like me" and cannot recognize the many differences that actually exist. This overinvolvement is different from genuine care and concern for the client, which is appropriate and necessary. When enmeshment occurs, therapists usually need the client to change in order to meet their own needs—for example, to shore up their own feelings of adequacy or to manage personal problems or feelings of their own that are similar to the client's. Recalling the concept of change from the inside out or an internal locus for change, the enmeshed therapist is trying to solve an internal problem of his own, externally, through the client.

Treatment progress and change usually slow to a halt at the point when therapists become overinvolved with a client. When they become enmeshed, therapists lose sight of the fact that clients have their own subjective worldview, shaped by many factors that actually differ from those of the therapist. They tend to become controlling of the client and to project their own problems and solutions onto the client. As we have seen, this response often recapitulates the client's developmental predicament in the therapeutic relationship. For example, many clients have grown up in families with overly controlling or intrusive parents. When the therapist's own experiential range is too limited to encourage the client's own autonomy, the therapist is responding in the same problematic way as caregivers did in formative relationships. This reenactment will prevent clients from being able to resolve their conflicts around intimacy and control in current relationships. Thus, one reason for paying attention to maintaining an optimal interpersonal balance, and always being receptive to the possibility of enmeshment as a form of client-induced or therapist-induced countertransference, is to avoid treating clients in the same ways that originally caused their problems.

Furthermore, when therapists become too close or overidentified, they lose their ability to be a participant-observer. That is, therapists cannot think objectively about the client or sufficiently step out of the interpersonal process they are enacting with their clients to consider what their interaction may mean. In parallel, their heightened emotional reactivity to the client will cloud their understanding and diminish their ability to provide a holding environment and remain a "steady presence" to contain the client's feelings and distress. And, instead of accurate empathy, enmeshed therapists will offer clients only a global, undifferentiated sympathy that does not consistently capture the essential or precise meaning that this particular experience holds for this client. Most important of all, perhaps, therapists will fail to discriminate the real differences between themselves and the client that do exist and will begin to see the client as being just like themselves. Although this overidentification may feel like closeness or caring and may be reassuring to clients at first, enduring change does not usually occur in an enmeshed relationship.

Disengagement

Just as most therapists will become enmeshed with clients at times, most therapists will hold themselves too far away from certain clients at times. Per client-induced countertransference, therapists will often observe themselves initially distancing from moving-against and other difficult clients who are angry, critical, or manipulative. The therapist is distancing herself just as others in the clients' lives often distance from them. And, rather than continue to enact this scenario, once therapists identify this interpersonal pattern they can begin to make process comments or offer interpersonal feedback to begin changing this maladaptive pattern.

Per therapist-induced countertransference, therapists also tend to disengage or distance from the client when the client arouses the therapists' own personal conflicts. For example, the therapist may disengage from the client as the client talks about the impending death of her aging parent. In this situation, the therapist may be activated by her own sadness about losses in her own life, by unresolved feelings about not being able to say good-bye to her own deceased parent, and so forth. The purpose of supervision is to help therapists recognize their own countertransference reaction, find a more appropriate way to manage their own personal feelings (for example, talk with a supervisor or colleague about them), and empower the therapist to be able to engage the client again with their own personal boundaries better in place.

Most important, however, some therapists will consistently be too removed from their clients because of their own enduring conflicts and coping styles. For many clients, this distance renders treatment ineffective. When therapists cannot be emotionally available or responsive to their clients, the working alliance is lost or does not ever form. Treatment loses its intensity and often becomes intellectualized when the therapist cannot be a "participant" in the real relationship with the client. Therapists need to be able to risk effecting and being effected by the client—nothing ventured, nothing gained.

Therapists also lose their intuition and creativity when they distance themselves from the client. When this detachment occurs, the client does not have the supportive relationship or holding environment necessary to explore affect-laden problems. Counseling is more than an intellectual discourse, and little enduring change will occur in an emotionally disengaged relationship. Without the impetus or real meaning provided by genuine personal involvement, a corrective emotional experience will not occur.

Finally, a therapeutic relationship in which the therapist is too distant or uninvolved will also recapitulate many clients' developmental problems. For example, for clients whose aloof, dismissive, self-centered, or authoritarian parents could not meet their childhood needs for emotional support, a distant therapist might be frustrating but ultimately safe. Their anxieties over their insecure attachments—and their conflicts associated with being close to others—will not be activated in this therapeutic relationship. Unfortunately, just as the clients' conflicts are not aroused in a disengaged relationship, the opportunity to resolve these conflicts is lost as well. Thus, when therapists defend themselves against their own anxieties that are aroused by an authentic relationship and genuine emotional involvement with clients, many clients will not be able to change. *Novice therapists who are consistently unwilling to risk genuine involvement with clients are less likely to be effective than those who are willing to risk a real relationship but tend to become overinvolved at times.*

Optimum Middle Ground of Effective Involvement

To be an effective participant-observer, the therapist is trying to maintain an optimum middle ground of involvement. For two distinct reasons, maintaining this middle ground of involvement is not easy. First, let's explore the therapist's side of this—therapist-induced countertransference reactions that get in the way. Second, we will explore client-induced countertransference further and see what clients sometimes do to make it harder for therapists to maintain an optimum interpersonal distance.

Therapist-Induced Countertransference Therapists are continually confronted with personal issues in their own lives by the sensitive, affect-laden material that clients present. For every therapist, it's just part of this work—it comes with the territory. As a basic element of their training, all therapists are encouraged to pay attention to their own countertransference propensities and become more aware of how they tend to respond when clients activate their own anxieties or personal issues. For example, when the client talks about an issue that is sensitive or difficult for you, what is your initial response tendency? Some therapists initially withdraw from the client and distance themselves by intellectualizing—for example, by interpreting the client's affect before the client fully experiences it or ineffectively "explaining" or interpreting what something means. Others become controlling and overly prescriptive, telling clients what they should do in a particular situation, usually without really listening to what the client thinks is wrong. Whereas some therapists begin to talk too much, many others become too quiet and unresponsive, leaving clients feeling that they aren't getting any help or don't have a real partner with them in the counseling room. With only a little effort and some nondefensiveness, however, therapists can begin to identify their own countertransference propensities and learn to hold them in check.

Becoming familiar with our own response tendencies becomes even more important with moving-against and other difficult clients who respond to the therapist with "negativity." As emphasized before, this is one of the most important but neglected issues in clinical training. Through eliciting maneuvers, testing behavior, and transference distortions, clients will respond in negative ways to therapists—critically and competitively, with anger or blame, contemptuously or manipulatively, and so forth. Although many clients are going to respond in these ways at times, most therapists have not been given the practice and preparation they need in order to respond nondefensively to these difficult reactions that are so unwanted by most therapists—whether novice or experienced. They have not role-played or rehearsed alternative responses that offer more neutrality and therefore are less likely to reenact the familiar but problematic responses that this client's negativity usually elicits from others. Too often, that is, new therapists have not been encouraged to anticipate and discuss how their own personal issues and response tendencies will be activated by different kinds of negativity from the client—such as criticism, power struggles, sexual innuendo, and so forth.

Whereas much is often said about "nice" things such as empathy, genuineness and warmth, or a working alliance, less training is usually provided to help beginning therapists respond to a demanding, passive aggressive or demeaning client (the one who comments, for example, "Have you been gaining a little weight lately?"). In this regard, researchers find that therapists do not detect or observe client's negative reactions toward the therapist very well—therapists tend to deny or avoid the unwanted remark. It seems that it is much easier for therapists to miss or gloss over negative/hostile responses from clients than positive responses (Hill, Thompson & Corbett, 1992). However, we don't want to tune out or miss these important messages from the client. Why? The negativity coming toward us is often at the heart of what is causing problems in the client's relationships with others. We need to make the most of the time we have with clients, and addressing clients' problems with others—especially as they are occurring right now with us—is usually the most powerful way to do this.

Thus, our intention is to approach negativity nondefensively, rather than avoid it as if something significant was not being said, even though this may be hard for us to do. Our goal is to work with it in an up-front way that helps clients learn about the problematic impact they are having on others and learn how to deal with negative feelings and interpersonal conflict in healthier ways:

CLIENT: Your office looks a little tired. This furniture and brown color went out back in the 1970s.

THERAPIST: (calmly) Uh huh. Things look a little out-of-date to you in here.

CLIENT: Yeah, you should do something to change it.

THERAPIST: No, I don't think I want to make any changes; I'm comfortable here. But maybe it would be helpful for us to talk about this for a minute.

I'm wondering how others usually respond to you when you tell them what's wrong like this or what they should do?

CLIENT: Well, I don't know, I guess they usually do it.

THERAPIST: I can see that—you do have good taste and express strong opinions in an authoritative voice. But I'm wondering how it affects your relationships. I don't think most people would like too much of this.

CLIENT: No, I don't think people at work or at my church like me very much—I guess they would say I'm "bossy."

THERAPIST: Yeah, I can imagine some people feeling intimidated and wanting to get away from you, and others feeling resentful and wanting to argue with you. What it's like for you to be seen as "bossy" in this way—how does that feel?

CLIENT: Well, when you say it like that, it makes me feel like crying—I'm taking over just like my mother always did. Sometimes I wonder if I'm going to ruin my marriage—just like she ruined hers. . . .

This therapist was effective because she remained nondefensive in the face of the client's criticism and control. She *approached* this sensitive issue—she named it and asked about it—rather than going on and pretending as if something important wasn't transpiring between them. In contrast, therapists often respond ineffectively to such negativity with their own personal or countertransference reactions:

CLIENT: Your office looks a little tired. This furniture and brown color went out back in the 1970s.

THERAPIST: (*moving toward and complying*) Oh, I'm sorry, maybe I should fix it up a little.
          (*moving away and avoiding*) So, what do you think we should work on today?
          (*moving against and chiding*) Maybe you should spend more time thinking about your own problems and less about everybody else's.

With practice, and by observing instructors model a range of effective responses, therapists can learn better how to *remain nondefensive and approach the negativity*—the key to responding effectively. This approach empowers the therapist to intervene effectively by providing interpersonal feedback, making process comments, or simply adopting a neutral stance to question, highlight, or explore this negativity. Although still poorly integrated into clinical training, researchers have been trying for decades to tell practitioners that negativity will occur with many of their clients and that therapists need to learn how to deal with it more effectively. It is an important part of the treatment process that can be especially difficult for new therapists. Unfortunately, even experienced, highly trained therapists tend to respond with the same fight-or-flight reactions that clients elicit from others and that cause problems in their lives (Strupp & Hadley, 1979; Binder & Strupp, 1997; Safran & Muran, 2003).

To help therapists better maintain an optimum interpersonal distance with clients, let's look further at what therapists can do to manage their countertransference reactions to clients' negativity. One useful exercise for therapists is to write down their initial reaction propensities in different anxiety-arousing situations—for instance, what you are most likely to say or do first when a client is demeaning or critical of you. Or, with a partner, trainees can role-play their responses to different clients in difficult situations. For example, one person can role-play being a competitive, intimidating, or insulting client, perhaps someone who is questioning whether the therapist is really capable enough to be of help. Taking turns, the other trainee practices responding to this provocative client. In small discussion groups of three or four, trainees can provide each other with feedback, or watch themselves being replayed on videotape, and share their observations of the other therapists' responses while role-playing these challenging situations. With practice and guidance from supervisors, new therapists can learn to approach these unwanted, negative situations and turn them into constructive relearning experiences for clients.

Client-Induced Countertransference: Eliciting Overinvolvement and Under-involvement   In addition to therapist-induced countertransference reactions, the balance between overinvolvement and underinvolvement can be difficult to maintain because *clients often try to move the therapist along this continuum of involvement as part of their interpersonal coping strategy.* This is client-induced countertransference. When the therapist has been responding effectively, some clients may become threatened by the anxiety-arousing material that is coming up or they are beginning to address. To defend against the difficult emotions that are being aroused by the therapist's effectiveness, some clients may try to create either enmeshment or disengagement in the therapeutic relationship. For example, some clients may try to bore the therapist by talking about issues they are not really concerned about. In contrast, others may try to elicit too much involvement from the therapist by exaggerating their distress and escalating their demands for help. In these ways, the client's eliciting behaviors may press the therapist away from an effective middle ground of involvement.

If therapists can maintain this middle ground of involvement—remain both a participant and an observer in the counseling relationship, therapists can provide a more even therapy with no stormy ups and downs or lengthy impasses. It is also safer for the client to be vulnerable and trust when the therapist is dependably available without swinging to either side and becoming over-reactive or unresponsive. By offering this steady presence in the face of the client's eliciting maneuvers, the therapist is providing a "secure base" for the client. More specifically, when therapists maintain an optimum interpersonal balance, the client does not have to be concerned about disappointing an over-involved therapist or eliciting the responsiveness or emotional accessibility of a distant technician. The attachment researchers have appreciated best the importance of the caregiver's *consistent* response in developing security and trust.

The essential balance is for therapists to maintain their own personal boundaries (separateness) while still being responsive to the client (relatedness). That is, therapists are striving to enter the client's subjective worldview—to be fully present and connected—yet still maintain appropriate self-boundaries and self/other differentiation. Regardless of the therapist's theoretical orientation, treatment is more likely to succeed when the therapist can maintain this optimum interpersonal involvement. Conversely, treatment usually reaches an impasse—and ends before the client's work is complete—when the therapist and client become enmeshed or do not emotionally connect sufficiently to establish a collaborative working alliance.

To help therapists recognize the interpersonal process they are enacting with their clients and assess their balance of involvement with clients, therapists can consider the following questions after every session:

1. What are my feelings and personal reactions toward this client?
2. How might my reactions parallel those of significant others in the client's life?
3. Are my feelings typical of my reactions to others, or are they more confined to this particular client?

By writing answers to these questions after each session, therapists ensure that they are attending to the process dimension. Asking themselves such questions will also help therapists distinguish their own personal reactions toward the client (therapist-induced countertransference) from the client's eliciting behaviors and interpersonal coping style that affect many people (client-induced countertransference). Building upon these three questions, therapists are encouraged to utilize the guidelines for writing process notes in Appendix A after each session.

As we have already noted, all therapists will sometimes become overinvolved with a client. Indeed, many theorists have described the therapeutic process as therapists' ability to become immersed in clients' modes of relating and then to work their way out (Gill & Muslin, 1976; Levenson, 1982; Mitchell, 1988). However, new therapists are especially prone to become overidentified with their clients and to misperceive their clients' problems as being the same as their own. This is one of the most common ways in which clients' maladaptive relational patterns come to be reenacted in the therapeutic process. New therapists often feel discouraged when their supervisors accurately point out that, here again, they are reenacting aspects of the client's problems in their interpersonal process. New therapists need to be patient with themselves and accept that it is difficult to maintain this effective middle ground of involvement that safeguards against reenacting the client's maladaptive relational patterns. By examining their own experiential range, anticipating their own countertransference propensities, and tracking the separateness–relatedness dialectic in the therapeutic relationship, new therapists will become increasingly able to maintain an optimum interpersonal balance.

When therapists realize that they have become over- or underinvolved with a client and find that they cannot realign the relationship on their own, it is best to consult with a colleague or supervisor. One of the most productive uses of supervision is to help therapists regain control of their own emotional reactions so that they can reestablish an effective degree of involvement with a client who has successfully pushed them away or drawn them in too much. Maintaining this optimum interpersonal distance is one of the best ways to keep from reenacting the familiar but problematic scenarios that occur with others.

Enduring Problems Tend to Be Paradoxical and Ambivalent

The Two Sides of Clients' Conflicts

Therapists can better understand their clients' problems when they recognize the paradoxical and ambivalent nature of many conflicts. Usually, there are two sides to the client's central conflict, and therapists need to respond to both sides of their dilemma before problems can be resolved. Often, because of their own countertransference, therapists may fail to recognize both sides of their client's conflict and only respond to one side of the issue. Therapists are more helpful to their clients when they can identify the two opposing feelings or competing beliefs that make up the double-binding conflict.

Colloquial expressions convey the two-sided nature of conflict. We often say that someone is "stuck between a rock and a hard place," that someone has grabbed "both horns of the dilemma," or that a difficult relationship is "like a double-edged sword." These expressions capture the push-pull nature of conflict—that you are "damned if you do and damned if you don't."

In parallel, therapists working within different theoretical perspectives have used differing terms to describe the same concept. Long ago, for example, the behavioral therapists Dollard and Miller (1950) highlighted how clients were often struggling with "approach-approach" or "avoid-avoid" conflicts. Communication theorists and early family therapists highlighted double-bind interactions and paradoxical communications (Bateson, 1972). Contemporary attachment researchers reveal the plight of children who are attached to a caretaker who also frightens or hurts them (Hesse & Main, 1999). To illustrate this "Category D" attachment style, imagine a young child who is growing up with a highly inconsistent or unpredictable caregiver. This attachment figure may have borderline or narcissistic features, may struggle with bipolar depression or be preoccupied with strong unresolved grief reactions, or may be dissociative at times as a result of the caregiver's own physical or sexual abuse in childhood. In a truly maddening paradox, this child receives genuine comfort and understanding from this caregiver at times. At other times, however, this same benevolent caretaker who meets their emotional needs is also the source of terror and shame. This occurs when the caretaker acts in strange or disturbing ways that the child cannot understand (for example, observing a drug-induced parent talking to themselves with exaggerated hand and facial expressions), is threatening or enraged toward the child ("I wish you were never born! You're ruining my life. I hate you"), and so forth. This child has an unsolvable dilemma: The source of comfort is simultaneously the source of danger. Aptly calling it "fear without resolution," the person who helps me is also the person who hurts me. To the child, seemingly, there is no solution to this double bind—"I can't walk away from this caregiver who I depend on yet, at the same time, I can't be close and need this person who frightens or hurts me so much." In sum, theorists from many theoretical perspectives have tried to help therapists recognize and respond to the double-binding, paradoxical, or two-sided nature of conflict.

There are many different facets to this issue of the two-sided nature of conflict. One important aspect is how therapists often fail to appreciate that many clients have concerns both about being hurt by the therapist and others, and about hurting others (Weiss, 1993). The counseling field has well-developed concepts regarding clients' expectations about again being hurt, misunderstood, betrayed, and so forth, by the therapist—as they have been in other important relationships. However, far less attention has been paid to clients' worry and guilt about hurting the therapist and others by doing well, such as succeeding at a goal and surpassing their parent, appropriately standing up for themselves and advocating on their own behalf, and so forth. When clients make progress or improve in treatment, or successfully pursue their own interests in life (for example, by graduating from college, receiving a promotion, or enjoying a good relationship), many will retreat from these healthy, success experiences. In various ways, they will sabotage themselves, be unable to enjoy or take pleasure in the success they achieve, or be unable to sustain their progress. Often, they will become depressed as a result of separation guilt and/or survivor guilt, or become anxious as success threatens their attachment ties or challenges their familial role.

To highlight the two-sided or double-binding nature of many clients' conflicts, let's return to a previous case study. As we will see, once the therapist is able to recognize the two-sided structure of the conflict and reflect the ambivalent feelings that result from the client's bind, the client's seemingly irrational or self-defeating behavior makes sense and becomes resolvable.

Recall Anna, the young adult client who was discussing her dream with her therapist at the end of Chapter 4. Anna wanted to end her chronic depression, become more involved with friends, and find a career path that she could pursue. Although she had expressed these wishes for many years, she could not follow through and realize them. She rarely initiated activities with peers and often declined the invitations she received from others. She repeatedly enrolled in college but would eventually "lose interest" and drop out after one or two semesters. Without recognizing the two-sided conflict that plagued Anna, her self-defeating behavior seems irrational.

On the one hand, Anna wanted to grow up and have her own independent life with her own friends, marriage and family, and career. At the same time, however, she felt a responsibility to remain close to home and support her mother—who acts as a guilt-inducing martyr whenever she starts to establish her own goals or pursue her own interests. The repetitive relational scenario is that whenever Anna begins to succeed with friends, in school, or with life in general, her mother will look hurt, withdraw physically and emotionally, and induce guilt. Unable to break this tie that binds, Anna continues this maladaptive relational pattern by repeatedly failing, and having to become dependent again and return home. Only upon her return home does her mother stop acting like a martyr and again become available and responsive to her. Having to sacrifice their own independence to maintain their attachment ties, clients like

Anna become depressed over their inability to have their own lives and lose any sense of agency or self-efficacy from such binding guilt.

After a few months, however, the same cycle will be reenacted again. Anna will start to act on her own healthy wish for more autonomy, seek peers and friends, and reenroll in school. However, at this point, her mother will withdraw and act hurt, and Anna will not be able to "individuate" and initiate her own adult life without her mother's support. In a few months, she will again return home, feeling hopeless and a failure. Completing this scenario, her mother will become comforting and supportive of Anna in her dependency and social failure. This type of maladaptive relational pattern can be reenacted for years, unless both sides of the client's conflict can be identified or made overt and worked through in treatment.

Without understanding these family dynamics (see Haley, 1980), the therapist and others might erroneously conclude that Anna is lazy, unmotivated, or receiving too many secondary gains from being home and depressed. However, once we understand the maladaptive relational pattern that immobilizes her, we see that such pejorative labels are unwarranted. Clients' contradictory or self-defeating behavior does make sense, but not until we understand the two-sided, push-pull, or double-binding nature of their conflict.

Exploring Ambivalent Feelings

In addition to understanding when the client's problem is a two-sided or double-binding conflict, *therapists will be more effective when they can help clients explore and sort through both sides of their ambivalence.* That is, therapists take a neutral, exploratory stance and invite clients to expand and elaborate their thoughts and feelings on both sides of their conflict. For example:

> THERAPIST: It sounds like one part of you wants to get married but another part of you doesn't. Tell me about both sides—the part of you that wants to marry him and the part of you that doesn't. I have no investment in what you decide; I'd just like to help you sort through the pros and cons on each side.

In this way, the therapist welcomes and invites both sides of the client's ambivalence. The therapist's intention is to respond affirmingly to the opposing or contradictory feelings that accompany both sides of the conflict. As clients find that they have the support of the therapist to explore all of their feelings and concerns about a particular decision, they are able to clarify their own preferences, take responsibility for their own choices, and act on their own decisions. In contrast, therapists will not be effective when they only respond to one half of the client's ambivalence or do not encourage the client to freely explore both sides of the issue. When this occurs, the therapist's own countertransference issues are usually in play, as the following case vignette illustrates.

Marie, a depressed, 25-year-old graduate student, entered time-limited (12-session) treatment at the Student Counseling Center on campus. During the first few sessions, Marie explored how angry she was at her mother. Her mother expected Marie to be perfect, insisted that Marie never had any problems, and always needed Marie to be "happy." In counseling, Marie began to realize how much she resented her mother for denying so many of her true feelings and for having so many expectations of how she "should" be. Recalling similar issues in her own childhood, Marie's therapist resonated with her anger and actively supported her protest.

At first, it was liberating for Marie to have her long-suppressed anger affirmed. She was both excited and relieved to be realizing that she no longer needed to keep fulfilling her mother's unrealistic expectations. This good feeling was short-lived, however. In the next few sessions, Marie's newfound freedom gave way to a growing despondency and return to her long-standing dysthymia. The therapist thought that Marie must be worried about the consequences of defying her mother's expectations, feeling disloyal to her family for complaining about her mother to the therapist, and feeling guilty about being angry with her mother.

While asking about and looking for opportunities to explore these reasonable possibilities, the therapist continued to draw out and encourage Marie's anger. However, moving in a different direction than the therapist, Marie began to share fond memories about her mother and recalled special times they had spent together baking and making cookies after school. The therapist was not very responsive to Marie's positive feelings toward her mother, however. The therapist thought that Marie's fond recollections were part of denying her anger toward her mother, avoiding the reality-based conflicts in their relationship, and was concerned that Marie would continue to be ruled by her mother's problematic expectations.

As so often happens in treatment, Marie's therapist only recognized one side of her ambivalence and failed to respond to the feelings on both sides of her conflict. Yes, Marie was angry with her mother, and these feelings needed to be explored and affirmed. At the same time, however, Marie knew that her mother had offered her many fine things as a parent and she did not want to forgo the good things they had shared. Marie's conflict was that she wanted to keep the good things she had experienced with her mother without having to take on or adopt the problematic aspects of their relationship as well. Thus, on the one hand, Marie wanted to reject the unrealistic demands to be perfect and happy that her mother placed on her. Yet, on the other, Marie became depressed when it seemed that the only way to do this was to give up her identification with her mother altogether. This had been the most important relationship in her life, and Marie did not want to risk losing the good parts of this relationship as well.

Realizing that time was short and treatment was stagnating, her therapist sought consultation from a colleague. The colleague readily recognized both sides of Marie's ambivalence and encouraged the therapist to allow Marie to simultaneously have contradictory feelings of appreciation and anger toward her mother. At first, the therapist was reluctant and argued with the colleague. Soon, however, she began to consider the possibility that her own countertransference may be operating, and that because of her own history, it was easier to support Marie's anger toward her mother than her appreciation. As a result of this helpful consultation, the therapist began to invite and affirm the other side of Marie's feelings as well:

> THERAPIST: You feel two different ways toward your mother at the same time. You're angry with her for expecting you to be perfect, but you also treasure many of the fond times you've had together. Tell me about both your angry and your loving feelings toward her.

The behavioral impact on Marie of this new approach was immediate. As the therapist encouraged Marie to explore both sides of her feelings and, in effect, gave her permission to love her mother and be angry with her at the same time, she felt enlivened and began to make real progress. Marie began to clarify what qualities of her mother she wanted to keep and make a part of herself (such as her mother's easy laughter and genuine warmth at times) and what aspects of her mother she wanted to discard (her mother's preoccupation with appearances and what other people might think). This process of *differentiation* also allowed Marie to clarify which issues she wanted to confront and try to change in her current relationship with her mother, and which issues she preferred to let go of and simply accommodate herself to.

Marie continued to make progress in treatment as the therapist stayed this course and continued to provide support for both sides of her ambivalence. At times, for example, the therapist would observe, "It really made you mad when she did that." However, this would soon be followed by the reflection, "I can see how important you two are to each other, and how much you love her even though these problems get in the way sometimes."

Therapy was successfully terminated at the 12th session. Marie was no longer depressed and had better integrated aspects of her own identity as a woman. In addition, Marie was initiating a rapprochement with her mother and checking out what could change and be better now in their relationship and what couldn't—exploring realistically how much authenticity and closeness they could share at this point in their lives.

To summarize, therapists help clients change when they recognize the two-sided structure of their clients' conflicts, and help clients explore and integrate both sides of their ambivalent feelings. Trained to look for pathology, therapists tend to miss personal strengths or positive aspects of conflicted relationships. However, clients will make far more progress when the therapist can actively encourage clients to keep the good parts of the parental (or spousal) relationship while they are realistically examining the problematic parts. In this regard, *the therapist can tell clients explicitly that they are not being disloyal if they talk about problems with a parent or spouse, and that the therapist knows that their complaints do not reflect all of their feelings about this person.* In this regard, many clients will not be able to engage deeply in treatment and work productively until the therapist recognizes and supports both sides of their ambivalence. For so many clients, a primary treatment goal will be to integrate their positive and negative feelings so they can be angry with someone they love, or still feel worthwhile even though they are disappointed with themselves about something.

Finally, just as therapists in individual practice are looking for opportunities to affirm both sides of the client's conflict or ambivalence, couple therapists are trying to support both partners equally in an interpersonal dispute. The primary task in marital therapy is to resist the couple's eliciting maneuvers and keep from taking sides by participating in the good guy/bad guy, overadequate/inadequate, healthy/neurotic, or other polarizations that couples often present. When couple counseling fails, one member of the couple usually stops treatment because she or he comes to believe that the therapist has taken sides against them in the marital conflict.

Closing

In this chapter, we examined three dimensions of clients' current interpersonal functioning: (1) how clients' problems with others are brought into the therapeutic relationship through eliciting maneuvers, testing behaviors, and transference reactions; (2) how therapists can establish an optimum degree of interpersonal relatedness and enact a corrective balance of separateness-relatedness; and (3) how therapists can understand the two-sided structure of clients' conflicts and support clients in exploring both sides of their ambivalent feelings. These interpersonal factors, together with the developmental/familial perspective and interpersonal model presented in the previous two chapters, will help therapists better understand and intervene with the problems clients present.

With these conceptual guidelines in mind, we now return to intervention strategies. The next chapter explores how therapists can use the interpersonal process they enact with clients to identify maladaptive relational patterns, begin resolving clients' problems in the current interaction with the therapist, and then help clients generalize this new behavior with the therapist to other important relationships in the client's life. Working in the here-and-now with the process dimension is essential to all forms of interpersonally oriented therapy. However, if therapists are to use process-oriented interventions and the therapeutic relationship to effect change, they need to be able to talk forthrightly with clients about their current interaction and what's going on between them. Initially, however, it often feels uncomfortable for new therapists to metacommunicate in this way and "talk about you and me." Such immediacy interventions, which include process comments, self-involving statements, and interpersonal feedback, are powerful yet challenging for most new therapists to adopt. In Chapter 9, we explore therapists' concerns about working with the process dimension and consider guidelines to help therapists intervene more effectively within the therapeutic relationship.

Suggestions for Further Reading

1. In a managed care milieu, writing effective treatment plans becomes an essential skill for clinical trainees. Illustrating the guidelines for writing case conceptualizations that are provided in Appendix B, Chapter 8 of the Student Workbook provides an in-depth case formulation of Linda in Arthur Miller's *Death of a Salesman*.

2. Readers are encouraged to learn more about clients' testing behavior by reading Chapters 4–6 of *The Psychoanalytic Process* by Weiss and Sampson (1986). Two popular press versions of this work, which also emphasize the role of unconscious guilt in many clients' symptoms and problems, may be found in *Hidden Guilt* by Engel and Ferguson (1990) and *How Psychotherapy Works* by Weiss (1993; see especially Chapter 4).

3. Chapter 2 of Jay Haley's (1980) *Leaving Home* elucidates the family dynamics and structural family relationships that impede individuation and emancipation for many young adult clients.

4. Family systems concepts of "cohesion and adaptability" can also be used to describe an optimal degree of interpersonal relatedness. Interested readers may examine *Families: What Makes Them Work* by Olson, McCubbin, Barnes, Larsen, Muxen, and Wilson (1983).

Resolution and Change

CHAPTER NINE

An Interpersonal Solution

Conceptual Overview

The previous three chapters provided information to help therapists conceptualize their clients' problems and formulate treatment plans; now we focus further on the process dimension and how change occurs. To begin, therapists need to help clients develop new narratives for their lives that allow them to better make sense of the influences and events that have shaped their development and who they have become. Unless they do so, it will be harder for them to guide where they are going in the future and to maintain changes after treatment has stopped. However, a consistent focus on historical relationships and what happened in the past will not produce change and often serves only to help clients avoid the anxiety of addressing current problems. Certainly, information on the developmental themes and patterns that clients experienced in the past does help both the therapist and clients to understand how faulty schemas have developed, and why current relationships are being framed and recreated in a particular way.

Going beyond mere insight or understanding, *change occurs when clients experience a behavioral or in vivo resolution of their problems in the current relationship with the therapist, and this experiential relearning is successfully generalized to clients' lives*. Said differently, the primary reason for exploring developmental and familial relationships is not to gain insight or catharsis, but to throw light on clients' faulty coping strategies and characteristic ways of responding to the therapist and others in current relationships. In this way, understanding formative relationships may be a relevant aspect of change for some clients yet is, in itself, usually insufficient. Instead, enduring change comes about when clients enact a resolution of their conflicts in their current interaction with the therapist, and the therapist is successful in helping clients apply this relearning to other arenas in their everyday lives where the same themes and patterns are causing problems.

If the primary focus of treatment and fulcrum of change is in the current interaction between the therapist and the client, rather than a psychodynamic exploration of the past, what can the therapist do to help clients change? Does the therapist need to take responsibility for finding solutions and telling clients what choices and decisions to make? No, therapists can't just prescribe solutions or direct clients what to do—telling clients how to live their lives only fosters their dependency and undermines their self-efficacy. Instead, therapists help by providing accurate empathy that discerns the key concerns or highlights the core issues in their current problems. This empathic understanding also allows the therapist to provide clients with a different and better response to their maladaptive relational patterns than they have received in the past and come to expect from others. Enduring change occurs as these new behavioral patterns with the therapist expands clients' cognitive schemas, alters their beliefs about themselves and expectations of others, and allows them to increase their interpersonal range with others.

This corrective emotional experience, which comes from the new ways of relating with the therapist that do not fit the familiar but unwanted responses they have learned to expect from others, creates *interpersonal safety*. This newly felt safety, in turn, frees clients to become more authentically connected to themselves and fosters greater choice and flexibility in how they respond to others. Clients are empowered to take more responsibility for their own choices, advocate on their own behalf, and better pursue their own interests and goals.

We have seen that clients' problems are not just talked about abstractly in treatment; in addition, they are brought into the therapeutic relationship and played out in the current interaction between the therapist and client as well. In order for clients to be able to make meaningful and sustainable changes with others in their lives, the therapist and client cannot continue to reenact the client's maladaptive relational patterns in their interpersonal process. Instead, they need to mutually work out a resolution to them in their real-life relationship. Highlighting the power of this experiential relearning, the attachment researchers put it best: you do not get an empathic child by teaching or admonishing the child to be empathic; you get an empathic child by being empathic with the child (Karen, 1998).

The therapist's intention is to make this corrective emotional experience an ongoing pattern of interaction in the therapeutic relationship—one trial learning is not sufficient. As this occurs, the focus of treatment shifts to helping the client successfully *generalize* this in vivo relearning to other arenas in the client's life where the same themes and patterns are being enacted. Conversely, if the therapist and client merely talk about important issues and behavioral options, but the therapist does not actively help the client link these new ways of relating with the therapist to other relationships where the same problematic themes and patterns are being disruptive, change will not occur. Thus, the purpose of this chapter is to further clarify the process dimension and illustrate how therapists working within different theoretical orientations can use the therapeutic relationship to help clients change.

Resolving Problems through the Interpersonal Process

In this section, we explore how the interpersonal process that therapists enact with clients can be used to resolve clients' problems. First, we will review the course of therapy up to this point in treatment and highlight how clients' problems are brought into the treatment setting and reenacted with the therapist. This review serves to introduce the next stage of treatment: providing a corrective emotional experience and helping clients apply this relearning to other relationships. Following this overview, we will look at five case vignettes that illustrate how therapists can use the process dimension to help clients resolve their problems and change.

Bringing Clients' Conflicts into the Therapeutic Relationship

Maladaptive relational patterns with caregivers give rise to developmental conflicts that clients have not been able to resolve on their own. Symptoms and problems develop as clients try to cope with difficult feelings and faulty beliefs by adopting fixed interpersonal coping strategies. Although these coping strategies once were necessary and adaptive, they are no longer necessary and create problems in many current relationships. The therapist's first task is to establish a collaborative, working alliance with the client by making accurately empathic connections. They then work together to identify and change the maladaptive relational patterns that are occurring with others and in the therapeutic relationship, as well as the conflicted feelings and pathogenic beliefs that accompany them. As this work proceeds, the therapist's intention is to watch for and recognize the pull from the client to reenact these relational themes in the therapeutic relationship. The therapist's goal, then, is to respond differently than others usually have and change the familiar but problematic scenario—providing a different type of relationship that resolves, rather than reenacts, the client's repetitive relational themes. By responding in more empathic and affirming ways, the therapist gives clients the opportunity to experience new ways of relating, and to evaluate the costs and gains of their habitual coping styles. In this process, clients are empowered to choose more flexibly how they want to respond in current relationships, and to develop more realistic and affirming self-concepts.

The therapist's primary role is not to give advice, explain, reassure, interpret, self-disclose, or focus on the behavior or motives of others, although each of these responses will be effective at times. Instead, the therapist encourages the client to take ownership of the treatment process by setting the direction for therapy. The therapist does this by helping the client identify the issues or concerns that the client feels are most pressing or important right now. While encouraging and following the client's lead in this way, the therapist participates in shaping the course of therapy and providing a *treatment focus* by listening for

- maladaptive relational patterns or problematic interpersonal scenarios that keep occurring,
- pathogenic beliefs about self and faulty expectations of others, and
- affective themes that recur throughout the material the client presents.

The therapist continues to focus the client inward, reflects the core affective messages, links the client's maladaptive relational patterns with others to their own current interaction, and maintains a collaborative, working alliance by ensuring that clients are active participants who feel *ownership* of the change process. Continually refining working hypotheses, the therapist will be able to conceptualize the core conflicts and problematic coping styles that each client presents.

This conceptual formulation provides a treatment focus or direction for the ongoing course of therapy. For example, the therapist will be able to do the following:

- Point out when and how clients employ their interpersonal coping strategy to rise above their conflicts.
- Help clients become aware of how they block their own needs and feelings by responding to themselves in the same problematic way that others originally responded to them.
- Help clients explore why they become anxious at a particular time, what types of responses they tend to elicit from the therapist and others to manage this anxiety, and how they systematically avoid certain interpersonal modes or feelings.

In these and other ways, the therapist helps clients become aware of their core conflicts and the interpersonal strategies they have adopted to cope with them. Although this conceptual awareness will be an important part of helping clients change, something more is needed.

At the same time as the therapist and client explore the content of the client's concerns, they also tend to play out these same relational themes in their interpersonal process. As we have seen, the client and therapist do not just talk about issues in therapy; they actually relive them in the therapeutic relationship. This occurs in three ways.

Transference Reactions   First, transference reactions bring the client's core conflicts into the current relationship with the therapist. As the client's conflicts begin to emerge in therapy, the client will become increasingly concerned that the therapist has been responding—or is going to respond—in the same unwanted ways that significant others have in the past. These fears and misperceptions follow from the client's cognitive schemas, and *they are most likely to occur when strong feelings have been evoked and the client feels distressed or* *vulnerable.* In some cases, treatment will reach an impasse until the therapist can help the client differentiate the therapist's actual intentions from the problematic responses that the client expects and has in fact received in other important relationships:

> THERAPIST: As you tell me about this, what do you think I might be feeling about you?
>
> CLIENT: Well, I guess I'm concerned that you might be a little disappointed in me.
>
> THERAPIST: Disappointed? Oh, no, not at all. Actually, I was thinking . . . .

Eliciting Behaviors   Second, clients will systematically *elicit* responses from the therapist that pull the therapist into reenacting their old scenarios or confirming their problematic expectations of others. When therapy stalls or ends prematurely, clients have usually reenacted their generic conflict in this way with the therapist. This unwanted reenactment occurs when clients elicit responses from the therapist that are thematically similar to those they have received from others in the past—responses that clients perceive, once again, as rejecting, idealizing, competitive, critical, and so forth. Clients elicit these familiar yet problematic responses from the therapist for different reasons. Sometimes defensively—to avoid the internal aspects of their conflicts—and sometimes more adaptively—by testing whether they can obtain a better response to their familiar relational scenario than they have come to expect from others:

> THERAPIST: How do others usually respond when you talk to them like you're talking to me right now?
>
> CLIENT: They probably don't like it much—you know, everybody just gets pissed with me.
>
> THERAPIST: I can see how that happens, and I have some ideas about how we could try to make that different here in therapy—in our relationship. I think if we could start talking about what's going on between us and change how this goes in here, that would be a very helpful step to start changing this with others in your life. How does that sound to you?

Testing Behaviors   Third, the manner or way in which the therapist responds to clients—the therapeutic process—may unwittingly reenact their conflicts. As we have emphasized, the clients' relationship with the therapist needs to provide a different or more helpful response to their conflict than they have received in the past. However, it is not always easy to provide this corrective emotional experience. Clients will expect, and often successfully elicit, the same problematic responses from the therapist that they have received from others in the past. The task for therapists, then, is *to identify how their interpersonal process with the client may be reenacting some aspect of the client's problem with others and, when this is occurring, to use a process comment to make this interaction overt as a topic for discussion.* The therapist then works together with clients (and with the help of a supervisor when the clients' issues have evoked the therapist's own matching or reciprocal conflicts) to establish a different pattern of interaction that does not repeat the old relational scenario. As we have seen, however, this idea is far easier to say than to put into practice:

THERAPIST: I just disagreed with you—I'm wondering how that was for you?

CLIENT: Well, kind of different, I guess. I'm not used to people speaking up and disagreeing with me much.

THERAPIST: Yes, I can see why they don't—you are so forceful when you speak. I hear you that you want to "communicate better" with your 15-year-old daughter, but it makes sense to me when she says that she feels "dominated" by you and reluctant to disagree or speak up. I'm a 45-year-old, educated professional, and you're a challenge even for me.

CLIENT: We need to work on this, don't we? I think this is what my wife has been trying to tell me for a long time.

Thus, if it is to lead to change, the therapeutic process needs to enact a resolution of clients' conflicts rather than a repetition of them. When this occurs, clients have received far more than just an explanation for their problems. They have experienced a meaningful relationship in which their old conflicts have been aroused but, this time, they have found a better outcome than the maladaptive relational pattern they have come to expect. They have been able to reveal themselves, to act more boldly, to express a need, to have someone constructively stand up to them, and so forth, without receiving the unwanted response they have come to expect (for example, being ignored, exploited, or diminished). When this occurs in the context of a meaningful relationship, such in vivo or experiential relearning is powerful, indeed.

Experiencing an interaction with the therapist that is incompatible with their early maladaptive schemas does not make up for the deprivations or disappointments that have shaped the client's life, of course. However, it does behaviorally demonstrate that change can occur—that at least some relationships, sometimes, can play out in a different and better way. This corrective emotional experience begins to broaden clients' schemas for what can occur between people in relationships, and their interpersonal modes of responding with others. At this critical juncture in treatment, when clients are having a meaningful experience of change with the therapist and the therapeutic process is providing a resolution of the clients' conflicts, two important things occur. First, *intervention techniques from varying theoretical orientations all become more effective.* That is, cognitive, interpretive, self-monitoring, educational, skill development, and other interventions all can be used more productively by clients because the useful new content in these interventions is congruent with the corrective process they are enacting. Second, as clients have repeated experiences of change with the therapist, the therapist can help clients begin to apply this in vivo relearning to other arenas in their lives where similar conflicts are being enacted. This next period of therapy—transferring this experiential relearning that has occurred with the therapist to other relationships—is called the *working-through* phase of treatment and will be discussed in the next chapter.

With this overview of the change process in mind, let's now examine more closely how therapists can use the interpersonal process they enact with their clients to effect change.

Using the Process Dimension to Facilitate Change

To effect change, the therapist tries to keep clients from reenacting their maladaptive relational patterns in the therapeutic relationship. Although such reenactments are inevitable at times, we do not want them to come to characterize the ongoing interaction with the therapist. To alter clients' schemas and expand their interpersonal range, clients need to have the real-life experience of a relationship that does not go down the same old problematic lines they have known before and come to expect. Providing this in vivo or experiential relearning is the core component of change in the interpersonal process approach.

Within this context, interventions from other theories can helpfully be integrated. Following Rogers, it is essential for the therapist to listen empathically and work collaboratively. Also, it is helpful with some clients to suggest psychodynamic interpretations and facilitate client insight about the sources or development of their problems. Behavioral and cognitive therapists significantly help clients by training new relaxation, desensitization, and assertiveness skills; cognitively reframing situations; and suggesting new, more adaptive behavioral responses for clients to try between sessions. Interventions from these and other theories will certainly be useful, but each can become more effective when the therapist is also attending to the interpersonal process that the therapist and client are enacting.

Clients benefit when they live out with the therapist a relationship in which their core conflicts are activated but the therapist does not respond in the problematic way that clients expect. This recommendation is easy to say but often so difficult to do. Why? Therapists often fail to recognize how they are metaphorically or thematically reenacting aspects of their clients' conflicts in their interpersonal process. Clients do not just talk about their problems in the abstract; they reenact the same relational themes with the therapist—in the three ways described earlier. When treatment fails, the working alliance usually has been ruptured because the therapeutic process has repeated some aspect of the same interpersonal problem that the client has been struggling with in other relationships, and that originally brought the client to treatment. Furthermore, *this often occurs when the client's eliciting maneuvers, testing behavior, or transference reactions have tapped into the therapist's own personal issues and created a mutual or shared conflict for both of them.* For example, perhaps the client's need to control others taps into the therapist's own concerns about being controlled by others, or the client's criticalness activates the therapist's own insecurity about always needing to be right, and so forth. As we will see, it can be personally challenging for the therapist to utilize the process dimension and enact a solution to the client's old scenario rather than a repetition of it. Next, we look at five different case examples to illustrate how the process dimension can be used to facilitate change.

Example 1: Reenacting the Problem—An Inability to Recognize the Interpersonal Process   Therapists usually find it easy to understand the process dimension conceptually, and relatively easy to observe when a client's maladaptive relational patterns are being reenacted with another therapist. In contrast, however, it is far more difficult for therapists to recognize how the interpersonal process they enact with their own clients may be reenacting the conflict rather than resolving it.

An alcoholic client began treatment by expressing how much pain he was in as a result of his drinking problem. His wife was threatening to leave him and he was on probation at work. His 10-year-old son had recently seen him in an embarrassing situation while intoxicated and had asked him not to drink anymore. The client was distraught that, even though the structure of his life was collapsing, he could not stop drinking. During their first session, the distressed client said, "Nothing ever works out for me. Every relationship I have seems to go bad and fall apart in the end." The therapist was moved by the client's plight but became overly invested in having him change. The therapist found it difficult to listen to how hopeless he sounded, and she began reassuring him that things would get better. The therapist disclosed that she had had a drinking problem of her own years ago and that she knew what he needed to do in order to stop drinking. The therapist and client quickly established a friendly relationship in which the client successfully elicited a great deal of nurturance from the responsive therapist. The client enjoyed this support immensely and sincerely tried to follow the therapist's advice.

Things went better for about a month, but then the client "fell off the wagon" and began drinking again. When he started arriving late for his therapy sessions, and then missed two of them, the therapist began to feel let down. It was the final straw for the therapist when the client arrived for therapy smelling of alcohol. The therapist felt angry and betrayed because she had extended herself to him more than she usually did with other clients. She felt that he was letting them both down and told him so. Unwittingly, the therapist became punitive toward the client in this way, and she induced guilt over the impact of his "irresponsible" drinking on his family. The client was contrite and tried, unsuccessfully, to elicit her sympathy and support again. The client did not show up for his next appointment, however, and did not return to therapy.

This interaction reenacted the client's conflict in two ways. First, the client initially elicited sympathy from the therapist, but the therapist eventually turned to criticism and control. This confirmed the client's pathogenic belief that nothing ever works out and relationships all go bad in the end—the specific relational pattern that he had forewarned the therapist about during their initial session. Second, the therapist benevolently tried to rescue this client, who was behaving as a victim. The caretaking, advice-giving response from the therapist inadvertently defined the client as incapable of managing his own life. This response paralleled a similarly belittling attitude that the client had received from his indulgent but undermining parent. The client's maladaptive relational pattern of a rescuer–victim relationship was reenacted in treatment and prevented the client from addressing his real concerns over shame-based feelings of inadequacy, which he defended against, in part, through drinking.

In reviewing this case several weeks later, the therapist reported, "The client just wasn't ready to stop drinking yet. He's going to have to get worse and really hit bottom before he'll be able to stop denying there's a problem and do something about it." The therapist did not recognize that the interpersonal process they enacted in treatment unwittingly reenacted the client's maladaptive relational patterns and prevented change from occurring. As is often the case when treatment fails, the therapist attended only to the content of what they talked about (his drinking) and not as well to how their process might be replaying the client's interpersonal problems (pattern of successfully getting others to rescue him when he acted like a victim).

All therapists, beginning and experienced alike, will reenact aspects of the client's patterns and problems in their interpersonal process at times—without realizing that this parallel is occurring. From an emotionally neutral vantage point—that of a therapist's supervisor or colleague looking at the session on videotape—it is relatively easy to see how the old pattern is being reenacted in the interpersonal process. In humbling contrast, it can be exasperatingly difficult for therapists to see their own interpersonal process while engaged in an intense, affect-laden relationship. This is complex work, and it is challenging to perceive and assess at both the content level and the process level simultaneously. In the first year or two of their clinical training, it is especially difficult for new therapists to attend both to the content of what is being discussed and to the process that is being enacted as the discussion proceeds. With more experience and guidance from a supportive supervisor, trainees will become more effective at observing the process level. The bigger challenge, however, is using or applying the new understanding they now possess. That is, trainees may have difficulty making process comments and intervening in the moment with the interpersonal process they now see occurring with their clients. For example:

> THERAPIST: (*internal dialogue*) OK, I can see that he's talking to me just like he does to everybody else. Great, I'm getting just as bored and frustrated as they do, but I sure don't want to bring this up. How can I possibly talk with him about what's going on without hurting his feelings?! I don't want to make him feel bad, or get mad—he might not come back, and then I'll have lost two clients. . . .

Clearly, therapists in training are facing a complex challenge in their professional development here, and they need to be patient with themselves. The ability first, to recognize, and then, to intervene, on the process level is usually acquired gradually over a period of a year or two. In their attempts to adopt a process-oriented approach, therapists will benefit from further reading on this topic (see the suggestions at the end of the chapter), rehearsal/practice with classmates, demonstration/role modeling from practicum instructors, and the guidance of a supportive supervisor. It is not easy in the beginning to work this way, and student therapists need help.

Example 2: The Role of Process Comments in Keeping Therapists from Repeating Maladaptive Relational Patterns   In the previous example, the therapeutic process reenacted the client's old scenario. In the next illustration, a similar dynamic begins but, this time, the therapist recognizes their interpersonal process and uses a process comment to effectively realign this problematic interaction.

Many clients enter treatment because of depression. They feel sad, believe they are bad, and see themselves as helpless to change their circumstances. In many cases, these clients communicate their very real suffering to the therapist with emotional pleas for help. However, whatever the therapist tries does not work. Often, the therapist's efforts to meet the client's request for help are met with "Yes, but. . . ." A problematic cycle begins: The client feels increasingly distressed and intensifies his plea for help. For example, clients might even exclaim that they "can't go on living" if things don't change. In response, the concerned therapist becomes more active and tries even harder to find some way to help. However, nothing the therapist does has any impact or provides any relief for the client.

As the client rejects the help that he actively has been eliciting, the therapist's own personal need to be helpful is frustrated. For example, the new therapist's own tenuous confidence or fledgling sense of adequacy as a helper may be threatened by this client's response. Or, other therapists who were parentified in their family of origin may feel guilty for failing to meet the emotional needs of others. When the client's eliciting behavior has tapped into the therapist's own personal issues in one of these ways, some therapists will become punitive and critical toward the client; others will withdraw and emotionally disengage. (Think about your own response tendencies for a moment, and consider how you would likely respond to this mixed message from the client.) When this occurs, the therapeutic process thematically repeats the client's maladaptive relational patterns.

Treatment will stall at an impasse or terminate prematurely when the therapist *remains* critical toward or disengaged from the client. The depressed client's conflict is reenacted, as the client again feels dependent on a relationship with someone who is critical, inconsistently available, or threatens to terminate the relationship. What is the alternative? In successful treatment, the therapist is able to make a process comment (metacommunicate) and begin a dialogue with the client about what might be going on in their present interaction. Rather than focus solely on the content of what they are talking about (for example, depression), *the therapist can wonder aloud or tentatively inquire about how they seem to be responding to each other.* This is often the best way to effectively change or alter the reenactment:

> THERAPIST: Let's talk about what is going on between us right now and see if we can understand what's happening in our relationship. It seems to me that you keep asking me for help, but you keep saying "Yes, but. . . ." Maybe I get frustrated then, feeling that you don't allow me to help. What do you see going on between us?

With a process comment of this type, the therapist is inviting the client to be a partner or collaborator and join in trying to understand their mutual interaction. The therapist is also providing interpersonal feedback and helping the client look at how his depression is expressed to and experienced by others.

As we will see, the process comment also allows the therapist to break the cycle, at least for the moment, of the client's escalating need and the therapist's increasing frustration.

How does this process comment help the client to change? In this moment, the client is experiencing a relationship with the therapist that provides a new and different response to his old relational pattern. The therapist does not "give up" on the client, as has happened in other relationships with friends and his wife. Instead, the client remains engaged with someone who finds other ways to relate. The client has not experienced this consistent availability, or the holding environment it provides, in past relationships. If this corrective emotional experience is repeated in many other ways—large and small—throughout treatment, the client is living out a resolution of his core conflict in the relationship with the therapist. When clients have this real-life experience of change, rather than passively or intellectually hearing advice, reassurance, or interpretations, they are empowered to begin making similar changes with others in their lives.

A key concept in the interpersonal process approach is that the capacity for change increases as the therapeutic relationship becomes more meaningful to the client (Gelso & Hayes, 1998). When an interpersonal solution is enacted in a valued relationship with the therapist, the therapeutic alliance is strengthened and change is facilitated in two ways. First, this new type of relationship provides interpersonal safety for the client. The interpersonal safety of this new context permits clients to come to terms with the core conflicted feelings and pathogenic beliefs that accompany their maladaptive relational patterns. In other words, clients have the supportive relationship necessary to experience and integrate those feelings and situations that previously needed to be disavowed or split off as they were too threatening to be dealt with. Second, the experience of change with the therapist *demonstrates* to clients, rather than merely tells them, that current relationships can be different than they have learned to expect. It does this by expanding their cognitive schemas for relationships, by altering their core beliefs about the possibilities that relationships hold for them—for example, "I do matter and can be cared about"; "I can say what I want and not be left"; "I do have the right to set limits and stop letting my children run over me." Next, with the therapist's assistance, the client can begin to transfer this relearning and establish better relationships with others along these more flexible and self-affirming lines. As an illustration, let's consider the depressed client's response to the therapist's process comment:

CLIENT: Yeah, I do keep saying "Yes, but . . ." to you—like my wife always says I say to her. So this is all just pointless. If I'm doing this with you, too, then there's no way I'm going to get better. Maybe we should just forget it and stop now.

THERAPIST: Oh, no—that's not what I'm suggesting at all! It makes sense that you are having the same problems in here with me that you have with other people. In a way, it is a problem, but in another way, it gives us the opportunity to resolve your problem right here in our relationship.

CLIENT: How?

THERAPIST: If you and I can work out a better way of doing this in our relationship—you know, find a way to keep this old pattern from repeating in our relationship—I think it will go a long way toward helping you change this with your wife and others.

CLIENT: Do you think we can do that now if we haven't been able to up to this point?

THERAPIST: Yes, I think it's very possible. In fact, I think we are breaking that pattern right now, just by talking about the way we interact together. Tell me, how is it to be talking with me about our relationship and the way we respond to each other?

CLIENT: It's different, but I like it.

THERAPIST: I like it, too. I feel like I'm working with you, rather than being pushed away as I was feeling before.

In this example, the therapist has used a process comment to effectively alter the current interaction with the client. For the moment, the process comment has kept their interaction from repeating the client's maladaptive relational patterns. Some variation on the old "Yes, but . . ." pattern will probably soon reappear, and the therapist will have to make another process comment and work through a similar cycle again. However, with the process comment, the therapist is temporarily providing the client with a different response that does not replay the client's old scenario. If the therapist continues to find ways to provide this response throughout treatment, the client will experience a corrective relationship and, as we will highlight in the next chapter, be able to change this pattern in other relationships as well. The next example illustrates a macroperspective that tracks the process dimension over the course of treatment.

Example 3: The Process Dimension as a Means to Bridge Differing Theoretical Orientations

There are many ways to help clients change. Therapists working from different theoretical orientations can all help clients change. Every theoretical approach clarifies some aspects of clients' problems and every theoretical approach has limitations. Within each theoretical orientation, the primary component of change is not the theory per se (for example, client centered versus psychodynamic) but how effectively individual therapists apply it in their therapeutic relationships. That is, some cognitive behavioral therapists are far more effective than other cognitive behavioral therapists and account for most of the combined treatment effect in psychotherapy outcome studies (Bergin & Garfield, 1994; Norcross, 2002; Teyber & McClure, 2000; Walborn, 1996). Long ago, Kiesler (1966) called this the "uniformity myth." He encouraged psychotherapy outcome researchers to study *within*-group differences (that is, more and less effective psychodynamic therapists, and more and less effective cognitive behavioral therapists), where large differences in effectiveness are found. This is more informative than continuing to focus on *between*-group differences (cognitive-behavioral versus psychodynamic treatment approaches) where only small differences in treatment effectiveness are found (Bergin, 1997; Garfield, 1997; Blatt, Sanislow, Zuroff, & Pilkonis, 1996). However, regardless of theoretical orientation or treatment approach, change is likely to occur when therapists can establish a strong working alliance and repair it when it is ruptured. Working within any theoretical modality or treatment length, therapists will find that change is also more likely to occur when the therapeutic relationship provides a resolution of the client's conflicts rather than a repetition of them.

After 2 years in analytically oriented therapy, Rachel still could not take charge of her life. She was always complying with her husband's demands, and it was impossible for her to get her children to do what she asked. She also complained of her stultifying daily routine as a homemaker, yet she was never able to do anything to improve it. At her husband's insistence, she would periodically enroll in a class or interview for an office job, but she never followed through on any of his suggestions.

In therapy, Rachel had spent many hours exploring her childhood. Her therapist was skillful in seeing unifying themes in the recollections that she shared with him. For example, her therapist astutely observed how she was subtly discouraged from initiating activities on her own as a child. Subtly, she was not allowed to feel good about her success experiences. Anytime she expressed an enthusiastic interest, tried out a new venture, or achieved something she was proud of as a child, her parents did not notice it—or, if they did notice, they somehow didn't seem very happy about it.

Rachel's therapist once explained to her, in a sensitive and nonjudgmental way, that she had a passive-dependent personality style. Rather than feeling labeled or put down, Rachel was impressed that her therapist seemed to understand her so well. He knew so many things about her without her even having to tell him. Although her problems had not changed much yet, she still believed her therapist would cure her. He was so bright and insightful, and he seemed genuinely to care about her. He didn't like to tell her what to do but, when things became too much, he could usually help by explaining what the problem really meant. It was comforting to be with him and Rachel did not know what she would do without him.

It was Rachel's husband, Frank, who finally ran out of patience with the slow course of treatment. After 2 years, he was fed up with the unending therapy bills and his wife's unremitting discontent. His wife's helpless dissatisfaction felt like an unspoken but unending demand for him to love her more, give her more, or somehow fill up her life. He was tired of these subtle, nagging demands, and he wanted a change.

A man of action, Frank obtained the name of a behaviorally oriented therapist on the faculty of a nearby college. A friend told Frank that this therapist was a "problem-solving realist" who could make things happen quickly. This sounded like just the right approach to him. Frank insisted that Rachel stop treatment with her present therapist and begin with the new therapist. Initially, Rachel was sickened at the thought of leaving her therapist but she sensed that Frank was truly at the end of his rope. Although she still believed in her therapist and felt loyal to him, she was afraid that Frank might actually leave her if she did not go along with this demand.

After her first two sessions with the new therapist, Rachel was surprised to think that perhaps Frank might have been right after all. The new therapist wasted no time in taking charge of the situation. It was actually encouraging to have the therapist outline a treatment plan with steps for her to follow. The therapist discussed a list of specific treatment goals with her, and they planned a set of graduated assignments for meeting these goals on a scheduled timetable.

In their first hour together, the therapist had Rachel role-play how she responded to her children when they disobeyed her. Then, with the therapist serving as a model and coach, they rehearsed more assertive responses that Rachel could try with her children. The therapist also had Rachel enroll in an assertiveness-training class that the therapist was running for some other female clients. Each week, Rachel was also to complete a homework assignment. For the second week, she was to call one new person she might like to get to know and ask her to lunch. She was supposed to report back to the therapist on this assignment at the beginning of their next appointment.

Frank was encouraged by this practical, problem-solving approach to his wife's problems. He began to think that something might change after all. Rachel was surprised to find that she actually felt hopeful, too. She felt reassured by her new therapist's goal-oriented, problem-solving approach. In fact, Rachel became determined to make this therapy work, even though she didn't want anything to do with it initially. She promised herself that she was going to try to do everything the therapist asked of her.

Therapy progressed well for the first few weeks, but things soon started to slow again. Without really knowing why, Rachel began to find it hard to muster the energy to attend the assertiveness class. She knew her therapist would be disappointed in her, but she just couldn't seem to help it. Although she felt guilty and confused about it, Rachel began to come late to her therapy sessions. Over the next month, she began to miss them altogether and soon faded out of treatment.

Both of these therapists failed to have a significant impact on Rachel's problems, even though the treatment approaches they used were seemingly very different. Theoretically, the analytically oriented therapist might attribute the unsuccessful outcome to the great difficulty in restructuring the passive dependency needs of a basic oral character. The behaviorist might note that Rachel was not sufficiently motivated to change because she was receiving too many secondary gains from her help-seeking behavior. If we look at the interpersonal process that transpired between both therapists and the client, however, a different picture emerges. These two therapists actually responded to Rachel in a very similar—and problematic—way.

Let's look carefully at their interpersonal process. With both therapists, Rachel reenacted the same maladaptive relational pattern that she had with her children and with her husband. Her presenting problem was that her children ran over her and she could not make them listen to her. Her disturbing degree of compliance with her take-charge husband was a profound example of the same problem. Therapy failed because her pattern of being passive and compliant was reenacted with both therapists. Unwittingly, both therapists provided a hierarchical relationship in which she remained the passive helpee led by an all-knowing helper. This interpersonal process reinforced her pathogenic belief that the source of strength and the ability to solve problems did not reside in her but in the therapist or others.

In order for Rachel to change, she needs to experience a therapeutic relationship that behaviorally affirms her own efficacy. This would be a relationship in which she is actively encouraged to initiate what she wants to explore in treatment, invited to set personal limits with and share control with the therapist, and more overtly supported by the therapist in making her own decisions and choices. The client's symptoms will visibly improve in the context of a new interpersonal process in which the therapist

- encourages her to act more independently *in their relationship* and
- joins with her in collaboratively exploring the anxiety and guilt that arises each time she tries to act more assertively or independently with the therapist and with others in her life.

This experience of change with the therapist behaviorally shows her that her own strengths are valued and can be used, in conjunction with the therapist's skill and understanding, in a more productive relationship.

Unfortunately, both of Rachel's therapists were comfortable with the hierarchical helper–helpee relational process and did not bring it up as a focus for treatment. Neither therapist offered the meta-communication that, in certain ways, their current interaction was reenacting the same maladaptive relational pattern that was problematic for Rachel in other relationships. Such a process comment, tactfully shared, would have allowed Rachel to begin looking at this maladaptive relational pattern in a supportive environment. More importantly, it would have given her permission to change her dependent coping style *with the therapist* and to begin expressing her own feelings, interests, and authentic voice *in their relationship*. The sequence is often important: This would likely occur first, in her relationship with the therapist and then, in her relationships with others.

What do these changes in the ways she interacts with the therapist have to do with solving the real-life problems with others that brought her to treatment? As long as the therapists were telling her what to do, Rachel could not set limits with her children, take a more assertive stance with her husband, become more aware of her own genuine interests and goals, or follow through and act on what she wanted to do. That is, the highly relevant content of what she discussed with both therapists (issues about autonomy, assertiveness, and her own identity) was not matched by the interpersonal process they enacted. *Unless Rachel has the actual experience of behaving as an active, equal participant in her relationship with the therapist, she will not be able to adopt this stronger stance in other areas of her life.* In other words, the process must be congruent with the content.

To enact a more egalitarian or collaborative relationship with someone like Rachel is not a simple task for the therapist, however. It requires therapeutic skill and thoughtful monitoring of the process dimension. In response to her life experiences, Rachel has become accomplished at getting her therapists, her husband, her children, and others to lead, direct, and control her. Both the behaviorally and analytically oriented therapists could have successfully used their differing theories and techniques to help Rachel if they had enacted a different process in their relationship. If they had encouraged her to initiate more in the session, and then focused together on exploring her reluctance to lead, Rachel's generic conflict would have emerged quite overtly in the therapeutic relationship. That is, as soon as either of the therapists invited her to follow her own agenda and bring up whatever she felt was most important to talk about, encouraged her to disagree with or express any dissatisfaction she may have with the therapist, or celebrated whenever she was acting stronger or behaving competently in the therapeutic relationship, Rachel would have become anxious. At that moment, either therapist could have focused her inward on this anxiety so that they could begin exploring together what the threat or danger was for her to step out of her compliant and help-seeking mode (such as the belief that others would leave her if she wasn't always pleasing and "nice"; that she would be acting like her dominating mother, whom she was always afraid of being like; and so forth).

Simultaneously, either therapist could have given her permission to act more assertively within the therapeutic relationship and responded to her in ways that facilitated this new behavior in their interaction together. The therapist could do this in two ways. First, the therapist could watch for and affirm this effective new stance whenever it emerged in their relationship or with others. Second, the therapist could respond to instances in which Rachel "undid" herself by retreating to the safer, nonassertive mode just after she had risked acting in a stronger way with the therapist. Given Rachel's life experiences and interpersonal coping style, she was likely to become apologetic or confused, or to act dependently and ask for direction, soon after disagreeing with the therapist, making an insightful connection on her own, or redirecting the session more toward her own interests or concerns at that moment. If therapists have formulated working hypotheses about her maladaptive relational patterns, they can track this potential reenactment in their interpersonal process. If so, they will be prepared to help her identify this problematic pattern as it is occurring, and then explore the threat or danger she feels when she has just retreated from her new, stronger stance with the therapist:

> THERAPIST: What do you think I might be feeling, and how are you afraid I might respond, after you have just acted more assertively by disagreeing with me like that?

With both of her therapists, however, Rachel only continued to defend against her anxiety over being more assertive, independent, and successful by retreating to her dependent, help-seeking role and, in their interpersonal process, merely reenacting this maladaptive relational pattern.

In most therapeutic relationships, the therapist and client will temporarily reenact the client's core conflict in their interpersonal process—it just happens.

However, in successful therapy of every theoretical orientation, the therapist and client do not *continue* to reenact the maladaptive relational pattern in an ongoing way. Instead, they are able to recognize this problematic pattern (and perhaps give this pattern a name to help them talk about it, such as "going along" for Rachel) and work out a different type of relationship that changes the course of the old familiar scenario. Once clients find that their conflicts can be activated or come in to play with the therapist, but do not have to result in the same hurtful or frustrating outcomes they have come to expect, their schemas and expectations expand and become more flexible. At this pivotal point in treatment, it is relatively easy for therapists to move to the next phase of treatment and help clients generalize this experiential relearning to other relationships in their lives.

Example 4: Resolving Conflicts by Working with the Process Dimension   Excessive performance demands often leave trainees feeling pressure to do something to make their clients change. Unfortunately, these internal pressures on the therapist are usually translated into interpersonal pressures on the client to change, often before either the therapist or the client know what's really wrong, let alone what they want to do about it. This is often evidenced by a premature emphasis on intervention techniques and usually at the expense of too little exploration and understanding. Unless the therapist has conceptualized what has gone awry for the client in other relationships and considered how these maladaptive relational patterns could be reenacted in their therapeutic process, intervention techniques will often fail. In contrast, once the therapist has several good working hypotheses about the interaction between the therapist and client—and the specific relational experiences that this client needs in order to change—it is usually easy to find effective ways to intervene. Therapists are encouraged to ask themselves repeatedly, "What does this mean?" rather than, "What should I do?" As we will see, the second question is usually answered by the first.

Therapists can employ a wide range of techniques from different theoretical modalities (Wachtel, 1997). The key, then, is to observe closely the client's response to what they have just said or done (Hill, 2004; Weiss, 1993). That is, the therapist evaluates the effectiveness of each successive intervention in terms of the client's ability to utilize this type of response to make progress in treatment. Based on this behavioral feedback from the client, *therapists need the personal flexibility to modify their interventions* to provide the responses that work best for this particular client (Teyber & McClure, 2000). Unfortunately, researchers find that many therapists do not flexibly alter their interventions to match the client's needs, but dogmatically stick to the same approach whether the client is finding it helpful or not (Najavits & Strupp, 1994).

To illustrate the need to use our understanding to guide our interventions, we examine two critical incidents in the treatment of an incest survivor. In these incidents, the therapist uses intervention techniques that have a highly significant impact on the client: validating the resistance and role-playing. It is not the intervention techniques in themselves that facilitate change, however. Both interventions are effective because they follow from the therapist's understanding of what is transpiring in their current interaction, and because the therapist then is able to work with these issues in the immediacy of the client–therapist relationship.

Early in treatment, Sandy had confided to her male therapist that she was an incest survivor. Grasping this profound betrayal, he anticipated that trust was likely to become a central issue in their relationship. Therapy had gotten off to a good start but, before long, progress began to slow as the material that Sandy presented became repetitive. About this time, Sandy recounted two different narratives in which the relational theme was feeling unsafe with men. Based on this and other material they had been talking about, the therapist hypothesized that, without being aware of it, Sandy was using these vignettes to broach the topic of trust in their relationship. As their relationship was becoming more important to Sandy, it seemed that her deep concerns about safety and betrayal were now being activated with him.

Responding to these "embedded messages" about their relationship, the therapist began to talk with Sandy about trust between them and asked about the different thoughts and feelings she was having toward him. Sandy genuinely liked the therapist and was finding him helpful. However, when the therapist explored the trust issue and asked specifically whether she felt safe with him, her affirmative response sounded half-hearted and unconvincing. It soon became clear to both of them that Sandy was emotionally removing herself from the therapist as they talked more directly about safety in their relationship.

The therapist responded to Sandy's concern by affirming both sides of her ambivalent feelings and, using immediacy, working with her concerns directly in terms of their relationship.

THERAPIST: I know that one part of you likes and trusts me, but it makes sense to me that another part of you doesn't feel safe. I think that both sides of your feelings toward me are valid and important for us to work with.

SANDY: (*sheepishly nodding and gesturing vaguely to indicate that this was true*)

Because they had been talking about issues of trust and betrayal for some time and this groundwork had been prepared, the therapist thought that this might be an opportunity to try and go further with these issues. Rather than trying to talk her out of these concerns or convince her of his trustworthiness, the therapist validated Sandy's distrust and actually went on to articulate or develop it more fully:

THERAPIST: If I violated your trust in some way after you had taken the risk to ask me to help you, it would be very bad for you. If I tried to approach you sexually or foster any other kind of relationship between us, it would hurt you very much. Maybe it would even hurt you so much that you might not be able to risk trusting or asking for help again.

SANDY: (*tearing, looked at the therapist and slowly nodded in agreement*)

The therapist continued to elaborate Sandy's concerns about mistrust and exploitation from men in general within the immediacy of their relationship:

> THERAPIST: If I took advantage of our relationship in some way, I think you would feel hopelessly betrayed. I think you might become very depressed again, enter into other relationships that would be hurtful or exploitative, and may even start thinking again that you do not want to be alive.

As the therapist spoke and made the concerns that they had been discussing with others overt in terms of their relationship, he observed that her entire demeanor changed. She remained tearful but became alert and present, nodding agreement with what he was saying. As they continued to talk together about this, the therapist went on to describe his own attitude, and reframe her distrust as a strength:

> THERAPIST: Yes, I can see how much it would hurt you if I betrayed your trust, and no part of me wants you to have that wounding experience again. In fact, I respect the cautious part of you that isn't sure about trusting me. That distrustful part of you is your ally—it's a strength of yours. We need her—because she is committed to not letting you get hurt again. That's why it's so important that we go at your speed in here, that you have control over what we talk about, and that you can say no to me and know that I will honor your limits.

After the therapist responded so affirmingly to Sandy's resistance, and worked with this previously unspoken but dreaded expectation of betrayal in terms of their relationship, important changes occurred. Whereas in the past Sandy had only alluded vaguely to her childhood abuse, she now chose to share it more explicitly. Over the next weeks, she recounted in painful detail, and with strong emotions, how she had been molested over a period of years by her stepbrother, who was 12 years older. When she first went to her mother for help, her mother was not supportive, denied that the abuse was occurring, and told Sandy never to talk about anything like that again. Sandy did not even consider seeking protection from her stepfather, who had always been distant and unresponsive. After failing to receive help, Sandy recalled sitting alone on the floor of her closet for long periods with the door closed, feeling afraid and ashamed. Sandy said that this was the point in her life when she concluded that she was "just always going to be alone." In response to this deep sharing, the therapist was compassionate, validated her experience, and began to help her with the many significant connections she began to make between this familial tragedy and the symptoms and problems she was struggling with in her current life.

The therapist had already learned in their work together that role playing was an effective intervention that Sandy enjoyed and found useful. During one of these sessions, the therapist talked with Sandy about using a role-playing technique to provide a different response to the abuse she had suffered, and Sandy welcomed the suggestion.

> THERAPIST: I wish someone could have been there to stop him and protect you. No one was there for you but, if I had been there, I would have walked into your bedroom when he was there, turned on the light, and in a loud

voice commanded, "Stop it! Get away from her and leave her alone right now! I see what you're doing and it's not fair. You're hurting her, and I won't allow it."

After speaking this forcefully, as if he were actually saying it to the perpetrator, the therapist rolled Sandy's coat into a ball. Using it as a little Sandy doll, he spoke to it reassuringly:

> THERAPIST: You're safe now, and he's gone for good. I'm going to call the police now and help protect you so you'll never have to worry about him hurting you again.

As the therapist metaphorically gave Sandy the protective response that she desperately longed for as a girl but did not get, the full intensity of her pain and shame was evoked. As these important feelings ran their course, Sandy became more composed and said, "I'm not all alone anymore, I'm going to be all right." The therapist, still holding the rolled-up jacket tenderly, asked Sandy if she could join in and help him take care of this little girl who still needed help. Sandy's affirmative response was almost joyous, and the therapist carefully tucked it in her arms.

> THERAPIST: This is the little girl you were. She needs you to open up your heart to her and give her a home. You need to take care of her—and not push her away anymore like they did then and you have done since. You need to hold her, talk to her, and listen to what she tells you. I want you to join me in taking care of this part of you, so this little girl is no longer alone behind the closet door.

Sandy readily accepted this responsibility and later bought a doll that she used to represent the part of her that needed to be cared for but had not been protected. She had always thought of herself as "ugly" but went to considerable effort to find and purchase a doll for herself that she felt was pretty.

Sandy entered therapy almost 25 years after the abuse, anxious, unassertive, and almost incapable of leaving home alone. Just weeks after this session, however, she got her driver's license renewed and began driving again for the first time in several years. She also got a job as a waitress—her first paid employment in 6 years. Although she had always foiled any type of success for herself, she enrolled in a local community college and began receiving A's in many of her classes. Whereas Sandy had felt helpless, had characteristically acted as a victim, and had been repeatedly taken advantage of by others, she increasingly became more assertive and expansive. All of Sandy's problems did not go away, of course, but long-standing symptoms were resolved and did not return.

What allowed Sandy to get stronger in these significant ways? Several factors made this role-play intervention effective. By using this technique, the therapist brought Sandy's conflict into the therapeutic relationship and provided her with the reparative experiences she needed. In sharp contrast to what occurred in her family, this time Sandy experienced protection, validation, appropriate boundaries, and a supportive holding environment. This corrective experience with the therapist disconfirmed her pathogenic belief that she does not matter enough to be protected or looked after. The therapist's compassionate response also helped Sandy shift from an identification with her parent's rejection of her—and her own resulting shame and self-hatred—to an identification with the therapist and his compassion for her. Eventually, this healthy new identification led her to be able to care better for herself and believe that she did matter and was not alone with her problems for the first time in her life. These far-reaching consequences were set in motion by the therapist's affirming response to Sandy's vulnerability. Even though it was enacted in role-play, the therapist provided Sandy with a corrective emotional experience that played an important part in allowing her to become stronger and more capable of protecting herself.

Example 5: Recognizing the Process Dimension in All Interpersonal Relations
As we have seen, the most likely reason for counseling to fail is that the therapist and client unwittingly reenact aspects of the client's problems in their interpersonal process. At first, therapists in training often find it challenging to track the process dimension, and to distinguish the process that is being enacted from the content that is discussed. For our final example, let's look at how the process dimension operates outside the therapy setting—at a sporting event.

In a championship basketball game, the defending champions confidently face an underdog team made up of talented but younger players who are far less experienced. The defending champions are heavily favored to win.

Late in the fourth quarter, the game is tied. The fans, on their feet, are cheering at the prospect of an upset. However, the coach of the young team can feel that the game is about to slip away. The screaming fans, the pressure of time, and the unrelenting play of their opponent is rattling his players. To help compose his less experienced players, the coach calls a 2-minute time-out. His team has played well up to this point but the seasoned coach knows that, in these pressure-packed closing minutes, the other team's experience is likely to make the difference.

Crouched on the sideline, the coach positions his five players in a semicircle in front of him. With the roaring crowd only a few feet away, the coach quietly makes eye contact with the first player on his left. Without speaking, he calmly holds the first player's gaze for about 10 seconds (a very long time amid such pandemonium). Having compelled the first player's attention, the coach turns to the second player and maintains eye contact with him for another 10 seconds, and so on with each of the five players. Each time, as he turns to engage the next player, the previous player's attention remains riveted on the coach. Finally, after making contact with each player in this way, the coach holds up a chalkboard with one word written on it: *POISE*.

For the remaining seconds of the time-out, the players and coach sit together silently. The referee blows his whistle, the players return to the court, and the underdogs defeat the defending champions. The players did not make mental errors in the closing minutes of the game. They did not turn the ball over to the other team; they made clutch-free throws in the final seconds, and they smoothly executed the plays they had practiced all year. In a word, they played with poise.

This coach astutely identified the key issue that could defeat his team in the final minutes: being intimidated by a more confident and experienced team, losing their composure, and making mental mistakes that would cost them the game. Having recognized the central issue, however, the coach did not just *tell* his team that they had to remain poised. Instead, he gave them the *experience* of composure during the time-out. The players were able to finish the game with such composure because the coach used the relationship he had developed with his players to give them the experience of composure during the final tense minutes. During the time-out, the players were able to generalize this experience to their performance on the court and play with poise.

Similarly, it is far more effective if the therapist–client interaction provides an experience of change than if the therapist merely tells the client what to do or explains what something means. It can be challenging to provide a relationship that enacts a resolution of clients' conflicts, however, and there are no simple formulas that tell the therapist how to do so. Every client really is unique. The therapist needs to be willing to take the personal risk of entering into a relationship that matters with each client, and have the flexibility to provide the new relational experiences that this particular client needs in order to change.

Providing a Corrective Emotional Experience

As we saw in Chapter 8, the first step in working with the process dimension is to attend to clients' maladaptive relational patterns. The therapist is trying to identify how the same problematic scenarios that are causing problems with others in the client's life are starting to be played out or reenacted in the therapist–client interaction as well. As the therapist and client begin to identify these faulty patterns with others, they can start to highlight or punctuate when similar patterns are beginning to occur between them. Working collaboratively, they begin to discuss what's going on between them and find new ways to interact that resolve, rather than reenact, the maladaptive relational patterns that keep occurring with others. In this way, the therapist is trying to give clients the real-life experience that clients do not have to be controlled, judged, ignored, idealized, and so forth, in this relationship as they often have been in the past. As clients live out this experience of change with the therapist, *the therapist can then help clients generalize or transfer this in vivo relearning beyond the therapy setting.* That is, based on this real-life experience of change with the therapist—not just an interpretation or explanation, clients' cognitive schemas for relationships change and their interpersonal range expands.

At this point in treatment, it is relatively easy for the therapist to help the client sort through relationships with others more realistically. By trying out new ways of responding to others in the client's life, the therapist and client together can begin to *discern* others with whom the problematic patterns and unwanted responses continue to occur, and others who are able to respond in more affirming, accepting, or rewarding ways. For example, the client may test the waters again and find out that his father still has to "be right" all the time, just as he did years ago. In contrast, he also finds that his mother is less rigid than his father and can now listen or hear him better than when he was growing up. At times, however, she still tries to demand that he just comply and do everything her way. In this reassessment, he also clarifies that his girl-friend really is different than his parents and can be very cooperative, but he needs to speak up more clearly and better communicate to her what he does and doesn't want.

The key point here is that *issues that have caused problems for the client in other relationships can be made overt and talked about—especially when they begin to occur in the therapeutic relationship—and resolved differently with the therapist.* The best way to begin working with the interpersonal process and providing this corrective emotional experience is by asking clients, in an open-ended way, how the problem that they are talking about in relation to others could be occurring in their relationship as well:

> THERAPIST: I'm wondering if this battle for control that you have been describing with your partner ever goes on between us here in counseling. Does it ever seem like I am trying to control you or that you are controlling me?

> OR

> THERAPIST: It sounds like others have often criticized or "judged" you, and it really hasn't felt safe for you to talk about important things. I'm wondering if you have ever felt criticized by me or been afraid that I will judge you too?

By bringing the client's conflicts with others directly into the therapeutic relationship in this way, the therapist creates the opportunity to disconfirm the pathogenic belief and provide a corrective emotional experience. In order to change with others, first the client needs to find that the therapist is neither trying to control nor willing to be controlled by the client, as occurs for the client in other important relationships. Only if the client has this real-life experience of change can the client resolve this conflict, find that sometimes control can be shared in close relationships, and learn that some relationships can be different from those in the past. This immediate, real-life experience of change is a powerful way to intervene. As we have emphasized, however, many new therapists find it challenging to make this process comment or link the client's problems with others to how the therapist and client may be interacting together right now. Thus, the purpose of the next section is to help therapists make process-oriented interventions and take the risk of using their real-life relationships with clients to facilitate change.

Interpersonal Process Interventions

The interventions we have discussed have repeatedly invited the therapist to work on the client's problems within the *immediacy* of the therapist–client relationship. The following examples review how process-oriented interventions link clients' problems with others to the here-and-now relationship between the client and therapist. At the beginning of treatment, for example, clinicians can speak directly to clients about their current interaction when they are trying to use accurate empathy to establish a *working alliance* (Chapter 2):

THERAPIST: As we talk about these things, does it feel like I am understanding what is most important for you here—am I getting it right?

Talking about what is going on between the therapist and client becomes even more important in working with *resistance* (Chapter 3):

THERAPIST: What's it been like for you to talk with me today? What's felt good, and what hasn't?

Working directly with the current interaction is also helpful when establishing an *internal focus* (Chapter 4):

THERAPIST: Maybe we are arm wrestling a bit here today. It seems to me that you keep talking about what others are doing, and I keep asking about what you are thinking or how you responded. What do you see going on between us?

Therapists also want to work with the *client's feelings* in a way that brings the full intensity of whatever feelings the client is experiencing into the immediacy of the therapist–client relationship (Chapter 5):

THERAPIST: I can see how much this has hurt you, and how sad you are feeling right now.

Moreover, therapists intervene in the here and now by providing clients with *interpersonal feedback* about the impact their interpersonal coping style is having on others (Chapter 7):

THERAPIST: Aaron, may I have permission to give you some feedback about how you come across to me at times—and maybe to others, too? I'm not sure you are aware of the effect you have on others sometimes, but I want you to know I'm bringing up this sensitive issue with good intentions. Outside of therapy, people usually don't talk together so directly, and I want to make sure this is a helpful conversation. Can I share my thoughts with you?

Finally, this focus on the current interaction between the therapist and the client is central to working with *transference reactions* (Chapter 8):

THERAPIST: How do you think I am going to react to you if you do that? What am I likely to be thinking or feeling inside?

The unifying theme in all of these process-oriented interventions is to *link the client's issues and concerns with others to what is currently going on between the therapist and the client*. In this way, the client's problems with others are not just talked about in abstract discussions about others; they are reexperienced and potentially resolved in the real-life relationship with the therapist. In turn, this in vivo or experiential relearning with the therapist empowers clients to begin changing how they are responding to similar problems with significant others in their lives.

To illustrate, let's return to the previous client whose basic relational theme is repeatedly getting locked in control battles with others. To begin resolving this client's problem with others, the therapist's intention is to find an effective way (that is, not blaming or confrontational but tactful) to make this control issue overt and put it on the table for discussion. The therapist does this by wondering aloud or tentatively inquiring about this possibility, so the therapist and client can begin to explore it directly in terms of their relationship.

> THERAPIST: Does this type of control battle that you have been describing with others ever get going between us, too? Does it ever seem like you and I are jockeying over control in our relationship?

Making such a process comment and bringing the client's concerns with others into the here and now may seem a dreadful prospect to new therapists. However, by judiciously taking this risk, the therapist can bring real-life *immediacy* to the therapist–client interaction. Problems with others are not just being talked about intellectually; instead, for the first time perhaps, they are being addressed honestly and constructively with someone *as they occur*. If the therapist waits for her own sense of good timing (for example, the therapist may say to herself, "I think this might work right now") and responds in a respectful manner or tone, the potential for change is at hand.

Let's consider a range of three progressively more challenging responses to the therapist's query about control issues in their relationship.

Response 1: The Least Challenging to the Therapist Is a Disconfirming Response

An example of a disconfirming response from the client might be as follows:

> CLIENT: No, I don't feel like we are in a control battle—that's not going on here.

The therapist can then use this response to begin exploring what is different about their relationship.

> THERAPIST: Good, I'm glad that's not a problem for us. Any ideas about what makes our relationship different?

The client's answer may provide useful information:

> CLIENT: You treat me with respect; that's what's different.

The therapist can follow up on this comment by exploring the client's concerns about not being respected in other relationships. The therapist also can build on it to establish that mutually respectful relationships can be developed with some others in his life as well. Further, the therapist is modeling for this client how he can find diplomatic ways to talk with others when he feels that an unwanted control battle is coming in to play.

Response 2: Clients May Avoid the Immediacy of Their Conflicts by Discounting the Therapist's Relevance

> CLIENT: We're not in a control battle because this isn't a real relationship. You're just my therapist.

Clients make far more significant gains in treatment when this distancing defense is resolved. At this point, it is essential to clarify that the therapist and the client really are two people who have been having a relationship for some time, even though there are constraints on their relationship. The therapist can then begin to explore the threats aroused for the client by more meaningful involvement, such as being controlled or not being respected. The therapist and the client can then agree to watch for these concerns and address them if they arise in their relationship. It is usually more effective if therapists do not bring up these relational patterns in the abstract but wait until they think the client might be experiencing them right now, in their current interaction, before addressing them.

Response 3: Clients May Respond to the Therapist's Question Affirmatively

> CLIENT: (*in an exasperated tone*) Of course you're in control! You insist that we stop at 10 to the hour, whether I'm finished or not. And you're subtle about it, but I see you trying to take control all the time and make things go where you want.

Most new therapists fear such an accusing response and eschew process comments to safeguard against such criticism. Too often, student therapists take the client's comment at face value and believe that they have done something wrong and actually been too controlling if the client disapproves of them in this way. Perhaps the therapist actually has been too controlling—this possibility must always be considered and owned if necessary. In most cases, however, the client's reactions have as much or more to do with the client's own cognitive schemas than they do with the reality of how the therapist has responded. Thus, rather than posing a problem, this affirmative response from the client actually provides an important window of opportunity.

Openly acknowledging that the same conflict the client has with others is also occurring right now with the therapist creates the opportunity to explore this issue directly and begin working together to try and resolve it in their relationship. Why do we keep returning to this same point? As long as the client is feeling controlled by the therapist, the therapist is not going to succeed in helping the client change this problem with others in his life either. However, by utilizing the therapeutic relationship as a social learning lab, the therapist can use their interaction to begin changing this maladaptive pattern or schema in the following ways:

1. By remaining nondefensive and accepting the validity of the client's concerns whenever possible:

> THERAPIST: Yes, we do have to stop at 10 to the hour, and that is an unnatural ending for you. I can see how the control issue is brought up by those time constraints.

2. By exploring the client's perceptions further and working with the client to understand them better:

> THERAPIST: What do you think is going on for me when I try to take some things we talk about in a certain direction? What does it seem like I am trying to do at those times—what might my intentions be?

3. By "differentiating" the therapist from other figures in the client's life:

> THERAPIST: Yes, I have had ideas about where we should go, but I have also been interested in following your lead, too. In fact, I may be different from some other people in your life because I genuinely liked it when you disagreed with me last week and told me what you thought. I like it when neither of us feels controlled and we can both say what we want. It makes our relationship feel more alive to me.

4. By offering to be sensitive to this concern in their relationship, expressing a willingness to handle the issue differently in the future, and inviting the client to tell the therapist right then whenever they think this conflict is occurring between them—so together they can talk it through and change it right then:

> THERAPIST: I don't want you to be controlled in our relationship as you have been in others. That's no good for you or anybody else. Let's try to do something about it. From now on, I will watch the clock and let you know when it's 5 minutes before we have to stop. And any time you feel like I am directing you away from where you want to go or being controlling in any way, tell me, and we'll stop right then, sort through what's happening, and change it. What else could we do to help with this? Tell me your ideas.

Adopting these approaches will work easily with some clients, whereas other clients will insist on reestablishing the same constricted relational patterns. However, *if therapists are willing to remain nondefensive and tolerate their own discomfort for a few moments*, most clients will become significantly more engaged in the therapeutic relationship and motivated to explore this and other related issues more fully. As clients repeatedly find that the expected but unwanted old scenario does not repeat with the therapist, they experience greater interpersonal safety than they have known before. As a result, important new issues, pathogenic beliefs, and previously threatening feelings can emerge, providing new material for exploration and clarifying the treatment focus. Routinely, many clients will make attempts to try out this new way of relating with others and test out what they have just experienced with the therapist. That is, clients often will return to the next session and report on their attempts to change these same types of patterns and respond in a different way to someone in their life. Of course, clients will have successes and failures in their attempts to change these problematic patterns with others, and the next chapter will explore closely what therapists can do to help at this pivotal new point in treatment.

Thus, the real-life experience of change within the therapeutic relationship is a powerful relearning experience for clients, and propels clients to begin exploring new ways of responding with others in their lives where the same problems and patterns are occurring. As we have emphasized, however, intervening so directly in the here and now may break the social rules and be uncomfortable for therapists in the beginning. To help with this, we now explore several reasons for such discomfort and offer some practical guidelines to help therapists intervene effectively with the process dimension.

Therapists' Initial Reluctance to Work with the Process Dimension

Karen, a first-year practicum student, was seeing her first client. Everything her client talked about seemed unimportant—she just rambled on and on about nothing. Karen felt bored and, even though she knew inside that the client also felt like this wasn't going anywhere, neither one wanted to say anything. A few sessions went by like this and Karen's awkward, bad feeling became worse. She stopped enjoying their sessions and soon began to dread them. Karen developed a knot in her stomach as she sat politely with the client and worried: "Being a nice, empathetic person isn't going anywhere"; "Maybe I'm in the wrong field"; "I don't think I can do this. . . ."

Karen's supervisor sensed her distress and tried to help. He suggested that Karen talk with the client about what she was feeling and check things out—maybe the client was feeling the same way, too? For example, the supervisor suggested that Karen reflect aloud on their interaction, and he tried to show her how by role playing things she might say, such as: "Sometimes I'm wondering if what you're talking about feels like what's really most important for you"; "How does it feel for you to talk with me—or others—about problems or personal things that are meaningful for you?"; or "What's it like for you to come each week and talk with me? What feels good and what isn't being helpful?" Karen knew that this is what she "should" do, but there was no way she could bring herself to say things like that. She couldn't break the social rules and be so forthright; she couldn't do that with her client or with anyone— she just couldn't. Karen had never talked this way in her family, and felt torn apart about by the possibility that she might hurt her client's feelings by being so direct.

After six sessions, the client told Karen how nice and what a good listener she was, but that she wanted to stop coming. Karen continued to be nice and act socially proper by agreeing but, inside, she felt devastated knowing that it could have been—should have been—so different. Her supervisor knew that this was a painful failure experience for Karen. Based on his experience with many other graduate student therapists, he reassured her that, in a year or two, these process comments could become natural ways of responding that felt like second-nature to her. Karen was skeptical but hoped it could be true.

In this section, we explore why some beginning therapists, like Karen, may be reluctant to make process-oriented interventions, talk forthrightly with clients about their current interaction together, and explore how aspects of the clients' problems with others may be occurring in the therapeutic relationship. We will examine six reasons why therapists in training may find it difficult to work in the here and now and create immediacy, provide clients with interpersonal feedback about the impact they are having on others, and utilize process comments that focus on the current interaction between the therapist and the client. We will provide guidelines to help therapists begin working in these initially challenging yet potent and rewarding ways.

Although the best vehicle for effecting change is often the therapeutic relationship, it can be anxiety arousing for beginning therapists to take the plunge and say:

NEW THERAPIST: Does that kind of problem ever go on here, too—you know, between us here in therapy?

Talking together so directly about issues between two people breaks the cultural norms and family rules that most therapists grew up with. As a result, and because of the immediacy that such forthright communication creates, some beginning therapists are reluctant to make process comments in the beginning and explore how clients may be experiencing parallel problems with the therapist that they are having with others.

A few beginning therapists find it easy to work with clients using a process approach. They have always been able to talk directly with others when problems came up in their relationships, and it makes sense to them that what repeatedly goes on for clients in other relationships will also come into play with the therapist at times. For these therapists who are more comfortable or familiar with approaching interpersonal conflict, it is enlivening to make potential problems or misunderstandings overt and talk about them openly with the client. The process approach simply gives them the permission they may need to work with clients in this engaging way, and they welcome the intensity of relating in such a deeply personal way.

For others, however, process comments are more difficult at first. Although it makes sense intellectually to link the problems that clients are having with others to the current interaction with the therapist, they find it harder to actually put this into practice. Because they have had so little direct clinical experience, most first- and second-year graduate students have little confidence. Understandably, they are not eager to have what little self-confidence they may possess shaken by stepping outside of familiar bounds. Furthermore, even if they think that this type of intervention might be useful with a particular client, most beginning therapists struggle to discern whether they are objectively observing the client's behavior or whether their perceptions simply reflect their own countertransference. For example:

THERAPIST: (*internally*) Is this just me, or does she make everybody feel this way?

Finally, even if therapists feel confident that it is indeed the client's maladaptive relational pattern and what the client tends to elicit from others as well, *they may be unsure of how best to address this reenactment and find a helpful way to make it overt—without making the client feel criticized or blamed.*

Still other therapists feel concerned that to be forthright with clients could be hurtful because it would draw out or reveal the clients' pain. Others have seen directness used hurtfully—in angry confrontations or blaming personal attacks that were intended only to induce shame or guilt. Therapists never want to do this, of course, or intrude on their clients and violate their personal boundaries in any way (Kiesler, 1988). Indeed, diminishing clients' self-esteem or sense of personal safety in these ways will impair significantly their ability to make progress in treatment. However, clinicians readily can learn to address the process dimension in respectful ways that actually reassure clients, rather than make them feel uncomfortable. Let's examine six common concerns therapists have about using the therapeutic relationship to intervene with clients' problems and suggest some solutions.

Uncertainty about When to Intervene   Because new therapists have not had much experience attending to the process dimension or tracking it with clients, they are often unsure when or how the client's conflicts are potentially being played out in the therapeutic relationship. As trainees gain more experience and begin to integrate all of the complex new information their clinical training presents, most will become more comfortable intervening with the process dimension. The usual sequence is, at first, therapists will find that they are beginning to recognize the process dimension and see when something important is going on between them. However, it is often weeks or months later until new therapists feel confident enough to begin speaking up and inquiring about this possibility with the client. It is something of a personal risk for therapists to suggest:

> THERAPIST: I'm noticing something that may be going on between us right now, and wondering if. . . .

Although therapists will be accurate with some of their process observations, the client will not resonate with others. Therapists have not failed in any way or made a mistake when the observation they have tentatively suggested is inaccurate; they are simply trying to understand, and clients usually appreciate these good intentions. As emphasized earlier, the therapist's aim is not to be "right." Instead, they are trying to initiate a mutual collaboration or dialogue with the client—so together they can consider and explore what may be going on between them:

> THERAPIST: (*nondefensively, in a friendly and welcoming tone*) OK, what I'm suggesting doesn't quite fit. Help me say it more accurately. What are your words for what might be going on between us here?

Nevertheless, novice therapists should not be in a hurry to make process comments or use other interventions that create immediacy until they feel ready to do so. Forcing such responses will not be good for the therapist or effective for the client. If therapists are disempowered because they do not feel in control of choosing how and when they intervene with their client, in turn, they will not be capable of empowering the client.

How can therapists discern when the client's conflicts are being reenacted between them? Let's follow a five-step sequence. First, it is helpful if the therapist has tried to identify the client's maladaptive relational patterns with others. By formulating working hypotheses about what goes wrong with others—and how those same patterns or similar themes could come about in their interaction—the therapist is better prepared to respond to what this particular client is likely to present in treatment. Thus when therapists can anticipate the types of expectations, distortions, and reenactments that this client is likely to enact with them, therapists do not have to "think on their feet" as much. They better can "see" what is going on between them or "hear" what the client is really saying—right now as the client is saying it, rather than "getting it" later (a trainee reviewing a tape recording of his session might comment, "She's telling me that *everybody* judges her. She's probably telling me that she's worried that

I'm judging her and being critical, too. Why didn't I get that right then when she was saying it? This always happens!").

Second, these working hypotheses will help therapists make sense of their own experience and recognize when they are starting to feel or respond as others do in the client's old scenario. For example, when the client is making the therapist feel bored, impatient, or overwhelmed, these working hypotheses will help therapists consider the possibility that they are beginning to react as others in the client's life often do. That is, that their feelings and reactions toward the client may be informing them that the same scenario that causes problems with others may be under way here, too.

Third, this is a good time for the therapist to consult with a supervisor and check out together what may be occurring in the therapeutic interaction. If it does seem as if a reenactment is underway, supervisees can explore any concerns they may have about making this interaction overt the next time it comes up and talking it through with the client. Looking for effective ways to explore this possibility with the client, therapists can also role-play or rehearse alternative responses with the supervisor. Again, it is not realistic for therapists in training to be able to successfully explore and try out these process-oriented interventions without guidance from a supportive supervisor.

Fourth, with this preparation, therapists can wait to broach this possibility with the client until the next time they feel this reenactment may be occurring. *It is more effective to intervene at the moment when the interaction is occurring* (immediacy) *rather than bringing it up in the abstract* (for example, at the beginning of the next session). Recall the therapist earlier who realized, after the session, that his client likely was feeling judged or criticized by him. He did not start the next session by bringing up this possibility and talking with the client about it—as he wanted to do. Instead, he listened to his supervisor's advice to wait until he thought this issue might be occurring between them again, which did come up again about 20 minutes into their next session, and successfully inquired about this possibility *as it was occurring*.

Fifth, beginning therapists are encouraged to wait until they feel that this intervention (or any other) is likely to work. We need to respect our own sense of timing and listen to our own feeling that, "This just isn't going to work right now" versus "This might be a good time to say this." As emphasized, it is important that therapists *choose* whether to make this or any other intervention, and when to try it—compliance is just as problematic for therapists as it is for clients.

It will often take a year or two before these responses feel natural and come easily—but they will. Talking about "you and me," about what is going on between us right now and how we are interacting together, is a challenging (and exciting) new way of responding that brings real intensity to the therapist–client relationship. For most beginning therapists, however, it often breaks cultural norms and familial rules. Good advice for the novice therapist is to be patient with yourself, practice with classmates, and ask your instructors/supervisors to role-play and demonstrate these immediacy interventions.

Fear of Offending the Client   Some beginning therapists are concerned that the client will feel they are being too blunt, personal, or confrontational if they ask about what may be going on between them:

THERAPIST: What's it like to talk with me about this?

Therapists might worry that the client might feel angry or hurt, not like them, or—worse yet—just stare blankly back at them! Of course, process comments, like any other interventions, can be made in blunt, accusatory, insensitive, demanding, or otherwise ineffective ways. Warmth, tact, and a good sense of humor can all go a long way toward making every intervention more effective.

Therapists may also be worried about trespassing on social norms and expectations. Speaking respectfully yet forthrightly about what may be going on between two people may break unspoken social rules, and most trainees are concerned that clients may be surprised or taken back by this. To help ease the transition toward this more genuine dialogue, therapists can offer *contextual remarks* that facilitate the bid for more open or authentic communication. For example, the first time the therapist addresses the process dimension with a client, the therapist can create safety for the client by offering an introductory remark that acknowledges the shift to another level of discourse:

- Can we be forthright with each other and speak directly about something?
- Let me break the social rules for a minute and ask about something that might be going on between us.
- I know that people don't usually talk together this way, but I think it would help if we could talk about. . . .

Therapists will not be too blunt, take clients off guard, or offend clients if they respond respectfully and provide transition comments such as these. These contextual remarks are very effective in helping clients understand the therapist's good intentions as they help clients shift from the socially polite to this more straightforward approach. Rather than being threatened by this invitation for more forthright communication, most clients want this more authentic and substantive discussion. Welcoming it from the beginning, they are reassured by the therapist's willingness to get down to business and talk in plain terms about what's really wrong in their lives—something they often wanted but have not been able to do with others.

Therapists' Own Insecurities and Countertransference Issues   To work with clients in the highly personal manner described here, therapists are challenged to take the risk of engaging in a genuine relationship with the client. In the interpersonal process approach, therapy is a two-way street where each participant is willing to be affected by the other. Process comments lose their impact when therapists are not willing to be emotionally present and engaged in the "real relationship" with the client (Gelso & Carter, 1994). For example, the client receives a contradictory or mixed message from the therapist that is highly problematic when the therapist makes a process comment, inviting a more genuine dialogue with the client, but does not follow through. This mixed message occurs when the therapist becomes defensive and distances himself from the client, is unwilling to consider the potential validity of the client's criticism, or honestly explore the client's dissatisfaction with something the therapist has done. For example:

THERAPIST: What do you think as I say that to you?

CLIENT: Well, I feel like you're criticizing me a little bit there, and I guess it doesn't feel very good.

THERAPIST: (*defensive*) Well, of course I'm not being critical—maybe you shouldn't be so sensitive. Do you overreact like this to other people, too?

In contrast, when therapists are willing to remain nondefensive and talk with the client about their participation in problems and what is occurring between them, sessions become more intense and productive. Therapists' willingness to be personally affected by the client and to look at their own contribution to their interpersonal process will at times be anxiety arousing for all therapists. We wouldn't want it to be otherwise—nothing ventured, nothing gained. A corrective emotional experience does not occur unless the relationship is significant and holds real meaning for *both* the client and the therapist.

By looking honestly at their own contribution to problems or misunderstandings in the relationship, therapists facilitate an egalitarian relationship that holds genuine meaning for both. Of course, this increasing mutuality will activate therapists' own personal problems or countertransference issues at times (Jordan, Kaplan, Miller, Striver, & Surrey, 1991; Mitchell, 1993). For example, in order to respond more authentically, therapists need to be able to relinquish hierarchical control over the relationship, which will arouse anxiety for some therapists who need to be the authority, in charge of what happens next, or one-up in relationships. Therapists may also be concerned that this genuine responsiveness or emotional presence with the client will lead to a loss of appropriate therapeutic boundaries and result in overinvolvement or acting out on the part of the client or therapist. However, by consulting with a supervisor and applying the guidelines on optimum interpersonal relatedness in Chapter 6, therapists will be able to recognize when their own countertransference issues are prompting them to become too close to—or too distant from—the client.

In making process comments, how do therapists know whether what is going on is due to their own personal issues or to the client's dynamics—that is, to client-induced or therapist-induced countertransference? Therapists have two useful methods of distinguishing whether the client's dynamics—eliciting maneuvers, testing behavior, transference distortions, and so forth—or the therapist's own personal issues are operating. Here again, to help sort out their own issues from the client's, new therapists need a supportive supervisor who helps track the therapeutic process. Especially in the beginning, it is unrealistic to expect trainees to be a participant-observer who can maintain a balance of emotional relatedness and objectivity without ongoing assistance. This is a normative, developmental issue in clinical training. New therapists should be thinking about this distinction and working hard to clarify what is the client's issue and what is their own, but they should not worry or feel badly if they are having trouble distinguishing this in the beginning. It is complex and ambiguous, so let your supervisor help you.

Another way beginning therapists can safeguard against confusing their own and clients' issues is by not jumping in and making observations about the therapist–client relationship as soon as they see something significant happening. Hold back and follow the five-step sequence suggested here. It is more effective for the therapist to generate working hypotheses first, and then wait to see whether the observation also applies to subsequent exchanges between therapist and client. *If the issue or concern is relevant for the client, it will repeat as a theme or pattern. If it does not keep coming up, then it is likely that the therapist's own issues are involved or that it is not an important concern for the client and this hypothesis should be discarded.*

Thus, therapists are not cavalier about making process comments. When in doubt about whether an issue is the therapist's or the client's, it is best for the therapist to wait, gather additional information about the therapeutic process, and/or consult with a supervisor before raising the issue with the client. Remember, too, that all process comments are simply observations. They are offered to the client *tentatively*, as possibilities for joint exploration and mutual clarification, not as truth or fact.

Across varying treatment approaches, the therapist's relationship with the client may be the most significant means of effecting change (Norcross, 2002; Seligman, 1995; Najavits & Strupp, 1994). At the same time, one of the most likely reasons for treatment to end prematurely is that the client's interpersonal coping strategies have activated the therapists' own conflicts (countertransference). That is, the therapist becomes invested in rescuing the client who presents as a victim, the therapist becomes defensive and competitive with the client who is challenging, the therapist emotionally disengages and gives up on the client who is aloof or distancing, and so forth. In this way, therapists need to remain open to the possibility that their own countertransference issues may have come into play, and should generate working hypotheses about how each client's conflicts could be reenacted in treatment (see Appendix B, Part 6). Beginning and experienced therapists alike honor the therapeutic enterprise when they

- make a lifelong commitment to exploring and remaining open to their own countertransference propensities,
- consult with supervisors and colleagues as an ongoing career activity, and
- seek their own personal therapy when countertransference issues persist.

Countertransference issues are most likely to create problems when clinicians disregard them. The red flag goes up when clinicians are not willing to consider their own participation or potential contribution to the conflict. Therapists who are aware that they are susceptible to countertransference and who discuss possible instances of countertransference with supervisors as they arise should feel unfettered about working with clients in a process-oriented manner.

Concern about Appearing Confrontational   Some new therapists misunderstand process comments, misconstruing them as *confrontations*. In the typical scenario, these therapists are reluctant to try making a process comment because they anticipate that the client will be mad about being "confronted" in this hostile way, and angrily will walk away and leave treatment. *There should be nothing confrontational about a process comment. A process comment is a tentative wondering aloud, an invitation for a dialogue with the client about what may be going on between them right now.* If you think that the process comment you are about to make is going to make the client feel accused or blamed, or that it will "put the client on the spot," don't make it—that's not the stance we are seeking. Instead, therapists can wait for another time that feels better or, preferably talk with the client about their reservations about sharing this observation. For example:

> THERAPIST: There's something I'm thinking about right, but I'm feeling unsure of talking about it with you. I guess I'm concerned that you might feel criticized or blamed, which is not my intention. Can we talk about this for a minute?

In a similar vein, some therapists are concerned that, if they address issues directly, the client will misconstrue their straightforwardness as an unwanted and intimidating confrontation. Again, such a confrontational stance is not our intent—most therapists do not wish to be confrontational in any way, just as most clients do not want to be "confronted." Process interventions can be presented in an aggressive confrontational manner, as any intervention can be carried out ineffectively, but there is nothing insensitive or disrespectful about simply communicating forthrightly. Again, if therapists think these (or any other) interventions may be perceived by the client as demanding, intrusive, or unwanted, they should not use them. As we have just seen, therapists can either wait for what feels like a good time to try this particular intervention, metacommunicate about their reluctance to share observations with the client, or simply respond in other ways. We are seeking honest, straightforward communication, not confrontations.

Similarly, some therapists may worry that by responding more forthrightly, accurately reflecting the key concern or core message, or capturing the central feeling in what the client just said, they may "hurt" the client in some way. These expectations of client vulnerability are usually inaccurate, and are more likely to be related to the therapist's own countertransference issues about becoming more effective or acting stronger. Virtually all clients will welcome the invitation for a more authentic, straightforward dialogue. As clients find that it is safe and helpful to speak with the therapist about what goes on between them at times, new therapists will be reassured that they can be both sensitive and direct at the same time.

Fear about hurting the client takes on more personal significance for some therapists who grew up with highly authoritarian parents. Such therapists may have been exposed to double-binding family communications in which there was a great discrepancy between what was being done (for example, the child routinely may have been threatened, humiliated, or hit) and what was being said (for example, "We are a close and loving family. There are no problems here, and everyone is happy"). The essential element in the double-bind is the clearly understood but unspoken family rule that the child cannot acknowledge the incongruency in any way. For example, the child in such a double-bind cannot make these contradictory messages overt by metacommunicating and saying, "You're telling me to clean my plate, but later you'll make fun of me for being fat. Stop it—you're driving me crazy!" Children who have been intimidated and disempowered ("undone") by growing up with these double binds are highly anxious about making them overt because further rejection, debasement, or abandonment has been threatened or implied.

Such double-binding communications occur in many dysfunctional families, and are routine in alcoholic and in physically and sexually abusive families. Therapists who have suffered such mistreatment themselves may find it threatening (or at other times liberating) to use process comments and make overt what is occurring. An enduring legacy of such painful developmental experiences is that these therapists may fear that if they speak directly about what is going on, the client will be threatened or hurt—as they once were. For some, their pathogenic belief in their own inherent badness or toxic destructiveness, instilled by the mistreatment they suffered, will be confirmed. Thus, old threats can be evoked—but also resolved—by breaking double-binds and no longer complying with or being ruled by unfair family rules about how family members must communicate. However, such countertransference issues are most appropriately resolved in the therapist's own counseling, rather than in the supervisory relationship.

Therapists' Fear of Revealing Their Own Inadequacies   Some therapists may fear that, if they bring out how the client's conflict is being replayed in the therapeutic relationship, their own mistakes or inability to respond effectively will be revealed. For example:

> THERAPIST: (*thinking to herself*) Yes, there it is, the same problem that he has with his wife is going on between us right now. It sure is getting in our way, but I don't have a clue about how to bring it up!

In that case, making their interpersonal process overt and trying to talk together about it may seem like the last thing in the world the therapist would want to do.

Whether the therapist chooses to address them or not, such reenactments will occur at times in most therapeutic relationships—for novice and experienced therapists alike. By making them overt, however, clinicians do not reveal their own inadequacies or mistakes, assume responsibility for causing the client's conflicts or pain because their effective responses have revealed them (a common pathogenic belief for many who enter the helping professions), or assume sole responsibility for changing them. Rather, the therapist is simply wondering aloud, in a tentative manner, about what may be happening in the relationship and giving clients the invitation to work together to change or rectify this maladaptive relational pattern if the client also sees it occurring.

To address potential reenactments effectively, therapists work together with the client to *clarify the repetitive sequence of interactions* that typically unfolds in problematic scenarios with others. Continuing this collaborative effort, therapist and client can explore whether and when these problematic patterns that have been occurring with others have ever come up in the therapeutic relationship as well. If they have, therapists can acknowledge their own participation in this shared conflict, express their willingness to change their part in the old scenario, and try to make this relationship come out better for the client. Still working collaboratively, the therapist and the client search for new ways and better ways of relating that do not continue the old relational pattern. By doing so, the client no longer feels invisible, controlled, responsible, and so forth, at least in this relationship, and is encouraged to begin constructing more rewarding relationships with others along these new lines.

When the therapist first begins to make these reenactments overt, clients may feel discouraged—which can evoke feelings of inadequacy or guilt in the therapist. Initially, the client may feel discouraged because this reenactment confirms the clients' expectations that relationships cannot be different than they have been in the past. Intellectually, the client may recognize that other ways of relating are possible. However, this possibility holds little meaning experientially when the current enactment with the therapist confirms the client's maladaptive schemas and follows the same, problematic scenario that she has experienced repeatedly. Therapists' feelings of failure will be resolved when therapists

- realize they are already responding adequately to the client by recognizing that this is how the client's relationships have gone in the past;
- understand that this has been an unsolvable and painful problem that has characterized the client's life;
- affirm the historical validity of the client's hopeless or discouraged feelings, based on the reality of what the client has actually experienced with others in the past; and
- help clarify how the therapist and client are changing this unwanted pattern right now by recognizing and altering it in their current interaction.

Thus, the therapist is *behaviorally demonstrating* that the relationship between the therapist and client is no longer following the same well-worn, problematic patterns and is giving clients the corrective relationship they need in order to begin changing with others.

By thoughtful preparation—mentally formulating and revising working hypotheses, keeping written process notes, and writing case conceptualizations—new therapists will be able to resolve legitimate concerns about their own adequacy and performance. In the beginning, most therapists cannot successfully track the process dimension just by thinking on their feet—there's too much going on. Instead, therapists can prepare themselves by considering in advance various hypotheses about what may be occurring in the therapeutic relationship along the process dimension. As we have already noted, therapists should not make a process comment until they have seen an issue occur several times and until they have some tentative understanding of what it may mean. When the client's coping strategy elicits anger, discouragement, or some other strong emotion in them, therapists are encouraged to wait to respond until they better understand their own personal or countertransference reaction and can reestablish their neutrality. Finally, therapists who have prepared themselves with working hypotheses will feel more confident about making process comments. They will be less defensive about feedback the client may give them, more able to hear whatever concerns their clients' present, and better able to adjust more flexibly to their clients' needs.

As developing therapists become more experienced and confident, it will be easier for them to explore more open-endedly what may be occurring in the client–therapist relationship. When feeling confused, for example, an experienced therapist may be able to inquire simply:

THERAPIST: Hey, how did we get into this, anyway?

In the beginning, however, therapists may want to wait until they have some working hypotheses in hand to help them understand what may be occurring before they venture asking about it with the client.

Concerns about Owning Personal Power   A pervasive but generally unacknowledged issue in clinical training is therapists' concerns about owning their own personal power. Many new therapists feel uncomfortable about allowing themselves to become someone who is important to clients and having a significant impact on their lives. Especially in the beginning, it is also anxiety arousing to make strong interventions that can influence clients profoundly and to accept the responsibility that comes with exercising such personal power. As a result, *many beginning therapists overqualify their comments and water down the impact of their interventions.*

The therapist's own effectiveness can arouse anxiety for many reasons. Some clinicians were parentified or aggrandized as children. For these therapists, legitimate competencies can be readily exaggerated into unrealistic all-powerful or all-responsible grandiosities. Initially, as a child, it may have been exciting to be special and powerful vis-à-vis their parent in this way, but it soon becomes lonely, intimidating, or burdensome. As a result, these therapists may undo, or quickly retreat from, strong interventions they make or may avoid them altogether.

More commonly, developing therapists are reluctant to have a strong impact on clients because of their own separation anxiety or separation guilt. These therapists' experience, in their families of origin, was that competent or independent functioning threatened their emotional ties to parental caregivers. As the young child began to explore independently and move toward greater autonomy, certain caregivers may have looked sad, acted indifferent, demanded more, ridiculed attempts at mastery, or otherwise paired anxiety with competence strivings. Therapists with these developmental experiences also tend to retreat from their own effective interventions, and often may be heard saying, "I don't know what to do," "I'm afraid of hurting the client," or "I'm so screwed up myself that I have no right to try to help somebody else." All therapists certainly have their own personal problems and limitations. However, when such comments persist, it may indicate that the therapist wants to avoid the anxiety evoked by clearly and effectively addressing the client's problems.

What can help trainees with these concerns? An affirming supervisory relationship is crucial if therapists are to embrace their own personal power and be as effective as they can be. The supervisor–supervisee relationship is most productive when supervisees

- feel personally supported by the supervisor,
- receive conceptual information and practical guidelines when needed,
- obtain non-judgmental assistance in sorting out their own conflicted reactions toward the client that have been triggered by the client's eliciting maneuvers or by the therapist's own personal issues, and
- are able to address and resolve the interpersonal conflicts that are likely to arise in the supervisor–supervisee relationship.

If the supervisee feels that the supervisor is solely directing the case, the interpersonal process between the supervisor and supervisee is problematic. Even though the supervisor is ultimately responsible for the client, supervisor and supervisee should be consulting together collaboratively in a way that allows the supervisee to feel ownership of the treatment process. Supervisees need to hold primary responsibility for what occurs in sessions and need to feel free to act on their own ideas, while taking into account the supervisor's input. Otherwise, supervisees will not be able to experience their successes and failures as their own and will not be able to learn from them.

Ideally, the supervisor will help the supervisee evaluate the effectiveness of interventions and consider alternatives without taking away the supervisee's own initiative. When the supervisor and supervisee become stuck on a conflict in their relationship, the same issues often carry over to the supervisee–client relationship and will be reenacted there. In other words, the interpersonal process between the supervisor and supervisee is often paralleled in the supervisee–client relationship. Often, the therapeutic relationship will not progress until the conflicts in the supervisory relationship are resolved. In contrast, when the supervisor and supervisee can maintain a collaborative relationship, the supervisee is better able to enact a working alliance with the client as well.

In sum, it may take several years before therapists can allow themselves to have as significant an impact on clients as possible and to fully possess their own personal power; this is a gradual developmental process. If student therapists do not progress along this feeling-of-adequacy dimension as they move through their training, they should discuss this issue with their supervisors and/or seek treatment for themselves.

Closing

This chapter has drawn primarily from process-oriented concepts in the group therapy and existential psychotherapy literature, and from the counseling literature on immediacy interventions.

Therapists will find that, as in individual therapy, working with the process dimension is also the central focus in marital counseling, group therapy, and family therapy. The interpersonal patterns and relational conflicts between couples, family members, and group members similarly begin to involve the therapist. As in individual therapy, process comments that describe the current interaction with the therapist, or what is transpiring between group members or family members, can be used to identify and change these maladaptive relational patterns. In couple counseling, for example, the therapist may repeatedly share observations about the way in which the couple seems to be interacting or talking together:

THERAPIST: You two are arguing about who takes out the garbage, but I'm wondering if the issue is really about who has the power to make decisions or the right to tell the other what to do. What do you two think—would it be helpful to talk about that issue more directly?

Finally, the underlying assumption in all interpersonally oriented therapies is that the therapeutic relationship will come to resemble other prototypic relationships in the client's life. Clients' problems will emerge in the therapeutic relationship, especially if the therapist tracks the process dimension and attends to the reactions that clients tend to elicit from the therapist and others. Therapists do not need to artificially construct situations to re-create clients' conflicts and make them accessible for treatment. In other words, therapists can respond to clients in genuine ways. They do not need to contrive or manipulate events in order to strategically recreate clients' conflicts and thereby create "therapeutic experiences" for clients (for example, start the session late on purpose in order to make the client feel mad or unimportant, and use this experience to bring up the client's conflicts about expressing anger or readiness to feel unwanted). This approach undermines the authenticity and integrity of the therapeutic relationship. Furthermore, this strategic process often recapitulates the client's developmental history as the client is again forced to comply with others' covert control or attempt to decipher incongruent metacommunications. In contrast, if therapists are willing to work with the natural expression of the client's conflicts in the therapeutic relationship, they can offer the client a meaningful, real-life experience of trust, empowerment, and change.

Suggestions for Further Reading

1. Drawn from the film *Ordinary People* (directed by Robert Redford, 1980), a case conceptualization is provided in Chapter 9 of the Student Workbook.

2. There are many helpful books available to teach new therapists how to use immediacy interventions effectively. Practical guidelines to help therapists learn how to metacommunicate and when to intervene with the process dimension are found in Chapter 15 of *Handbook of Interpersonal Psychotherapy* (Anchin & Kiesler, 1982; see especially pp. 285–294). Cashdan's (1988) *Object Relations Therapy* provides informative illustrations of the interpersonal process, helps therapists recognize and respond to metacommunicative messages, and demonstrates how the therapeutic relationship can be used to effect change; see especially Chapters 3 and 5.

In addition, Ivey and Ivey (1999) helps therapists work "in the moment" and talk in the present tense about what is going on in the counseling relationship rather than in the past tense about others.

3. Useful illustrations of how therapists can integrate other theoretical modalities with the interpersonal process approach are provided by Wachtel (1997) in *Psychoanalysis, Behavior Therapy, and the Relational World*. Readers are also encouraged to examine *Therapeutic Communication: Principles and Effective Practice* (Wachtel, 1993).

CHAPTER TEN

Working Through
and Termination

Conceptual Overview

Working through and *termination* are two distinct phases of treatment. As clients continue to change maladaptive relational patterns with the therapist and find that familiar but problematic scenarios do not reoccur in this relationship, they begin to generalize this experience of change beyond the therapy setting. Finding that conflicts can be resolved in the therapeutic relationship, clients begin to explore how they can make the same types of changes in other relationships as well. This working-through phase of treatment is an exciting period of growth and change as clients try out with others the emotional relearning that has occurred with the therapist. As the therapist helps clients resolve with others the conflicts that they have already been able to resolve with the therapist, treatment evolves to a natural close. The termination phase provides one last opportunity to reexperience and resolve old conflicts, internalize the helping relationship with the therapist, and successfully terminate treatment.

Working Through
The Course of Client Change: An Overview

New therapists usually lack a conceptual overview of how client change comes about. Because they have not worked with many clients or may not have experienced their own successful therapy, most beginning therapists do not have a sense of the ordered sequence in which change often occurs. In the previous chapter, we reviewed the sequence of therapist activities over the course of treatment. We begin this section by extending this framework for conceptualizing the course of therapy. To get an overview of the change process, we look at when and how clients tend to resolve their presenting problems and adopt new, more adaptive responses with others. Clients often change in a predictable sequence of steps, and this unfolding course of change is reviewed from the beginning of treatment to the final working-through and termination phases.

For some clients, change begins as soon as they decide to enter treatment. That decision involves acknowledging that there really is a problem and they are no longer going to deny or avoid it. It is also an acknowledgment that clients cannot resolve the problem alone and need help. Some clients recognize that their resolution to seek help is a healthy step forward, and they feel good about making this decision. Sadly, others may experience it as a failure experience, evidence of their inadequacy, disloyalty to their family, or a violation of cultural beliefs. When clients can allow themselves to feel good about the decision to seek help, or when the therapist can help clients reframe their need for help with more self-empathy, some initial relief of symptoms—such as anxiety or depression—may result. This internal commitment to seek help is a crucial first step.

Some changes in feeling and behavior may also occur during clients' initial sessions with the therapist. Clients are reassured when the therapist

- invites them to express their concerns directly and fully;
- listens intently with respect and concern for their distress;
- is able to enter their subjective worldview and grasp the core meaning that these concerns hold for them; and
- demonstrates a practical ability to help solve at least some small part of the client's immediate problem (for example, by providing practical information about discipline/childrearing, teaching relaxation techniques for anxiety, identifying thought processes or behavior patterns that are maladaptive, role-playing new behaviors that will help with an upcoming situation, and so forth).

As clients find that the therapist can understand their experience and respond compassionately to them, a working alliance is established that also reduces anxiety and depression. Such validation and support do not resolve most clients' conflicts but do engender hope and help to relieve their distress. Finding this "secure base" in the therapeutic relationship, clients are no longer alone in their problems—they have the benevolent ally they need who can see and understand what's wrong. The essence of a secure attachment is that *the child is secure in the expectation that when she is distressed, the caregiver will see or register the distress and try to help the child solve her problem* (thus, a secure attachment has nothing to do with common misconceptions about warmth, being nice or "bonding"). This is our role as therapists, too, as life changes for clients when they are no longer facing their problems alone. Taken together, these initial responses—which in fact are complex interventions that require skillful application—comprise the essence of crisis intervention treatment. They also form the bedrock of therapeutic effectiveness for therapists working within every theoretical orientation and treatment modality.

More substantial changes begin when the therapist succeeds in focusing clients inward on their own thoughts, feelings, and response patterns and away from their preoccupation with the problematic behavior of others. When clients examine their own internal and interpersonal reactions, they often recognize how their behavior contributes to an interpersonal conflict and may be able to change their participation in it. Adopting an internal focus for change is also an important way to reframe how clients think about their problems. Redefining the conflict with the other person as, in part, an internal problem usually reveals a wider array of new and more adaptive alternatives that clients can begin to try out with others.

Next, as the therapist focuses clients inward on their own experience, the conflicted emotions that accompany most clients' problems will emerge. The pace of change accelerates as clients begin to express each sequential feeling in their affective constellations. Clients have the opportunity to resolve their problems when they stop defending against their core conflicted emotions, feel the safety to begin sharing them, and receive a more affirming response from the therapist than they have come to expect from others. Far-reaching changes are set in motion as clients begin to integrate their affective constellations through the "holding environment" provided by the therapist's acceptance and understanding.

As clients' conflicted emotions emerge, the therapist is also better able to conceptualize clients' relational schemas and the pathogenic beliefs that accompany them. The therapist then focuses treatment along these conceptual lines and clarifies how clients are reexperiencing the same relational patterns, faulty assumptions, and affective themes throughout the various issues and concerns they present. Some further change may occur as the therapist helps clients identify when each of these three patterns occur in their current interaction, and begin to recognize when and how they are occurring with others. This new awareness often contributes to behavior change, and it facilitates the next and most significant point of change.

As described in the previous chapter, the same conflicts that originally led clients to seek treatment, and that they have been discussing with the therapist, will usually be reenacted in the therapeutic relationship, especially along the process dimension—that is, the way in which the therapist and client interact. This reenactment in the therapeutic relationship provides both an opportunity for therapy to fail—by reenacting the client's conflict—and an opportunity for therapy to succeed—by resolving the conflict that has emerged in the therapeutic relationship. *Treatment reaches a critical juncture when clients feel that the therapist is responding in the same problematic manner as significant others have done in the past.* This usually occurs in one of three ways: transference, client-induced countertransference, and therapist-induced countertransference.

With his delightful sense of humor, the marital therapist Harville Hendrix (2001) refers to this as "the 3 P's." Clients *pick* others who keep presenting them with old patterns and problems, they *provoke* others to respond in unwanted but expected ways, and they *perceive* others as responding in the same problematic ways—even when they haven't! However this reenactment comes about, most clients cannot make significant progress in resolving their conflicts with others until this replay with the therapist is resolved. Only when the therapist and client achieve such resolution does the client have *a real-life experience of change.* This experience of change, as opposed to insight, cognitive reframing, or challenging faulty assumptions, is often the pivotal step that shapes whether enduring change can occur. One episode of such in vivo or experiential relearning will not usually be sufficient, however, as clients repeatedly need to find that their maladaptive relational patterns continue to unfold differently with the therapist than they have with others.

When the client is finding that the therapeutic relationship is different and problematic expectations are not confirmed, the door opens to change with others in two ways. First, the therapist begins in earnest to actively transfer the client's here-and-now relearning to other relationships:

> THERAPIST: Good, it's getting clear between us that I'm really not judging or criticizing you. So let's start sorting this through better with others in your life. Who does consistently do this to you, and who in your life doesn't?

Second, under their own initiative, many clients will come back the next session, following a corrective emotional experience with the therapist, and relate successes and failures in their attempts to relate in this new way with others in their life:

> CLIENT: I tried speaking up with my boyfriend this week, you know, like I did with you last time. In a way it felt good, but in a way it didn't. When he said. . . .

At this point in the change process, the therapist often acts as a "coach" to help clients deal with the positive and negative responses of others to their new ways of relating. Through behavioral rehearsal or role-playing, and direct guidance or teaching, the therapist helps the client transfer the new ways of interacting that have occurred with the therapist to others in their lives.

With this introduction in mind, let's now enter the working-through phase of therapy more closely, and see how the changes that occur with the therapist can be generalized to others.

The Working-Through Process

When clients change, it often begins in the therapeutic relationship. With just the therapist's encouragement and advice, some clients who function well can successfully adopt new behavior. When the new behavior is integrally linked to their core conflict, however, most clients will need to practice this new response in the therapeutic relationship. Thus, *therapists actively encourage clients to try out new ways of responding with them in the therapy setting* and reassure clients that they intend to greet clients' new responses in affirming ways, rather than in the problematic ways that clients have grown to expect. If therapists give this permission verbally *and* enact it behaviorally, clients will soon test the therapist and try out anxiety-arousing new responses (for example, by being more assertive or disagreeing with the therapist). If the therapist passes the clients' test—responds affirmingly to the new behavior, rather than unwittingly repeating unwanted but familiar responses—clients have the real-life experience that change can occur.

This corrective emotional experience is a powerful agent for change but, as we have noted, one trial learning is not sufficient to effect change. Most clients will need to reenact this and other reparative relational themes over and over again in the therapeutic relationship. As a rule of thumb, the more they have been hurt, the more often they will need to reexperience a new and safer response that doesn't fit the old schema or template. However, once clients have seen that relationships can be different, the therapist can actively engage clients in generalizing this new experience of change beyond the treatment setting. Thus, working-through is not about gaining insight or exploring the past—it is about linking the changes that have occurred with the therapist to significant others in their current lives. Typically—but by no means always—clients try out new ways of responding with others in the following progression.

First, change occurs in the client's relationship with the therapist. Clients then change with acquaintances they do not know well or with others who are not especially important to them. This is often followed by change with supportive others who are important to the client, such as caring friends, teachers, or mentors. Next, clients often change old response patterns with the historical figures with whom the conflicts originally arose, such as caregivers and important family members. Finally, clients change their behavior with primary others with whom the conflict is currently being lived out, such as spouses or children. This final arena is the most challenging because the consequences have so much greater import.

During the working-through stage of therapy, some clients will rapidly assimilate the new ways of responding and readily apply them throughout their lives. Other clients, who have more pervasive conflicts or who have been traumatized or deprived more severely, will work through their problems more slowly. These clients need to confront the same fears and expectations over and over again. Each time, the therapist's aim is to clarify how, here again, the same relational patterns, affective themes, or faulty beliefs are being played out in this situation, and help clients find a better way to respond to this particular manifestation or expression of their core conflict. In this regard, the core conflict will be repeatedly expressed in four areas of client functioning:

• In the current interaction or interpersonal process that is transpiring with the therapist;
• When reviewing crisis events in the client's life or the crisis events that originally prompted the client to seek treatment;
• In current relationships with friends and significant others in which the client's emotional problems are being activated;
• In developmental relationships with family members.

For many therapists, working-through is the most rewarding phase of treatment. The therapist often serves as a cheering squad who takes pleasure in clients' newly obtained mastery and celebrates with them as they successfully adopt new responses in progressively more challenging situations. During this stage when change is occurring, the therapist can also become more actively involved in answering direct questions and providing suggestions and information to help clients enact new behavior or solve problems. For example:

THERAPIST: Maybe you could try saying something like this to her. . . .

Therapists also provide more interpersonal feedback about how the client comes across to the therapist and others (Brammer & MacDonald, 1996):

THERAPIST: Hey, that wasn't that "small voice" we've been talking about. You sounded strong and clear as you said that—it held real conviction. It's great to see you like this. Have you been acting stronger like this with others, too?

Many therapists find that cognitive interventions and behavioral procedures for rehearsing alternative responses are especially helpful during this phase. Once the interpersonal process is enacting a corrective experience, the client becomes more responsive to other types of intervention, such as assertiveness training and parenting education, self-monitoring and self-instructional training, role-playing and other modeling techniques, and educational inputs or readings. The nature of the therapeutic relationship also changes during this period. As clients improve through the working-through period, they will increasingly perceive the therapist in more realistic terms. As this occurs—that is, once the transference projections and eliciting maneuvers have been jointly identified and are being resolved—the therapist can disclose more personal information to the client and enjoy further mutuality with the client. However, the therapist is still prepared to respond to the reenactments, maladaptive relational patterns, or resistance the client will present.

At this point in treatment, clients are living out a new and different relationship with the therapist—a relationship that is resolving rather than reenacting their maladaptive relational patterns. As clients find with the therapist that new ways of relating with others are possible, they often will begin—on their own—to try out these more adaptive responses with others in their lives. This is a critical juncture in treatment, and the *therapist's task is to actively help clients anticipate and negotiate the successes and failures that are likely to follow with others.* As schemas change and expand, clients need to be informed that, although some individuals will respond positively to the clients' new changes, others will not. It will be exciting when others respond affirmingly to clients' changes. However, therapists must also help clients discern who is likely to respond poorly when, for example, they step out of their old caretaking role and say no or ask for what they want. Therapists need to help clients anticipate realistically how it will feel, and how they are likely to react, if the other person responds in the old, unwanted way to their new response. This is especially important in primary relationships with spouses and parents, where it can be so deeply discouraging when clients, yet again, receive the same unwanted but familiar responses:

THERAPIST: It's great that your husband has been able to listen to you better, and take your concerns more seriously now that you are expressing them more directly. But I'm wondering how this will go with your mother when she visits this weekend. What if you speak up with her, as you have been doing with others, but she keeps turning everything back around to her need and what she wants? Let's think this through before you see her.

In sum, most clients will have experiences with others that both affirm the new ways of being and other experiences that painfully reenact the old relational patterns or expectations. Thus, we must examine what therapists can do to help clients anticipate these disappointments, which are an inevitable part of the change process, and how therapists can prepare clients to respond more effectively to such unwanted responses than they have been able to do in the past.

As clients find that they can safely respond in new ways with the therapist and then make successful changes with certain individuals in their current lives, their expectations of change in some other relationships may become unrealistically high. For example, suppose an adult survivor of childhood abuse has been deeply understood and validated by an effective therapist. Following this success with the therapist, the client risks disclosing this shame-laden secret to her husband, and then to her best friend, and they both respond affirmingly as well. Next, the client's lifelong wish that the abusing parent can now acknowledge the reality of what occurred long ago is evoked, or that the nonabusing parent, who could not be protective at the time, will now hear or believe the client. With expectations buoyed by successes with others, this client may address parents or other family members, only to encounter the same invalidation, scapegoating, and threats of ostracization that she received decades ago. Or, some clients may be profoundly discouraged to find that, unlike the therapist and some others in their lives, their marital partner cannot change or respond positively to this new behavior (for example, being more independent, assertive or limit setting). In this way, it is important for therapists to help clients examine their expectations of others *before* they try out new responses. In addition, the therapist's goal is to prepare clients by anticipating (1) how others are likely to respond to their new behavior and (2) how the clients are likely to respond (what they have said and done in the past) if they receive the same unwanted responses they have received before.

Thus, in order to manage these disappointments that are an inevitable part of the change process, therapists can prepare clients in the following ways:

1. By helping clients *realistically anticipate* how each new person is likely to respond to their changes. For example, if the client is a married woman, what will her husband probably say and do when she acts more assertively with him for the first time?

    CLIENT: He's likely to smile at me like I'm a cute child or something, and then just change the topic as if I didn't even say anything.

2. By helping clients spell out in detail how they are likely to feel, and what they are likely to say and do, if they receive the old unwanted response to their healthier new behavior.

    CLIENT: I think I'd feel ashamed. You know, feel stupid for trying to stand up for myself, and just give up and withdraw inside—like I've always done before.

3. By role-playing and providing new, more adaptive responses that, after rehearsed in therapy, clients can use at this discouraging moment, instead of repeating what they have usually done in such circumstances in the past. Recalling the concept of "change from the inside out," the key concept is trying to help clients change their own response in problematic interactions, rather than focusing on the often futile attempt to get others to respond in new or better ways.

> CLIENT: I'm trying to help our marriage and tell you something important right now, but you're not taking me seriously. I don't like feeling dismissed like this. Are you willing to take this more seriously or should we just stop this conversation now?

Realistically, clients will continue to have experiences with others that reenact maladaptive relational patterns and confirm their pathogenic beliefs. *However, by helping clients walk through these three steps, the therapist can help them find ways of changing their own internal and interpersonal responses—even when others are unable to change.* For example, clients can learn that they are not to blame for, or deserving of, the way they were mistreated—even though family members still cannot affirm them. The married woman in the prior example learns that she can still advocate for herself. She does not have to go along with or comply with her husband's dismissal, and she can establish new relationships with others where her limits or opinions can be respected, even though her spouse cannot do this. In this process, clients are empowered to learn that they can change their own internal and interpersonal responses, even when significant others in their lives are not supportive.

Thus, the familiar but unwanted reactions that clients sometimes receive as they try out new ways of responding with others can be deeply disappointing, yet they also provide an opportunity to further work through and resolve the conflict. Through each setback with a significant other, the therapist and client gain further awareness of the client's maladaptive relational patterns and deeper access to the painful feelings and faulty assumptions that accompany them. Working together, *therapists and clients can recognize how the client's old response patterns are being evoked again and formulate new responses that clients might try out the next time they are in that situation.* More specifically, the therapist might explore each successive step with the client:

- What were you thinking and feeling inside when he dismissed you like that?
- Tell me what you did when he said that to you.
- What would you like to be able to say to him instead the next time this comes up?

In this way, therapists can help clients recognize what they say and do to contribute to these prototypical problematic interactions. With this awareness, it is especially helpful to role-play or practice new responses that would change their role in the maladaptive relational pattern. Again, clients feel empowered by learning that they can alter the unwanted scenario by changing their own responses, even when the others cannot change. In addition, clients can begin to establish new, more affirming relationships with others that accept or even welcome these important changes. In other words, *clients' resolutions do not rest on changes in the parent, spouse, or others, as most clients inaccurately believe when they enter treatment, but on changing how they respond to themselves and others in current interactions.*

It is not always easy for therapists to provide a corrective emotional experience, however. Although clients are improving during the working-through period, they will still be struggling with their core or generic conflicts. Eliciting behavior, testing behavior, and transference distortions that narrowly slot the therapist into the old relational templates or role will still have to be negotiated. Especially when clients are distressed or feel vulnerable, therapists need to work through the distortions that stem from the old schemas. For example, misperceptions that the therapist "doesn't really like seeing me," "isn't strong enough to handle me," "feels burdened by my problems," "is judging me," "doesn't respect me," and so forth.

In addition, most clients will become discouraged at times by the repetitious working-through process. Seeing that the same old conflict is confronting them again, clients may feel that "nothing has changed," that it is futile to keep trying, and that they should consider dropping out of treatment. Therapists can make no guarantees of change, of course, and cannot assume responsibility for the clients' motivation to continue. At these critical junctures, however, therapists want to actively reach out and extend themselves to their clients. Only as therapists come to grasp the profound influence of schemas in shaping our experience of "reality" can they be empathic with the clients' discouragement. Although clients may realize it intellectually, experientially it really does seem to them as if this is the only way that relationships can be (Client: "Why do I always have to be the responsible one who takes care of everything!").

The client needs the therapist to hold steadfast that there can be another way—some relationships can be different some of the time. Therapists can point out possible instances where such other ways have occurred and convey that they remain committed to working with the client, even though they appreciate how frustrated or discouraged the client is feeling right now. It is therapists' resolve to effect change in their relationship, and their personal commitment to helping this particular client change with others, that pulls the client through these expectable crisis points in the working-through phase.

As an illustration of the working-through phase, we now consider a case example in which the client changes first in her relationship with the therapist and then, with the therapist's assistance, is able to extend that change to other relationships in her life. The episode in which clients' change first emerges— the *critical incident*—may be an overt, interpersonal conflict with the therapist or a compelling, transference-laden misunderstanding. More likely, however, as in this example, it is a subtle aspect of the interaction with the therapist that could easily go unnoticed unless the therapist, informed by working hypotheses, has been prepared to look for this relational theme or pattern.

Our case example concerns Tracy, a moving-toward client who habitually responded to others in a compliant, pleasing manner. She usually went along with others, acting as if her wishes and feelings didn't matter. Tracy said that she didn't believe that others would be very interested in listening to her or doing what she wanted to do—she long had felt that either she had to "go along" or be alone. Her therapist linked Tracy's compliant behavior to her dysthymia, especially her presenting symptom of crying spells, and went on to explore Tracy's low self-esteem and feelings of worthlessness.

In addition, the therapist actively encouraged Tracy to respond differently in their relationship. The therapist behaviorally demonstrated that she cared about what Tracy had to say and invited her to express her own opinions in counseling—even if that meant disagreeing with the therapist. During their next session, Tracy tested to see whether the therapist meant what she said and risked this significant new behavior in the therapeutic relationship:

TRACY: You're right. I guess I never have felt very good about myself.

THERAPIST: From other things you've said, it seems to me that your mother just wasn't there for you as a child. You didn't get the support you needed from her. Maybe we should explore that further.

TRACY: No, I don't really want to talk about her. I remember her more fondly, but my father was pretty harsh with us.

THERAPIST: (*exclaiming happily*) You just said no, you don't want to talk about your mother and that I was wrong about her!

TRACY: I'm sorry; sure, we can talk about her. What do you want to know?

THERAPIST: Forget your mother; you just disagreed with me! You just told me you didn't want to do what I wanted to do and suggested what you thought would be better. That's great!

TRACY: What?

THERAPIST: You just did what we have been talking about. You expressed your own opinion, said what you thought, and disagreed with me. I am so happy for you!

TRACY: Aren't you mad at me? Didn't that hurt your feelings?

THERAPIST: Oh, no, we can disagree and still be close. That was a brave thing you just did.

TRACY: So . . . I guess what you're saying is that maybe it is OK for me to say what I think sometimes?

THERAPIST: Sure it is. I care about what you think. I want to know what you feel, and I want to do things your way, too.

TRACY: (*tearing*) But you're safe. It's easy to do that with you. You're not like other people.

THERAPIST: Yes, I am "safe," and it is easier to do things like that with me than with other people. But you have changed to be able to do that with me; you've grown. You couldn't do that before, you know.

TRACY: Yeah, that's right.

THERAPIST: And if you can value yourself enough to express what you think in here with me, then you can begin to do that out there with other people as well.

TRACY: Do you really think so?

THERAPIST: Yes, I'm sure you can. If you can do it with me, you can do it with them.

TRACY: Oh, I would love to be able to do this with my boyfriend.

THERAPIST: Tell me about how this usually plays out with your boyfriend.

TRACY: (*describes the usual scenario*)

THERAPIST: OK, what do you want to be able to say to him instead?

TRACY: (*describes an alternative scenario*)

THERAPIST: Would it be helpful to you if we role-play that together? We might be able to work out together where the problems might develop if you try that with him.

Why is the therapist so excited about such a seemingly small change? Is this single manifestation of a new behavior all that is necessary for Tracy to resolve her problems? No, but this highly significant event brings her to the working-through phase of treatment. Within the safe orbit of the therapeutic relationship, Tracy was able to try out a new behavior that taps into the most significant issue in her life and arouses intense anxiety. For Tracy, that is, adopting such a seemingly insignificant self-assertive behavior arouses lifelong feelings of worthlessness. And, at the same time, she is discarding the primary means she has developed to protect herself from these painful feelings—that is, to "move toward" and go along with others. In order to resolve her painful conflict, Tracy will have to repeatedly experience the same sequence:

1. Her old relational pattern is activated—having to "go along" and act as if everything is OK even when things are painfully wrong.
2. She tries out a new and more adaptive response with the therapist, which is to assert her own preference.
3. She receives a different and more satisfying response from the therapist— the therapist affirms Tracy's initiative.
4. She generalizes the new behavior to others beyond the therapy setting by considering how she typically behaves with significant others in her life and, specifically, what she might want to say and do instead the next time she is in this situation with her boyfriend. This repetitious reworking of the new behavior and the old conflict is called *working through*.

When Tracy returned for her next session, she made no mention of this incident with the therapist and said nothing about what went on with her boyfriend. Recognizing this as resistance, the therapist waited for the opportunity to acknowledge what had occurred between them the week before and to reassure Tracy that she was supportive of her new assertive response.

THERAPIST: We haven't talked yet today about the important new way that you responded to me last week. Did you have any thoughts about that during the week?

TRACY: No, not really.

THERAPIST: Fine, but I just wanted to say again how happy I was for you. It felt very good to me to see you express your own opinion and be able to disagree with me.

Evidently, this reassurance must have given Tracy the permission she needed. The following week, she was able to adopt the same type of assertive behavior with her boyfriend for the first time. This successful experience, in turn, encouraged her to confront the same issue in another more threatening relationship—with her father.

TRACY: (*beaming*) Guess what? I did it! I've been dying to tell you all week!

THERAPIST: (*enthusiastically*) Ha! What did you do?

TRACY: I did what we talked about with my boyfriend. He was talking about something he thought we should do, and I disagreed with him and told him what I thought we should do instead. It felt great, and I don't think he minded, either.

THERAPIST: Good for you! You're on your way.

TRACY: I can't believe how easy it was. (*pause*) I wish it could be that easy with my father. He's always putting me down when I say something.

THERAPIST: Maybe you're ready to start changing your relationship with your father as well. How do you respond to him when he puts you down?

TRACY: I get real quiet when he does that. I feel like crying, but I don't. I wish I could just tell him that I don't like it and I want him to stop.

THERAPIST: Yeah, confronting him at the time he does that and setting limits with him would certainly change your relationship. It would be saying to him—and to yourself—that you value who you are and what you have to say and you're not going to let him hurt you like that anymore.

TRACY: But I could never do that. I'd like to, but I just couldn't.

THERAPIST: You haven't been able to do that in the past, but you have been changing in here with me, and now with your boyfriend, too. So maybe you can do something different with your father as well. What holds you back from setting limits with him? What are you most afraid of?

TRACY: He wouldn't take me seriously; he'd just laugh at me.

THERAPIST: How would that make you feel?

TRACY: Worthless. (*begins crying*)

THERAPIST: So that's where that awful feeling comes from.

TRACY: Yeah. (*long pause*) It makes me mad, too.

THERAPIST: Of course it does! It hurts you very much when he diminishes you like that, and you have every right to be angry.

TRACY: It's not fair! I don't want to let him do that anymore. It's not good for me.

THERAPIST: That's right, it's not good for you, and you don't deserve it. Right now, you are making the transition from feeling worthless to feeling worthy— that you do matter. Now that you are seeing this so clearly, maybe you don't have to go along with it the way you used to.

TRACY: What can I do?

THERAPIST: What would you like to do the next time he does that?

TRACY: Well, maybe I can just tell him to stop it—tell him that I don't want to be treated like that anymore.

THERAPIST: Yes! That would be a strong and appropriate response on your part. But before you try it, let's explore further how that conversation might play out. Maybe we can identify the trouble spots and role-play some responses to help you through them.

In ever-widening circles such as these, clients confront and work through the same relational patterns in the different spheres of their lives.

Notice that the critical incident in this case was very subtle: When the therapist suggested that Tracy talk about her mother, Tracy said she didn't want to. Therapists could easily miss such slight shifts in behavior unless they have formulated working hypotheses about how the client's conflicts with others are likely to be reenacted in the therapeutic relationship. If Tracy's therapist hadn't *anticipated* that Tracy would tend to comply with her, and if the therapist hadn't been alert for any budding sign of self-direction or assertiveness from Tracy, the opportunity for change would most likely have been lost. That is, the therapist might have responded by focusing on the content (by saying, for example, "But it's important to look at your relationship with your mother because . . .") rather than on the process (Tracy's self-assertion). In that case, Tracy readily would have complied with the therapist and dutifully talked about her relationship with her mother. Then, the interpersonal process between Tracy and her therapist would be recapitulating her maladaptive relational pattern, and little significant change with others would result.

When working with a client who seems unable to change, the therapist may eventually grow discouraged and become critical toward, or disengaged from, the client. In those circumstances, the therapist should consider whether the therapeutic interaction is subtly reenacting the client's maladaptive relational patterns. That would explain why the client is unable to change, even though the therapist and client continue to talk about adopting more assertive behaviors with others. The therapist sincerely supports and encourages this, and the client learns useful skills that should facilitate the new behavior. Such subtle reenactments along the process dimension are expectable. If therapists utilize process notes (Appendix A) and generate working hypotheses (Appendix B), they can get an idea of how each particular client's conflicts are likely to be expressed or played out in the therapeutic relationship. *By anticipating what to listen and watch for, therapists will be able to recognize such reenactments—in the moment as they are occurring,* and be prepared to respond more effectively to them.

Again, as clients are trying to generalize change to relationships outside the therapy setting, they will regularly receive the familiar but unwanted responses from some that confirm faulty beliefs and evoke maladaptive relational patterns. As we will see later, for example, Tracy's father did not respond well to her more assertive voice, as her boyfriend did. Therapists can use this deeply disappointing experience as an opportunity to help clients explore and discern which relationships in their life can be different and better and which cannot change.

Going from Current Problems, through Family-of-Origin Work, and on to "the Dream"

In the working-through phase, the therapeutic action is primarily in the present. That is, many of the significant events that take place in treatment are current interactions with the therapist that enact resolutions of maladaptive relational patterns. As these reparative events occur, they also lead to internal reworkings of clients' schemas as new behavior comes into play. These cognitive changes are necessary to help clients assimilate and maintain these new and more effective ways of relating after treatment ends. The next step in the change process occurs as these corrective experiences and expanding schemas are generalized beyond the therapy setting and clients try out with others the new ways of relating they have found with the therapist. As we have seen, clients will have both successes and disappointments; some individuals will welcome the clients' changes, whereas others insist on continuing along well-worn, problematic lines.

As clients explore the potential for change in current relationships and come to terms more realistically with the possibilities and limits, two new subphases in the working through process may emerge. First, clients, on their own, often start to look back in time in order to better understand the formative experiences that originally shaped these problems they now are resolving. In the first part of this section, we examine therapeutic guidelines to help with this "family-of-origin work." Second, as clients develop a more realistic narrative that helps them better make sense of their life experiences and how they have become who they are, they also begin to look ahead. That is, they begin to think more about what they want their life to be in the future, and to reformulate life plans to better fit the person they are becoming. This subphase, often ushered in by the emergence of "the Dream," will be explored second.

Family-of-Origin Work   It can be liberating for clients to explore the familial interactions and developmental experiences that shaped their current conflicts. As they gain understanding of the family rules, roles, and childhood dilemmas that shaped how they learned to cope and adapt, they become more accepting of the choices, compromises, and adaptations they have had to fashion in their lives. Therapists can also begin to help clients feel more empathy—both for themselves, and for the personal limitations or difficult life circumstances that led their caregivers to respond in problematic ways. Clearly, this developmental perspective can be profoundly enriching. In most cases, however, it is not productive for therapists to lead clients back in time, to make historical interpretations, or try to make links between current relational patterns and formative family relationships. Although these developmental interpretations and historical connections may indeed be highly accurate, few clients will find them helpful with their current problems. Per client response specificity, some clients may be able to utilize historical connections—and therapists can assess whether they are helpful to this particular client. However, many clients will not find these developmental interpretations relevant—especially early in treatment, even though they may indeed be accurate. For example:

> THERAPIST: I keep hearing the same theme: that what you are complaining about with your spouse seems in some ways to be similar to what used to occur with your mother.
>
> CLIENT: My mother! What's my mother got to do with anything?! I'm 41 years old, and she's been dead for 8 years. Why are you talking about my mother?

In this way, many clients will not be able to use this type of historical connection to make progress in treatment. Although the developmental suggestion may be highly accurate, often it is too far from their current experience to hold meaning for them. Instead, a *mutual exploration* of familial and developmental experiences can teach therapists (and clients) a great deal about the faulty schemas and patterns that clients originally learned, how these are shaping problems in current relationships, and the corrective experiences that clients need to find now in treatment with the therapist. Thus, this developmental understanding may not help clients much with their current problems, but it does offer therapists a great deal of help in generating working hypotheses about the themes and patterns that are likely to go on between them in their interpersonal process. Toward this different purpose, for example, the therapist might ask:

> THERAPIST: What was it like to be a child growing up in your family?

> OR

> THERAPIST: Tell me about your parents' marriage. Can you bring it to life for me?

The therapeutic dialogue might go something like this:

> THERAPIST: Tell me how your parents responded, what they would say and do, when you were successful or felt proud of an accomplishment.
>
> CLIENT: I'm not sure what you mean.
>
> THERAPIST: What was the look on your mother's face when you showed her that spelling award years ago?
>
> CLIENT: (*pensively*) Oh, I don't think it was a very happy face. I don't know why, but she almost seemed sort of sad or hurt.
>
> THERAPIST: Uh huh, your success seemed to hurt her in some way, almost as if she felt it as a loss. . . . That sure would make it hard for a girl to feel good about her success experiences. How did your father respond when you told him about winning?
>
> CLIENT: There was nothing subtle about that—he was mad and told me to stop bragging in front of my brother.

To recap, the developmental information gained from this type of mutual exploration, as opposed to historical interpretations, will inform therapists about the formative experiences that shaped this client's cognitive schemas. This provides therapists with invaluable guidelines to begin formulating the kinds of therapeutic responses that are likely to reenact problematic relational patterns for this client, and the types of therapeutic responses that are likely to be reparative or resolving for this client. For example, the client just cited needs to find that this therapist takes unambiguous pleasure in the client's success experiences and personal strengths. Early in treatment, however, most clients will not be able to make meaningful bridges between current and past relationships from this developmental exploration. When the therapist highlights seemingly obvious connections, the relational patterns may not come alive for clients until they first have experienced a corrective response to these old relational patterns with the therapist. Following this corrective experience, many clients will spontaneously lead the therapist back to the developmental experiences that originally shaped some of their problems. These client-initiated explorations, *which usually occur immediately after clients have a corrective emotional experience with the therapist or have successfully adopted a new response with others*, continue throughout the working-through period. In this sequence, they are not sterile interpretations or abstract concepts. Instead, they are enlivening for clients, feel relevant and informative, and readily contribute to productive new behavior change with others.

Let's pause for a moment and highlight the important sequence of change that is being emphasized here. In contrast to other counseling theories, the interpersonal process approach does not suggest that behavior change leads to insight or, conversely, that insight leads to behavior change. Although both of these change processes will often occur, a different mechanism of change is proposed here. *Both meaningful insight and sustainable behavior change follow or result from clients' new or reparative experience with the therapist* (Wachtel, 1987).

As clients make their own meaningful links between formative and current relationships, they often ask the therapist what they should do about problems and patterns that are continuing in their current relationships with family members. We now examine two broad guidelines to help clients with this family-of-origin work. First, recall the "internal focus for change" described in Chapter 4. We will again see how clients can learn to change how *they* respond to maladaptive relational patterns in their current interactions with family members—rather than futile attempts to try and get parents or others to be different and change how they have always responded. Second, we look at the concept of "grief work," in which clients mourn and come to terms with what they have missed developmentally. Working in both of these areas results in an interpersonal and internal resolution of family dynamics.

Clients who struggle with more pervasive or long-standing problems usually have family-of-origin work to do. Some clients do not wish to discuss problems from the past with caregivers or family members, and therapists should not press clients to do so. However, some clients will want to acknowledge or address past conflicts with parents and other family members—as a way to stop participating in them or to better resolve them. It is enormously gratifying—and a powerful impetus for change—when such rapprochements succeed.

As we have been emphasizing, however, such current resolutions of historical problems often do not occur. Sometimes caregivers have grown and changed and can talk openly or nondefensively for the first time about hurtful interactions that once occurred. However, in many cases, and especially with more serious problems, the client will often receive the same invalidating, blaming, or hurtful responses as they did years ago in childhood. This is especially painful when the client's motivation for talking with the parent is the unrecognized wish to get the parental approval, affirmation or protection that the client has always longed to receive. If the therapist has not realistically prepared the client for the possibility that this disappointment may occur, the client may despair and feel hopeless about change occurring in any relationship. Oftentimes, the client will be breaking unspoken but strongly held family rules against addressing conflicts with parents directly and, in many cases, punitive homeostatic mechanisms will be set in motion. For example, when Tracy confronted her father about always putting her down, he again disparaged her—just as he had for the past 20 years. As if that doesn't make change difficult enough, look what happens next in the family system. Pressuring Tracy to just "go along" with this mistreatment, *her mother threatened to disown her for being "disrespectful" and "ungrateful," and, the next day, her depressed and obese sister telephoned from another city and tried to make her feel guilty for "hurting Daddy" and "stirring up trouble."* No, for clients like Tracy, change is not easy and problems are not simple.

As we see here, the problem is not simply the parent's mistreating the child. Instead, more serious consequences—such as the self-hatred, bulimia, and dysthymia that Tracy suffered—result because the parent and extended family system typically make the adult offspring again feel responsible for or deserving of the parent's original and current mistreatment.

Clients usually feel hopeless and blame themselves when parental caregivers and the broader family system cannot change. Although these clients may go on to act helpless in other current relationships, in actuality they are not. Their interpersonal resolution does not rest on the parent changing, but on how they change and respond in current interactions with living parents or, equally important, in their ongoing internal relationship with deceased caregivers. In order for clients to make enduring changes in deep-seated relational conflicts, they can change their own responses in these prototypical interactions. For example, Tracy's father was too limited to be able to hear her concern and talk with her about the problem. However, Tracy stopped going along with her father's disparagement of her. Instead, she began setting limits with her father and bravely metacommunicated with him:

> TRACY: I keep asking you to stop it when you put me down, and you laugh at me and keep doing it. I can't make you stop, but I don't have to pretend it's OK anymore. It hurts, and I don't like it.

Although her father stayed in the same disparaging mode and again made fun of her, things did change profoundly for Tracy when she changed how she responded to him. As she was able to sustain this stronger stance toward him and the family system over the next few months, she felt less insecure and intimidated with them and others in her life than she had ever been. Her long-standing feelings of worthlessness and depression significantly improved, and her intermittent battles with bulimia largely disappeared. These far-reaching changes occurred for Tracy even though her father—as well as her mother and sister, who pressured her to retreat and "go along" again—could not change, because she could successfully change her participation in these problematic interactions.

In this way, clients make progress by changing how they respond in current interactions with significant others in their lives. Clients also change by doing grief work—coming to terms with the feelings left in them by hurtful parent–child interactions. Clients do not need to discuss with caregivers problems in historical relationships. However, they do need to (1) stop disavowing and acknowledge to themselves what was legitimately wrong, (2) stop their own participation in any ongoing mistreatment, and (3) grieve for the support or validation they wanted but have missed. Regardless of whether formative attachment figures can change—and whether they are living or deceased—clients can resolve these problems through their own internal work.

Exploring the potential for better relationships with family members is often important, but it may or may not lead to improved relations. Often, aging caregivers cannot change, so therapists need to prepare clients that when they introduce healthier new behavior into their family systems it will often be met with significant resistance. Even if the family system has improved, however, grown offspring will still function better in current relationships as they can more realistically acknowledge what was wrong (and what was right) in the past and mourn what they missed developmentally. The same model applies again: The therapist provides a supportive holding environment that allows clients to come to terms with the sadness, anger, shame, and other feelings that were too threatening or unacceptable to contain before. However, to succeed in this work, therapists are encouraged to remember that clients will be safe to discuss the "bad" or problematic part of the caregiver (or spouse) only after being reassured that therapists also appreciate and affirm the "good" or well-intended parts of these relationships as well. The therapist's goal here is to resolve or integrate the ambivalence, without resorting to the splitting defenses or triangulation that the client has relied on in the past.

In another way, the therapist must also help clients come to terms with both the good news and the bad news about their current situations. The good news is that, through grief work, clients can disconfirm the pathogenic belief that they were in some way to blame for their rejection, parentification, and so forth. The bad news is that clients' developmental needs—such as the need for secure attachment ties—were not adequately met then, and these original developmental needs cannot now be met in current adult relationships (for example, with a spouse, friend, or therapist). Only after such losses or deprivation can be acknowledged as real, that is, mourned rather than disavowed, can clients go on for the first time to get appropriate and realistic adult versions of their emotional needs met in current relationships. As long as developmental needs are operating unacknowledged, clients are impaired in their ability to get legitimate adult needs met in current relationships. Finally, once clients have accepted both the "good news" and the "bad news" associated with their caregivers, they are then free to accept some of those same characteristics in themselves, and relinquish their own rigidity and perfectionism.

Two countertransference propensities can prevent therapists from helping clients achieve resolution of such family-of-origin work. First, because of their own splitting defenses, therapists sometimes do not have the breadth to help clients accept both what was good and what was problematic in their development. Therapists' countertransference propensities lead them to make one of two errors. On the one hand, some therapists characteristically want to downplay or minimize the extent and continuing influence of painful developmental experiences:

> THERAPIST: That was then. You need to forget about what happened in the past. Let's try to do something about the problems you're having now.

On the other hand, in contrast to this denial, other therapists only want to blame the caregivers, make them "bad," and ignore the strengths and positive contributions that also were present in the relationship:

> THERAPIST: (*judgmental tone*) Your father was sick to do that! What's wrong with that idiot?

Instead, therapists need to affirm the full reality of what went wrong in formative relations, while appreciating the (internalized) child's need to maintain whatever ties can be preserved. Except in extreme circumstances, therapists should discourage the client from breaking off all contact with caregivers. When a client cuts off all communication with family members, splitting defenses are usually operating and lasting change is unlikely. In such cases, the therapist will be idealized and clients will be compelled to recreate the split-off, bad side of their conflicts in other relationships or to make themselves "bad." Therapists will be far more effective—and clients will be better able to work with this guilt-inducing material—when therapists can acknowledge realistically what was wrong without making the parental figures "bad" and taking them away:

> THERAPIST: I can see how much it hurt you when he did that. He did go way over the top sometimes. I wonder what was going on for him when he lost it like that—you know, what led him to such a desperate place?

> VERSUS

> THERAPIST: I can't believe he did that to you. What a toxic, destructive self-centered, SOB!

The second countertransference propensity that interferes with family-of-origin work is that therapists often want to bypass the work of mourning. As clients become able to acknowledge realistically to themselves the ways in which they were hurt developmentally or missed what they needed, painful feelings are aroused. Therapists may find it difficult to help clients come to terms with those feelings. Most clients do not want to end the hope—or fantasy—of getting what they missed; some may even become angry when therapists acknowledge these limitations. Therapists' reluctance to address these issues will be reinforced if, as routinely occurs, such material arouses therapists' own unfinished family-of-origin and grief work. However, if the therapist tries to bypass the work of mourning and move too quickly to a problem-solving approach, clients will not be able to act on or sustain these practical suggestions— even though they may be useful ideas. Only after original losses have been acknowledged and integrated (that is, grieved for in a supportive holding environment with the therapist) can clients open up their adult emotional needs and feel sustained by new relationships in a way they have not been able to risk before. As clients improve in these significant ways, they become able to leave the past behind, to turn the page, and begin looking ahead to a different future.

The Dream   Clients make behavioral changes and feel better as they work through their problems. As they improve, the focus of therapy moves away from developmental conflicts and, to some extent, beyond current interpersonal problems. In a very positive transition, the therapeutic dialogue moves toward future plans that reflect life-enhancing aspirations and goals for the client. Referencing the literature on the psychology of hope, for example, therapists can help clients explore how they would like to be—"imagining possible selves," and clarifying what they might like to become (Snyder, 1994). Similarly, therapists can work in an existential tradition with issues regarding personal choice and explore: How can clients create more meaning in their lives, live more authentically, and make the most of the time and relationships they have? This exciting transition from solving problems to leading a fuller and more meaningful life is often signaled by the emergence of "the Dream."

In his classic books, *The Seasons of a Man's Life* and *The Seasons of a Woman's Life*, Levinson (1978, 2000) describes the profound influence of the Dream in shaping the structure of adult life and the course of personality development. As Levinson points out, the Dream does not refer to casual waking or sleeping dreams but, in the largest sense, to the kind of life that one wants to lead. At first, the Dream may be poorly articulated and tenuously connected to reality, but it holds imagined possibilities of self in the adult world that generate vitality. It is the central issue in Martin Luther King's historic "I Have a Dream" speech or in Delmore Schwartz's story "In Dreams Begin Responsibilities."

The Dream has roots in the grandiose and unrealistic hero fantasies of adolescence, but it is more than that. In early adulthood, the Dream still has the quality of a vision—of who and how we would like to be in the world. In this way, the Dream is inspiring and sustaining for the individual, even though it may be mundane to others. For example, individuals have Dreams to be a good mother or responsible husband, a respected community leader, an ethical attorney, a skilled craftsperson, a successful but honest businessperson, an artist, or a spiritual leader. *If the individual is to have purpose and a sense of being alive, occupational and marital choices need to incorporate some aspects of the Dream. When the Dream has been abandoned or set aside, life will not be infused with vitality and meaning, even though the person may be successful.*

Up until the working-through phase of therapy, therapists are usually responding to the despair of broken dreams, the disillusionment of unfulfilling dreams, and the cynicism of abandoned dreams. In many cases, the life-infusing Dream that Levinson has articulated has not yet been addressed in treatment. In the working-through phase, however, clients are resolving their problems and emerging as healthier individuals. At this point, therapists can help clients to articulate and renew their Dream and to find ways of incorporating aspects of their Dream in their everyday lives. To achieve this, therapists first need to be able to differentiate the Dream from the broken dreams that contribute to clients' presenting symptoms. These broken dreams often reflect failed attempts to rise above their conflicts and become special, as discussed in Chapter 7.

In the process of working through, therapists repeatedly help clients recognize their compensatory strivings to be special and rise above their conflicts by pleasing others (moving toward), achieving success and power (moving against), or becoming safely aloof and cynically superior (moving away). Therapists help clients progressively relinquish these inflexible coping styles and adopt a wider interpersonal range. In doing so, clients stop blocking their generic conflict, which presents the opportunity for clients' core conflicts to be expressed and resolved within the therapeutic relationship. However, therapists can anticipate that when clients relinquish these attempts to rise above their conflicts, they may experience a sense of failure or loss of self-esteem. Thus, one important component of helping clients resolve their problems is to replace these defensive, grandiose strivings with clients' own attainable, yet sustaining Dream.

Arthur Miller's (1949) Pulitzer Prize–winning play, *Death of a Salesman*, poignantly illustrates how the neurotic striving to rise above conflicts inevitably fails, how it can lock clients in their generic conflicts for a lifetime, and how it stifles the Dream. The protagonist, Willy Loman, has adopted a moving-toward interpersonal style to cope with his profound feelings of inadequacy and shame. With his smile and his freshly shined shoes, Willy pathetically strives not just to be liked but to be "well liked." In a brilliant exposition of multigenerational family roles and relationships, Miller shows how Willy's grandiose strivings to rise above his shame-based sense of self by being well liked are also acted out through his oldest son, Biff. Willy exaggerates Biff's accomplishments as "magnificent," excuses or ignores his shortcomings, and never responds to the reality of his son's actual feelings, needs, or wishes.

Near the end of the play, both father's and son's defensive strivings to rise above fail. Willy is fired from his job—a crushing humiliation that shatters his myth of being well liked and his lifelong coping strategy of winning approval. That same day, Biff fails to secure an important business opportunity that his father has encouraged. With this setback, Biff realizes that he has been living out his father's myth that he would become a fabulously successful entrepreneur. In the closing scene, father and son respond very differently to their respective crises.

Biff is able to relinquish the unrealistic script of superachievement and success that his father has demanded of him. In the last scene, Biff begs his father to release him from this grandiose role, to "speak the truth" in the family for the first time, to acknowledge that he is just another regular guy—nothing more, nothing less—and to let that be good enough. However, Willy cannot relinquish his own rising-above defense and, therefore, cannot grant Biff's plea to "let me off the hook." Instead, Willy alternately rejects Biff and idealizes him as magnificent.

Biff finally recognizes the maladaptive relational pattern that he and his father have been enacting all of their lives. By addressing it with his father, Biff is able to change his part of their conflict. Although this honest communication does not lead to an interpersonal resolution with Willy, it does lead to an internal resolution for Biff. Soon afterward, Biff decides to leave home. He says that he now knows who he is and what he wants to do with his life. He is going to pursue his own Dream: to leave the city, move out west, and work outdoors where he can feel the sun on his back and smell the grass. Whereas Biff could relinquish the neurotic strivings to be special and allow his own attainable Dream to emerge, Willy could not.

In one last attempt to change their relationship and have a more authentic connection, Biff reaches out and puts his arms around his father. Horrified, Willy recoils in fear and disgust. Although his lifelong obsession has been to win the approval and respect of others, he rigidly rejects exactly what he has always wanted from his son. To accept the love that Biff was offering—to get what he wants—would reveal Willy's own generic conflict—his unmet needs to be respected and wanted, his profound sense of being unlovable, and his intense feelings of inadequacy and shame. With no understanding of what these disavowed feelings are about, and no supportive relationship to help him contain this flooding, his unacceptable feelings would be overwhelming. Thus, Willy is now trapped in his conflict: He is no longer able to maintain his own (or Biff's) defensive strivings to rise above, and he is unable to let his true feelings and concerns emerge. Desperately bound in that unresolvable conflict, Willy's tragic solution is to end his own life.

As clients' neurotic strivings to be special and rise above their conflicts fail, their generic conflicts are revealed. Without a holding environment to help them contain their feelings, some individuals—like Willy—feel hopeless and a few may see suicide as the only way out. In contrast, if the therapist can provide a different and more satisfying response to the conflict, clients can begin to come to terms with these feelings in a better way and change. As clients begin to improve during the working-through phase, the Dream will often emerge.

To help clients change, therapists can support and encourage the Dream in many ways. First, the therapist is listening actively for the issues that genuinely enliven clients, engage their intrinsic interest, and bring them pleasure. By attending closely to what clients really want to do and inquiring about what feels right for them, therapists can help clients discover and articulate their Dream. More specifically, the therapist's intention is to ask questions that will help clients clarify what they like, find interesting, or want in order to make a better life for themselves. For example, clients may respond:

- I want to stop losing my temper, and my children's respect.
- I want to make peace with my father and bury the hatchet.

- I don't want to keep driving to this same dead-end job. I want to finish school and get my credential.

For other clients, their Dream is completely undeveloped and they cannot articulate what they want in any meaningful way because their caregivers never accepted their feelings or responded to their interests. These clients, with a less developed sense of self, are unaware not only of what they want to do in life but, more basically, even of what they like or dislike. Therapists can help these clients discover and successively articulate their feelings, interests, values, and ultimately their Dream by repeatedly

- asking clients to attend to what they are experiencing at that moment;
- orienting clients to listen to themselves and simply be aware of what they want to do in different situations—even if they cannot act on their wishes (for example, to realize "I'm not enjoying this right now");
- being interested in and entering into clients' subjective experience;
- calling attention to, and trying to participate in, what seems to interest or hold meaning for clients;
- acknowledging and taking overt pleasure in what clients do well;
- encouraging clients to act on their own feelings or preferences when possible (for example, even something as simple as saying, "I'm sleepy—I'm going to take a nap").

Although many clients have a more developed self than this, some will need to begin attending to their own internal experience in these very basic ways. It is rewarding for the therapist to help clients differentiate a self by responding—perhaps for the first time in their lives—to their own feelings and interests. Clinicians working within many theoretical orientations have emphasized this process, especially the concept of "mirroring" in self psychology (Kohut, 1977) or "experiencing" in the client-centered tradition (Carkhuff, 1999).

As clients begin to experience their inner life more fully, and the therapist responds affirmingly, their Dream will start to emerge. At this point, the therapist can do several things to help clients bring aspects of their Dream into their current lives. For example, the therapist can provide a "practicing sphere" in which clients can discuss and explore various possibilities of their Dream without having to become committed prematurely to any action. The therapist can also help clients tailor their Dream to reality and find ways to express aspects of their Dream in their everyday lives. The therapist can also help clients identify training or education routes that will facilitate realization of their Dream, and need to give clients permission to develop close relationships with others who can act as mentors in their chosen fields and pursuits.

Finally, clients will confront the same core conflicts in exploring their Dream that they have experienced in other aspects of their lives. Thus, the therapist will also have to help clients work through their old conflicts that are aroused by pursuing their Dream. For example, the therapist will have to help clients differentiate between their Dream and their interpersonal coping style and defensive strivings to rise above. For example, strivings to feel secure by being loved by everyone, to gain a sense of personal adequacy by becoming wealthy and powerful, or to escape criticism or rejection by being safely aloof and superior. Progress usually will not be simple or straightforward, however. For many clients, *actively pursuing what they really want to do, or successfully attaining what they want in life, often threatens attachment ties and arouses both separation anxiety and separation guilt—leading clients to retreat from progress or undo their success in treatment.* However, as clients successively work through these conflicts and find that they do have a right to their own life, they can follow through on realistic plans for attaining aspects of their Dream. With this success, treatment evolves to a natural conclusion.

Termination

Termination is an important and distinct phase of therapy that needs to be negotiated thoughtfully. Ending the therapeutic relationship will often be of great significance to clients. For some, this may be the first positive ending of a relationship they have ever experienced. The way in which this separation experience is resolved is so important that it often influences how well clients will be able to resolve future losses and endings in their lives. It also helps determine whether clients leave therapy with a greater sense of self-efficacy and the ability to successfully manage their own lives more independently. Most beginning therapists underestimate what an important experience a successful termination can be for clients. It holds the potential either to undo the changes that have come about in treatment or to help clients maintain behavior change after treatment has ended. Therefore, therapists want to be prepared to utilize the further potential for change that becomes available as therapists and clients work together to end their relationship.

Early in their training, therapists often ask, "How will I know when it's time to end therapy?" It is time to end when clients have achieved significant relief from their symptoms, can respond more flexibly in current situations rather than slotting all experience into predisposed categories, and have begun to take steps toward promising new directions in their lives. Therapists know that clients are ready to terminate when they have converging reports of client change from three different sources:

- when clients report that they consistently feel better and can respond in more adaptive ways to old conflict situations;
- when clients can consistently respond to the therapist in new ways that expand their old coping styles and do not reenact their maladaptive relational patterns; and
- when significant others in clients' lives give them feedback that they are different or make comments such as, "You never used to do that before."

With this convergence of perceptions that important changes have occurred, it is time for treatment to end. Before going on to discuss successful termination, we must distinguish between two types of endings: natural and unnatural.

The termination sequence just described is a *natural* ending because the client's work is finished. One of the therapist's primary goals in these natural endings is to affirm both sides of clients' feelings about ending: to take pleasure in clients' independence and actively support their movement out on their own; but also to let clients know that the therapist will accept their need for help or contact in the future if the need arises. Clients' old schemas, transference distortions, and faulty expectations will often be activated toward the therapist at this time. To correct these transference distortions, therapists can clarify that they will not be disappointed or burdened (and so forth, as clients inaccurately expect) if clients want to send a letter to stay in touch, call to share important life events such as marriages and births, or to recontact the therapist for help if they have problems in the future.

The key issue here is that so many clients did not have permission to be both "separate and related" at the same time. For example, growing up, becoming more independent, and differentiating from their caregiver by pursuing their own interests, goals, and relationships, often meant losing their emotional support and attachment ties. In contrast, staying connected or emotionally supported meant remaining dependent or not having their own voice, views, or feelings distinct from those of the people they were with. The termination phase provides these clients with an important opportunity to further resolve the separateness–relatedness dialectic.

Therapists offer many clients a potent corrective emotional experience when clients are assured of the therapist's support for both sides of their feelings. That is, many clients benefit greatly from finding that the therapist takes pleasure in their success and welcomes their independence, yet is still available and interested in their well-being if the client needs to recontact them (or another therapist) in the future. With this support for both sides of the separateness–relatedness dialectic, clients can further internalize the therapist and keep all that she has offered, solidify their own sense of self and personal efficacy, and successfully end the relationship.

Unfortunately, most graduate student therapists do not have the satisfaction of seeing many cases evolve to a natural close. Instead, therapy ends before the client is finished because the graduate student must move on to another placement, the school year ends, or the training clinic provides only time-limited services. Sometimes the therapist and client have made substantial gains at the point when external constraints demand termination; at other times, the treatment process may be only in midstream. In either case, significant therapeutic gains can still be made during the termination phase. To accomplish this more difficult task, however, it is essential that the therapist and client address and work through the complex issues that an unnatural or externally imposed ending arouses. Because graduate student therapists often have to initiate terminations before clients are ready, we will focus on unnatural endings in the discussion that follows. *It is necessary to examine unnatural or imposed endings closely because they routinely evoke angry, blaming, and distancing reactions in the client which, in turn, activate guilt, defensiveness, and other countertransference reactions in the therapist that can undo the significant gains that have been made.*

Accepting That the Relationship Must End

It is often difficult for both the therapist and the client to end their relationship. In many cases, beginning to talk about termination will bring up emotional reactions that are far more intense and significant than either the therapist or the client has anticipated. As a result, *therapists and clients regularly collude to deny the reality of the impending ending and to avoid the difficult feelings it arouses for both of them.* When therapists and clients stumble in this way and do not address the ending squarely, their interpersonal process will often recapitulate clients' central conflicts and prevent clients from resolving their problems as fully as they could have. Clearly, this becomes especially important with clients who have a history of painful interpersonal losses. Thus, the single most important guideline for negotiating a successful termination is to unambiguously acknowledge the reality of the ending.

> THERAPIST: We have three more sessions left before the school year ends and we have to stop. Let's talk about what this ending is going to mean for both of us.

The therapist and client can then discuss all of the client's positive and negative reactions about the ending, especially the client's emotional reactions toward the therapist. Although the therapist and client may both want to avoid this topic, the therapist cannot let that happen. Once the therapist and client mutually decide that the client is ready to terminate, or that the therapist must terminate because of outside constraints, they need to establish the specific date for the final session. For example:

> THERAPIST: Our last session will be in two weeks, on Thursday, December 9, at 4:00.

Because the therapist and client need time to work through their ending, the final session should be scheduled at least two weeks hence.

To have a successful ending, the therapist and client mutually establish a final date and then explore together the client's reactions to ending their relationship. In some cases, it will be difficult for the therapist to set a specific termination date and discuss the ending because this arouses the therapist's own conflicts over this ending—or more commonly, endings in general. In addition to the therapist's own countertransference propensities, which are especially likely to be evoked around terminations, setting a clear termination date will often bring up the client's presenting symptoms and core conflicts again. Some clients may need to work through again, in the termination phase, aspects of the same issues that have been dealt with before. In other words, when termination becomes a reality, some clients will temporarily retreat from or undo the changes they have made. For all of these reasons, therapists need to acknowledge the termination forthrightly, count down the final sessions, and repeatedly invite clients to discuss their reactions to the termination:

- After today, we will have only three sessions left. How is it for you to hear me say that?

- We have only two sessions left now, and I think it's important that we talk together about our ending. What comes up for you when I acknowledge that soon we are going to have to stop working together?
- Next week will be our last session. We've accomplished some things together, but problems remain, and our work is unfinished. What do you find yourself thinking about our relationship and the time we have spent together?
- This is the last time we are going to be able to meet. I'm feeling sad about that, and I'm wondering what kind of feelings you're having.

This type of countdown precludes any ambiguity that clients may have about the ending. For example, it will prevent the client from asking, at the end of the last session:

CLIENT: So, will I be seeing you next week?

This approach also keeps the therapist from acting out and avoiding the separation as well—for example, by saying at the end of the last session:

THERAPIST: Oh, I see our time is up. Well, I guess that's it.

When therapists do not address the termination directly they are often acting out their own separation anxieties or guilt over letting someone else down. In particular, their own unresolved feelings from past problematic endings in their own lives—parental divorces, deaths, and other unanticipated or unwanted leavings—are being reevoked. Such countertransference reactions are especially problematic because they often dovetail with clients' issues. For example, many clients are also struggling with their own unresolved feelings over past painful endings. Other clients, for example, those who have been parentified, are avoiding the topic because they sense it is difficult for the therapist. Unresolved conflicts over ending relationships may be the most common countertransference issue that therapists bring to treatment. When this occurs, therapists can reassure themselves that this is a common and understandable problem to have, but they also need to do something about it by seeking consultation from a supervisor or colleague.

Because both clients' transference distortions and therapists' countertransference reactions are especially likely to occur around unnatural terminations, *and at the end of time-limited treatments*, we need to explore these issues further. Both therapists and clients often have experienced painful endings with significant others as "just happening to them." In many cases, they

- were not prepared in advance for the separation,
- did not understand when or why this particular ending was occurring, or
- were not able to participate in the leave taking by discussing it with the departing person or by saying good-bye.

Such problematic endings have left many therapists and clients feeling powerless in regard to some of the most important experiences in their lives. In contrast, the approach suggested here enhances clients' efficacy by allowing them to be active, informed participants in the ending. Clients know when the ending is occurring; have the safety to express their sadness, anger, or disappointment about it; and have the opportunity to say good-bye. Thus, to achieve a successful leave taking, the therapist can

- acknowledge whatever therapeutic work remains unfinished;
- remain affirming or nondefensive in the face of the client's disappointment, angry feelings, or accusations of betrayal and other guilt-inducing responses that may occur; and
- explore how this termination may be evoking other unwanted or unexpected endings in the client's past.

Why is this so important? Especially with unnatural endings, it often will activate early maladaptive schemas and seem to the client that this relationship has just failed *in the same way* that others have in the past. Let's explore this expectable problem more fully.

As the old saying goes, children are supposed to grow up and put their feet under their own table, but it's hard to get up from the table when you're still hungry. The emotional deprivation that many clients bring to therapy is activated when treatment ends prematurely or before they feel ready to stop. Routinely, *clients' anger and disappointment about what they have wanted but missed in past relationships—even their sense of abandonment or betrayal— will be directed toward the therapist*, no matter how clearly the treatment parameters were explained at the beginning of treatment. Clients' guilt-inducing accusations can be difficult to hear:

- You're just like my mother. You're abandoning me, too.
- Our relationship never meant much to me, anyway.
- Nothing changed for me in therapy. This was just a waste of time.

The success of treatment in these cases depends in part on therapists' ability to remain *nondefensive* and tolerate clients' protests—rather than feel guilty, become defensive, and make ineffective attempts to talk clients out of their feelings:

> THERAPIST: But you *have* changed. You're not as depressed as you used to be, and now you can do _____, and before you couldn't do that.

Instead of these futile attempts, therapists first need to acknowledge clients' feelings:

> THERAPIST: You're really angry at me as our relationship is ending, and I respect those important feelings. Right now, it just seems like nothing meaningful has occurred and that nothing has gotten better for you.

By accepting clients' angry protests in these ways—without agreeing that their factual content is true—therapists give clients a reparative response that doesn't match their schemas or previous experiences. In turn, this affirmation of their current frustration or disappointment often allows clients to connect their strong emotional reactions to the developmental figures who originally disappointed or left them. Thus, *what therapists often find hardest about unnatural endings—and about time-limited therapy—is managing their own guilt or defensiveness in the face of clients' anger over the ending.* However, by drawing on the guidelines outlined here, the therapist can help clients see how this ending is in fact different in some important ways from other unwanted endings in their past—even though it feels as if the same unwanted scenario is happening to them again. Therapists can help make the termination of therapy different from past problematic endings and, in doing so, help resolve this unfinished grief work from the past in the following ways:

- Talking with clients about the ending in advance;
- Inviting clients to share their angry, disappointed, or sad feelings and being able to accept those feelings without becoming defensive;
- Talking with clients about the meaning the termination holds for them, and clarifying how it is similar to and different from other important endings in their lives;
- Discussing the meaning this relationship has held for the therapist and sharing some of the therapist's own feelings about ending this relationship;
- Validating clients' experience by acknowledging the ways that this ending may evoke other painful or unwanted endings with others;
- Ensuring that therapists and clients can say good-bye to each other.

In other previous endings that have been problematic, clients were not able to have such experiences with the departing person. Although most clients will have difficulty recognizing these differences initially, the therapist offers clients an opportunity to resolve long-standing problems by helping them differentiate this type of mutually acknowledged ending from the incomplete or unsatisfying endings with which they have had to cope in the past. In this way, successfully managing the termination also becomes the prototype for learning how to cope effectively with future endings and losses. One of the pioneers in short-term therapy, Mann (1973), writes informatively about this and the strong feelings of loss that are activated for so many clients during termination.

In the face of clients' withdrawal or blaming accusations that their termination is "just like" prior losses and disappointments, therapists, too, may lose sight of the real differences that exist between this termination and clients' problematic past endings. In that case, therapists may accept the clients' blame, feel guilty, and avoid the ending—or invalidate the clients' sad, disappointed, or angry feelings by trying to talk clients out of them. In that case, paradoxically, therapists make the clients' accusations come true: They metaphorically reenact the clients' past conflicted endings. Therapists do better by acknowledging the similarity of the clients' feelings at termination and at past separations, and then exploring with clients what they can do together to make this termination different and better from such problematic endings in the past.

During the termination period, clients often have a range of sad, hopeful, grateful, and other feelings toward the therapist. As therapists track clients' reactions to the ending, they will often see how the clients' core conflicts are simultaneously being aroused in two contexts. First, the clients' primary feelings about termination (such as guilt over feeling better and no longer needing the therapist, feeling rejected, abandoned or betrayed by the therapist, and so forth) may match the clients' feelings about the formative relationship in which the core conflict originally arose. Second, clients' reaction toward termination may also match the feelings that were aroused in the crisis situation that originally brought them to treatment. By responding affirmingly to the clients' feelings, and clarifying their connection to these two sources when appropriate, the therapist is helping them further resolve their core conflicts.

Finally, to help with all of this challenging material, Marx and Gelso (1987) emphasize three main steps in effective terminations:

1. looking back and reviewing what has changed;
2. looking ahead and making realistic plans for coping with the problems that are likely to come up; and
3. saying good bye.

Exploring effective terminations further, a useful intervention for the termination phase is to go through a *review-predict-practice* sequence with all clients. Therapists *review* progress, accomplishments, successful changes and transitions, and unfinished issues with clients. Especially important, therapists help clients *predict* the challenging events or anxiety-arousing situations that are likely to activate or bring up their dysfunctional patterns in the future. To help clients maintain the changes they have achieved, it is essential to anticipate these potentially problematic situations or relational scenarios. Therapists want to identify these activating circumstances or triggers as specifically as possible. As illustrated with Tracy, earlier in this chapter, the therapist might ask:

THERAPIST: What is your dad likely to say if you bring this up? How is that going to make you feel, and what are you likely to do at that moment?

Finally, *practice* in dealing with these triggering issues or familiar patterns also prepares clients to succeed on their own. Therapists best can provide this by role-playing effective and ineffective responses to the threatening or unwanted interactions that clients are likely to face in the future.

To sum up, the therapist's goal is to address the client's feelings about ending the relationship and help the client work them through. This is difficult to do, however, when both the therapist and the client may wish to avoid the ending—as so often occurs. The therapist and client have shared much in their work together and the relationship holds real meaning for both of them. To avoid the conflicts that are activated by the termination, clients—and therapists, too—often deny or try to diminish the importance of the ending. Some clients may try to devalue the significance of their relationship with the therapist or to diminish the importance of the work that has been accomplished. Other clients will become symptomatic again and anxiously communicate that they are still too troubled to terminate and make it on their own. A few clients will ask the therapist if they can continue the relationship by becoming friends. These common reactions during the termination phase keep clients from facing current and past losses, preclude clients from receiving a new and more reparative response from the therapist, and keep clients from being able to internalize the therapist and carry forth with others the gains they have made in treatment.

Ending the Relationship

Although some clients will reexperience their presenting problems at the time of termination, many will not. Both clients and therapists still need to talk about the ending, however. For example, therapists can talk with the client about the different ways in which they have seen the client change over the course of treatment. Therapists can also acknowledge the limitations of their work together and the unfinished issues that clients will need to continue to work with on their own or perhaps with another therapist at some point in the future. Fearing that they are being disloyal, many clients will need explicit permission from the therapist to work with another therapist if they have problems in the future. The therapist needs to actively reassure many clients that the therapist would be happy to know that they are taking good care of themselves and are getting the help they need. Recollections of close moments, awkward misunderstandings, risks ventured, and humorous incidents also can be shared. This also is a time when therapists can share their own feelings about the client:

- what they have learned about themselves and life from the client,
- how the client has enriched or influenced them,
- ways in which therapists have cared about or have enjoyed the client, and
- how therapists will remember the client.

Finally, in natural and unnatural endings, we have seen that termination brings the therapist and client back to the separateness–relatedness dialectic. In the case of natural endings, the therapist's intention is to give clients permission to leave. Clients need to know unambiguously that the therapist enjoys their success, is pleased by their independence, and takes pleasure in seeing them become committed to new relationships and activities. Per client response specificity, some clients will find it helpful to be reassured that they can have future contact with the therapist if the need arises; others will not need or want this. Knowing that the therapist is available if needed but also supports their decision to leave, clients are able to keep and apply the help the therapist has provided. This type of natural termination supports clients' own strivings for individuation in a supportive relational context, and empowers clients to further claim their own authentic voice.

Many clinicians, and graduate student therapists in particular, often cannot be available in this way because endings with their clients are of necessity unnatural. In these cases, the finality of the ending must be honestly acknowledged. Therapists, with the help of a supportive supervisor, want to anticipate their own personal reactions to ending with this particular client, and be prepared to manage their own countertransference reactions effectively. In particular, many new therapists will need assistance from their supervisors in order to respond acceptingly, rather than defensively overreacting, to the client's angry, disappointed, dismissing, and other defensive reactions. The unfinished business that remains also needs to be realistically acknowledged, and the therapist may need to help the client find a new therapist to finish the work that was just begun. When the therapist can do this, clients will be able to accept the limitations of the relationship without having to reject the therapist or undo the work that was accomplished. Although often incomplete, when clients can live out with others what they have experienced with the therapist, and therapists can appreciate what they have learned from and given to the client, both have been enriched by this successful relationship.

Closing

This book has followed the course of therapy from beginning to end. Just as clients still have unresolved problems when treatment ends, many questions about therapy still remain unanswered for therapists-in-training. Challenging personal issues and complex interpersonal processes have been introduced here, and therapists will need further reading, supervision, and experience to help with the myriad exceptions that occur in every therapeutic relationship. Despite these limitations, the reader has learned essential guidelines for working with the process dimension and intervening in the therapeutic relationship. The conceptual framework presented here will help new therapists work more effectively with their clients and, hopefully, better integrate training from a variety of theoretical perspectives.

In closing, the theme of this text is that enduring change occurs within the context of significant relationships. Therapists working within cognitive and behavioral approaches, interpersonal and psychodynamic frameworks, and family systems and other theoretical traditions can utilize the interpersonal process approach to make their own work more effective. Accordingly, therapists are encouraged to use themselves, and the relationships they provide, to help clients change.

Suggestions for Further Reading

1. At the end of Chapter 4, readers were encouraged to complete the self-assessment scale, "Interpersonal Process Skills: Assessing Competencies," in the Student Workbook. Having completed this text, students may wish to assess themselves again on these essential clinical skills. In addition, readers are encouraged to read the case study of Dorothy, from the *Wizard of Oz*, in Chapter 10 of the Student Workbook.

2. Student therapists who have found the ideas presented here to be helpful are also likely to enjoy *Between Therapist and Client* (Kahn, 1997) and *Helping Skills: Facilitating Exploration, Insight, and Change* (Hill, 2004). For more experienced therapists, *Relational Concepts in Psychoanalysis* (Mitchell, 1988) and *Coping with Conflict: Supervising Counselors and Psychotherapists* (Mueller & Kell, 1972) both provide an informative presentation of interpersonal therapy.

3. Useful information about identifying clients' core/generic conflicts and how they can be reactivated during the termination phase of treatment can be found in Mann's (1973) classic primer, *Time-Limited Psy-*

chotherapy, and in Chapter 9 of Strupp and Binder's (1984) *Psychotherapy in a New Key.*

4. The Dream is an important psychological dimension in people's lives that has received too little attention. The reader is encouraged to read Daniel Levinson's (1978) groundbreaking book, *The Seasons of a Man's Life*; see especially pages 91–109. *The Seasons of a Woman's Life* (2000), published soon after Levinson's death, tracks developmental stages and transitions in women's lives. The Dream, as it relates to women, is given special attention on pages 238–239.

5. Much useful information about successful terminations can be found in Wachtel (2002), "Termination of Therapy: An Effort at Integration." In addition, therapists are encouraged to read more about how therapists can help clients establish the support systems they need in order to maintain changes after treatment ends (McWhirter, 1996).

PROCESS NOTES

Client _____ Therapist _____ Session Number & Date _____

1. Issues presented: What are the key concerns expressed by the client or noted by you?

_____

_____

_____

2. Themes: Describe any repetitive themes, issues, or patterns observed.

_____

_____

_____

3. Relationship and resistance: Note any overt statements or covert references (embedded messages) that the client made about you or your interaction together. Did the client express any ambivalence about attending sessions or concerns about any aspect of being in treatment?

_____

_____

_____

4. Interpersonal process: What can you learn about your client's problematic interactions with others from your own personal reactions toward the client (such as boredom, frustration, protectiveness)? Were there ways in which your interaction may parallel or reenact what tends to go wrong with others in the client's life?

_____

_____

_____

5. Transference and countertransference: In what ways does the client tend to distort or misperceive you, and respond to you as she or he does to significant others? Are there ways in which your own personal history or current life circumstances are activated by this client and influence how you respond to him or her?

_____

_____

_____

6. Evaluating interventions: Assess the client's reactions to your interventions—what was effective and ineffective with this client? In retrospect, what might you have done differently?

_____

_____

_____

7. Treatment focus: What do you see as the key issues or concerns that are central to this client's problems? What do you want to focus on and explore further?

_____

_____

_____

8. Critical incidents: Note any potential legal, ethical, or reporting concerns for consultation with your supervisor.

_____

_____

_____

CASE FORMULATION GUIDELINES

Although writing a case formulation can be an anxiety-evoking process, completing one is important because it helps therapists provide more focused treatments and helps them respond to treatment planning requests from HMOs and other third-party insurers.

1. Formulation of the problems   Summarize the client's presenting problems and suggest an initial diagnosis. Why is the client seeking treatment now, and what has the client done in the past to address these problems? How does the client feel about him- or herself (for instance, worthless, superior, burdened), and how is this feeling manifested with others and with you? What organizing themes or patterns characterize the client's most important experiences with others? How have familial and sociocultural factors (race, religion, socioeconomic status, gender, sexual orientation) affected the client and shaped his or her subjective worldview?

2. Treatment focus   What has treatment focused on to date? Identify the maladaptive patterns (including style of relating to others, pathogenic beliefs, cognitive schemas, and emotional conflicts) that recur for the client and contribute to his or her presenting problems. For example, thinking of the client's internal working models or cognitive schemas, ask yourself the following questions: (1) What does the client want from others (for example, to be cared for, protected)? (2) What does the client expect from others (to be criticized, rejected)? (3) What is the client's experience of self in relationship to others (a sense of being disconnected, burdensome)? (4) What conflicted feelings are typical for this client (fear, shame, anger)? (5) What interpersonal coping strategies does this client typically use (such as comply and go along, withdraw, try to control)? (6) What do these coping strategies tend to elicit from the therapist and others (rejection, advice giving)? (7) What stressors, crises, and social supports does the client have? Use your responses to these questions to identify two or three treatment focuses that clarify what's wrong and needs to change.

3. Developmental context   How did the client's problems originally come about? Identify developmental experiences (individual, familial, cultural) and stressors, crises, and/or social supports that shaped the client's subjective world-view and current conflicts. What were the client's experiences in the family (for example, how did the family nurture, discipline, establish generational boundaries and roles, engage in abuse or neglect)? What cultural factors (religion, socioeconomic status, neighborhood safety) were evident in this client's life? In other words, how did the client's early maladaptive schemas develop?

4. Therapeutic process   Describe the quality of the working alliance and how you and the client have interacted together. In what ways is your interaction together parallel to the client's experiences with others (reenacting), and how is it different (corrective)? What automatic thoughts, pathogenic beliefs, distressing emotional states, and maladaptive relational patterns are evident in this client's life, and how have these been expressed in the therapeutic relationship and addressed in your work together? In what ways does the client engage in "tests" with you to confirm or disconfirm past hurtful experiences that have led to his or her current problems? When the client is distressed, how does she or he tend to distort or misperceive you? Have you been able to address potential problems and misunderstandings between you and restore ruptures in the working alliance?

5. Goals and interventions   As you focus on the two to three core treatment issues identified in item 2, indicate how you plan to address these. Articulate as clearly as possible the specific experiences (affective, cognitive, behavioral) that this client needs in order to meet the treatment goals identified. Evaluate the appropriateness of the goals set and interventions proposed given the client's personal, familial, and cultural context. Assess also the appropriateness of referrals (such as group counseling, possible need for medication, parenting education). In other words, based on where you want to go with this client, how do you plan to get there?

6. Impediments to change   What client characteristics are likely to contribute to the effectiveness or ineffectiveness of treatment (pathogenic beliefs, typical emotional reactions, coping styles or strategies, social supports, and resiliency factors)? If the client terminates prematurely or treatment is not successful, anticipate the factors that could be contributing. Identify the personal, familial, or cultural factors that may make it difficult for this client to acknowledge having problems or to ask for help. Are you able to talk with the client regularly, and in depth, about the way you interact and work together, his or her reactions toward you and the treatment process, and what you could do to make treatment more helpful? Are you able to help the client identify interpersonal styles that are no longer adaptive, and become more flexible in his or her current coping strategies? If not, what factors are contributing to the impasse? Most important, suggest how the client's maladaptive relational patterns with others could be subtly reenacted with you in the therapeutic relationship, especially when the client's problems or interpersonal style activates your own counter-transference issues.

REFERENCES

Alexander, F., & French, T. M. (1980). *Psychoanalytic therapy: Principles and applications.* Lincoln: University of Nebraska Press.

Anchin, J., & Kiesler, D. (1982). *Handbook of interpersonal psychotherapy.* New York: Pergamon.

Anderson, C. M., & Steward, S. (1983). *Mastering resistance: A practical guide to family therapy.* New York: Guilford.

Bachelor, A. (1995). Clients' perception of the therapeutic alliance: A qualitative analysis. *Journal of Counseling Psychology, 42,* 323–327.

Balcom, D., Lee, R., & Tager, J. (1995). The systemic treatment of shame in couples. *Journal of Marital and Family Therapy, 21*(1), 55–65.

Bandura, A. (1977). Self-efficacy: Toward a unifying theory of behavioral change. *Psychological Review, 84*(2), 191–215.

Bandura, A. (1982). Self-efficacy mechanisms in human agency. *American Psychologist, 37,* 122–147.

Bandura, A. (1997). *Self-efficacy: The exercise of control.* New York: Freeman.

Barkham, M., & Shapiro, D. (1986). Counselor verbal response modes and experienced empathy. *Journal of Counseling Psychology, 33*(1), 3–10.

Bateson, G. (1972). *Steps to an ecology of mind.* New York: Dutton.

Baumrind, D. (1967). Child care practices anteceding three patterns of preschool behavior. *Genetic Psychology Monographs, 75,* 43–88.

Baumrind, D. (1971). Current patterns of parental authority. *Developmental Psychology Monograph, 4* (Whole No. 1, Pt. 2).

Baumrind, D. (1983). Familial antecedents of social competence in young children. *Psychological Bulletin, 94*(1), 132–142.

Baumrind, D. (1991). The influences of parenting style on adolescent competence and substance use. *Journal of Early Adolescence, 11,* 56–95.

Beavers, W. R. (1982). Healthy, midrange and severely dysfunctional families. In F. Walsh (Ed.), *Normal family processes* (pp. 45–66). New York: Guilford.

Beavers, W. R., & Hampson, R. B. (1993). Measuring family competence: The Beavers Systems Model. In F. Walsh (Ed.), *Normal family processes* (2nd ed.). New York: Guilford.

Beck, A. (1967). *Depression: Causes and treatment*. Philadelphia: University of Pennsylvania Press.

Beck, A. (1976). *Cognitive therapy and the emotional disorders*. New York: International Universities Press.

Beck, A. (1995). *Cognitive therapy: Basics and beyond*. New York: Guilford.

Beck, A., Freeman, A., & Davis, D. (2003). *Cognitive therapy of personality disorders* (2nd ed.). New York: Guilford.

Benjamin, L. (2003). *Interpersonal reconstructive therapy*. New York: Guilford.

Bergin, A. E. (1997). Neglect of the therapist and the human dimensions of change: A commentary. *Clinical Psychology: Science and Practice, 4*, 83–89.

Bergin, A., & Garfield, S. (Eds.). (1994). *Handbook of psychotherapy and behavior change* (4th ed.). New York: Wiley.

Beutler, L. E. (1997). The psychotherapist as a neglected variable in psychotherapy: An illustration by reference to the role of therapist experience and training. *Clinical Psychology: Science and Practice, 4*, 44–52.

Binder, J., & Strupp, H. (1997). "Negative process": A recurrently discovered and underestimated facet of therapeutic process and outcome in the individual psychotherapy of adults. *Clinical Psychology: Science and Practice, 4*, 121–139.

Bischoff, M. M., & Tracey, T. J. (1995). Client resistance as predicted by therapist behavior: A study of sequential dependence. *Journal of Counseling Psychology, 42*, 487–495.

Blatt, S. J., Sanislow, C. A., Zuroff, D., & Pilkonis, P. (1996). Characteristics of the effective therapist: Further analysis of the data from the NIMH TDCRP. *Journal of Consulting and Clinical Psychology, 64*, 1276–1284.

Bloom, M., Fischer, J., & Orme, J. G. (2003). *Evaluating practice: Guidelines for the accountable professional* (5th ed.). Englewood Cliffs, NJ: Prentice Hall.

Boszormenyi-Nagy, I., Grunebaum, J., & Ulrich, D. (1991). Contextual therapy. In A. S. Gurman & D. P. Knishern (Eds.), *Handbook of family therapy* (Vol. 51). New York: Brunner/Mazel.

Boszormenyi-Nagy, I., & Spark, G. (1973). *Invisible loyalties: Reciprocity in intergenerational family therapy*. New York: Harper & Row.

Bowen, M. (1966). The use of family theory in clinical practice. *Comprehensive Psychiatry, 7*, 345–376.

Bowen, M. (1978). *Family therapy in clinical practice*. New York: Aronson.

Bowlby, J. (1988). *A secure base*. New York: Basic Books.

Brammer, L. M., & MacDonald, G. (1996). *The helping relationship: Process and skills* (6th ed.). Boston: Allyn & Bacon.

Brisch, K. H. (2002). *Treating attachment disorders: From theory to therapy*. New York: Guilford.

Burns, D. D., & Auerbach, A. (1996) Therapeutic empathy in cognitive-behavioral therapy: Does it really make a difference. In P. M. Salkovskis (Ed.), *Frontiers of cognitive therapy* (pp. 135–164). New York: Guilford.

Carkhuff, R. R. (1999). *The art of helping in the 21st century* (8th ed). Amherst, MA: Human Resource Development Press.

Cashdan, S. (1988). *Object relations therapy*. New York: Norton.

Cassidy, J., & Shaver, P. R. (2002). *Handbook of attachment; Theory, research, and clinical applications*. New York: Guilford.

Cervone, D. (2000). Thinking about self-efficacy. *Behavior modification, 24*, 30–56.

Colin, V. (1996). *Human attachment*. New York: McGraw-Hill.

Crits-Christoph, P., & Barber, J. P. (1991). (Eds.). *Handbook of short-term dynamic psychotherapy*. New York: Basic Books.

Crits-Christoph, P., & Gibbons, B. B. (2002). Relational interpretations. In J. C. Norcross (Ed.), *Psychotherapy relationships that work: Therapist contributions and responsiveness to patients* (pp. 285–300). New York: Oxford University Press.

Davanloo, H. (Ed.). (1980). *Short-term dynamic psychotherapy*. New York: Aronson.

Dollard, J., & Miller, N. (1950). *Personality and psychotherapy: An analysis in terms of learning, thinking, and culture*. New York: McGraw-Hill.

Dutton, D.G., (1995). *The batterer: A psychological profile*. New York: Basic Books.

Ecker, B., & Hulley, L. (1996). *Depth-oriented brief therapy*. San Francisco: Jossey-Bass.

Egan, G. (2002). *The skilled helper* (7th ed.). Pacific Grove, CA: Brooks/Cole.

Elliott, R., Shapiro, D. A., Firth-Cozens, J., Stiles, W. B., Hardy, G. E., Llewelyn, S. P., & Margison, F. R. (1994). Comprehensive process analysis of insight events in cognitive-behavioral and psychodynamic-interpersonal psychotherapies. *Journal of Counseling Psychology, 41*, 449–463.

Engel, L., & Ferguson, T. (1990). *Hidden guilt*. New York: Pocket Books.

Erikson, E. (1968). *Childhood and society* (2nd ed.). New York: Norton.

Eron, L. D., & Huesmann, L. R. (1990). The stability of aggressive behavior—even into the third generation. In M. Lewis & S. M. Miller (Eds.), *Handbook of developmental psychopathology* (pp. 147–155). New York: Plenum.

Farber, B., & Geller, J. (1994). Gender and representation in psychotherapy. *Psychotherapy, 31*, 318–326.

Farley, J. (1979, January). Family separation-individuation tolerance: A developmental conceptualization of the nuclear family. *Journal of Marital and Family Therapy*, 61–67.

Field, T. (1996). Attachment and separation in young children. *Annual Review of Psychology, 47*, 541–561.

Fonagy, P. (2002). *What works for whom? A critical review of treatments for children and adolescents*. New York: Guilford.

Frailberg, S., Adelson, E., & Shapiro, V. (1975). Ghosts in the nursery: A psychoanalytical approach to the problems of impaired mother–infant relationships. *Journal of the American Academy of Child Psychiatry, 14*, 387–421.

Fromm, E. (1982). *Escape from freedom*. New York: Avon.

Fromm-Reichmann, F. (1960). *Principles of intensive psychotherapy*. Chicago: University of Chicago Press.

Galassi, J. P., & Bruch, M. A. (1992). Counseling with social interaction problems: Assertion and social anxiety. In S. D. Brown & R. W. Lent (Eds.), *Handbook of counseling psychology* (pp. 753–791). New York: Wiley.

Garfield, L. (1997). The therapist as a neglected variable in psychotherapy research. *Clinical Psychology: Science and Practice, 4*, 40–43.

Gaston, L. (1990). The concept of alliance and its role in psychotherapy. In A. E. Bergin & S. L. Garfield (Eds.), *Handbook of psychotherapy and behavior change* (4th ed., pp. 190–228). New York: Wiley.

Gelso, C., & Carter, J. (1994). Components of the psychotherapy relationship: Their interaction and unfolding during treatment. *Journal of Counseling Psychology, 41*(3) 296–306.

Gelso, C., & Hayes, H. (1998). *The psychotherapy relationship: Theory, research, and practice*. New York: Wiley.

Gendlin, E. T. (1998). *Focusing-oriented psychotherapy*. New York: Guilford.

Gill, M., & Muslin, H. (1976). Early interpretations of transference. *Journal of the American Psychoanalytic Association, 24*, 779–794.

Gilligan, C. (1982). *In a different voice: Psychological theory and women's development*. Cambridge, MA: Harvard University Press.

Goldenberg, I., & Goldenberg, H. (2004). *Family therapy: An overview* (6th ed.). Pacific Grove, CA: Brooks/Cole.

Greenberg, J., & Mitchell, S. (1983). *Object relations in psychoanalytic theory*. Cambridge, MA: Harvard University Press.

Greenberg, L. S. (2002). *Emotion-focused therapy: Coaching clients to work through their feelings*. Washington, D.C.: American Psychological Association.

Greenberg, L., & Paivio, S. (2003). *Working with emotions in psychotherapy*. New York: Guilford.

Greenson, R. (1967). *The technique and practice of psychoanalysis* (Vol. 1). New York: International Universities Press.

Guy, J. (1987). *The personal life of the psychotherapist*. New York: Wiley.

Haley, J. (1967). Toward a theory of pathological systems. In G. H. Zuk & I. Boszormenyi-Nagy (Eds.), *Family therapy and disturbed families*. Palo Alto, CA: Science and Behavior Books.

Haley, J. (1980). *Leaving home: Therapy with disturbed young people*. New York: McGraw-Hill.

Haley, J. (1996). *Learning and teaching therapy*. New York: Guilford.

Hardy, G., & Shapiro, D. (1987). Therapist verbal response mode in prescriptive vs. exploratory psychotherapy. *British Journal of Clinical Psychology, 24*, 235–245.

Hartman, E. (1978, October). Using eco-maps and genograms in family therapy. *Social Casework*, 464–476.

Hendrix, H. (2001). *Getting the love you want: A guide for couples*. New York: Owl Books.

Henry, W., Schacht, T., & Strupp, H. (1990). Patient and therapist introject, interpersonal process, and differential psychotherapy outcome. *Journal of Consulting and Clinical Psychology, 58*, 768–774.

Henry, W., Strupp, H., Schacht, T., & Gaston, L. (1994). Psychodynamic approaches. In A. E. Bergin & S. L. Garfield (Eds.), *Handbook of psychotherapy and behavior change* (4th ed., pp. 467–508). New York: Wiley.

Hesse, E., & Main, M. (1999). Second generation effects of unresolved trauma in non-maltreating parents: Dissociated, frightened and threatening parental behavior. *Psychoanalytic Inquiry, 19*, 481–540.

Hill, C. (2004). *Helping skills: Facilitating exploration, insight, and action* (2nd ed.). Washington, D.C.: American Psychological Association.

Hill, C. E., & O'Brien, K. M. (1999). *Helping skills: Facilitation exploration, insight, and action*. Washington, D.C.: American Psychological Association.

Hill, C. E., & O'Grady, K. (1985). List of therapist intentions illustrated in a case study and with therapists of varying theoretical orientations. *Journal of Counseling Psychology, 32*(1), 3–22.

Hill, C. E., Thompson, B., J., Cogar, M. M., & Denman, D. W. (1993). Beneath the surface of long-term therapy: Client and therapist report of their own and each other's covert processes. *Journal of Counseling Psychology, 40*, 278–288.

Hill, C. E., Thompson, B. J., & Corbett, M. M. (1992). The impact of therapist ability to perceive displayed and hidden client reactions on immediate outcome in first sessions of brief therapy. *Psychotherapy Research, 2*, 143–155.

Hobson, R. F. (1985). *Forms of feeling: The heart of psychotherapy*. London: Tavistock.

Horney, K. (1966). *Our inner conflicts*. New York: Norton.

Horney, K. (1970). *Neurosis and human growth*. New York: Norton.

Horowitz, M., Marmar, C., Krupnick, J., Wilner, N., Kaltreider, N., & Wallerstein, R. (1984). *Personality styles and brief psychotherapy*. New York: Basic Books.

Hovarth, A., Gaston, L., & Luborsky, L. (1993). The therapeutic alliance and its measures. In N. E. Miles, L. Luborsky, J. P. Barber, & J. P. Docherty (Eds.), *Psychodynamic treatment research* (pp. 247–273). New York: Basic Books.

Hovarth, A. O. (2000). The therapeutic relationship: From transference to alliance. *Journal of Clinical Psychology, 56*, 163–173.

Hovarth, A. O., & Bedi, R. P. (2002). The alliance. In J. C. Norcross (Ed.), *Psychotherapy relationships that work: Therapist contributions and responsiveness to patients* (pp. 37–70). New York: Oxford University Press.

Hovarth, A. O., & Greenberg, L. S. (1994). *The working alliance: Theory, research, and practice*. New York: Wiley.

Ivey, A. E., & Ivey, M. B. (1999). *Intentional interviewing and counseling* (4th ed.). Pacific Grove: Brooks/Cole.

James, R. K., & Guilliland, B. (2000). *Crisis intervention strategies* (4th ed.). Monterey, CA: Wadsworth.

Johnson, B., Taylor, E., D'Elia, J., Tzanetos, T., Rhodes. R., & Geller, J. D. (1995). The emotional consequence of therapeutic misunderstandings. *Psychotherapy Bulletin, 30*, 139–149.

Jones, A. S., & Gelso, C. J. (1988). Differential effects of style of interpretation: Another look. *Journal of Counseling Psychology, 35*, 363–369.

Jordan, J., Kaplan, A., Miller, J., Striver, P., & Surrey, J. (1991). *Women's growth in connection* (pp. 126–184). New York: Guilford.

Kafka, F. (1966). *Letter to his father*. New York: Schocken.

Kahn, M. (1997). *Between therapist and client*. New York: Freeman.

Karen, R. (1998). *Becoming attached*. New York: Oxford University Press.

Karen, R. (1992, February). Shame. *Atlantic Monthly*, pp. 40–70.

Kaufman, G. (1989). *The psychology of shame*. New York: Springer.

Kelly, A. E. (1998). Clients' secret keeping in outpatient therapy. *Journal of Counseling Psychology, 45*, 50–57.

Kelly, G. (1963). *The psychology of personal constructs*. New York: Norton.

Kerr, M. E. & Bowen, M. (1988). *Family evaluation: An approach based on Bowen theory*. New York: Norton.

Kiesler, D. (1988). *Therapeutic metacommunication: Therapist impact disclosure as feedback in psychotherapy*. Palo Alto, CA: Consulting Psychologists Press.

Kiesler, D. (1996). *Contemporary interpersonal theory and research. Personality, psychopathology, and psychotherapy*. New York: Wiley.

Kiesler, D. J. (1966). Some myths of psychotherapy research and the search for a paradigm. *Psychological Bulletin, 65*, 110–136.

Kiesler, D., & Van Denburg, T. (1993). Therapeutic impact disclosure: A last taboo in psychoanalytic theory and practice. *Clinical Psychology & Psychotherapy, 1*(1), 3–13.

Kiesler, D., & Watkins, L., (1989). Interpersonal complementarity and the therapeutic alliance: A study of relationship in psychotherapy. *Psychotherapy, 26*(2), 183–194.

Klerman, G. L., Weissman, M., Rounsaville, B., & Chevron, E. (1984). *Interpersonal psychotherapy of depression*. New York: Basic Books.

Kohut, H. (1971). *The analysis of the self*. New York: International Universities Press.

Kohut, H. (1977). *The restoration of the self*. New York: International Universities Press.

Lafferty, P., Beutler, L. F., & Crago, M. (1991). Differences between more and less effective psychotherapists: A study of select therapist variables. *Journal of Consulting and Clinical Psychology, 57*, 76–80.

Laing, R. D., & Esterson, A. (1970). *Sanity, madness and the family*. Middlesex, England: Penguin.

Lazarus, A. O. (1973). Multimodal behavior therapy: Treating the BASIC I.D. *Journal of Nervous and Mental Disease, 156*, 404–411.

Lazarus, A. O. (1989). *The practice of multimodal therapy*. Baltimore: Johns Hopkins University Press.

Lazarus, A. O. (1993). Tailoring the therapeutic relationship, or being an authentic chameleon. *Psychotherapy, 30*, 404–407.

Lazarus, A. O., Lazarus, C. & Fey, A. (1993). *Don't believe it for a minute!: Forty toxic ideas that are driving you crazy*. San Luis Obispo, CA: Impact.

Levenson, E. (1982). Language and healing. In S. Slip (Ed.), *Curative factors in dynamic psychotherapy*. New York: McGraw-Hill.

Levenson, H. (1995). *Time-limited dynamic therapy*. New York: Basic Books.

Levinson, D. (1978). *The seasons of a man's life*. New York: Ballantine.

Levinson, D. (2000). *The seasons of a woman's life*. New York: Knopf.

Lewis, H. (1971). *Shame and guilt in neurosis*. New York: International Universities Press.

Luborsky, L., & DeRubeis, R. (1984). The use of psychotherapy treatment manuals: A small revolution in psychotherapy research style. *Clinical Psychology Review, 4*, 5–14.

Luborsky, L., et al. (1986). The nonspecific hypothesis of therapeutic effectiveness: A current assessment. *American Journal of Orthopsychiatry, 56*, 501–512.

Luborsky, L., & Marks, D. (1991). Short-term supportive-expressive psychoanalytic psychotherapy. In P. Crits-Cristoph & J. Barber (Eds.), *Handbook of short-term dynamic psychotherapy* (pp. 110–136). New York: Basic Books.

Mahalik, J. R. (1994). Development of the Client Resistance Scale. *Journal of Counseling Psychology, 41,* 58–68.

Malan, D. H. (1976). *The frontier of brief psychotherapy: An example of the convergence of research and clinical practice.* New York: Plenum.

Mann, J. (1973). *Time-limited psychotherapy.* Cambridge, MA: Harvard University Press.

Mann, J., & Goldman, R. (1982). *A casebook in time-limited psychotherapy.* New York: McGraw-Hill.

Marx, S. A., & Gelso, C. J. (1987). Termination of individual counseling in a university counseling center. *Journal of Counseling Psychology, 34,* 3–9.

Mash, E., & Hunsley, J. (1993). Assessment considerations in the identification of failing psychotherapy: Bringing the negatives out of the darkroom. *Psychological Assessment, 5,* 292–301.

Masterson, J. (1972). *Treatment of the borderline adolescent: A developmental approach.* New York: Wiley.

Masterson, J. (1976). *Psychotherapy of the borderline adult: A developmental approach.* New York: Brunner/Mazel.

May, R. (1977). *The meaning of anxiety.* New York: Norton.

McCarthy, P. (1982). Differential effects of counselor self-referent responses and counselor status. *Journal of Counseling Psychology, 29,* 125–131.

McClure, F., & Teyber, E. (2003). *Casebook in child and adolescent treatment: Cultural and familial contexts.* Pacific Grove, CA: Wadsworth.

McGoldrick, M., Gerson, R., & Shellenberger, S. (1999). *Genograms: Assessment and intervention.* New York: Norton.

McWhirter, E.H. (1996). *Counseling for empowerment.* Alexandria, VA: American Counseling Association.

McWilliams, N. (1999). *Psychoanalytic case formulation.* New York: Guilford.

Miller, A. (1949). *Death of a salesman.* New York: Penguin.

Miller, A. (1984). *Prisoners of childhood.* New York: Basic Books.

Miller, W., & Rollnick, S. (2002). *Motivational interviewing* (2nd ed.). New York: Guilford.

Millon, T. (1981). *Disorders of personality: DSM-III, Axis 11.* New York: Wiley.

Mills, J., Bauer, G., & Miars, R. (1989). Use of transference in short-term dynamic therapy. *Psychotherapy, 26,* 338–343.

Minuchin, S. (1974). *Families and family therapy.* Cambridge, MA: Harvard University Press.

Minuchin, S. (1984). *Family kaleidoscope.* Cambridge, MA: Harvard University Press.

Minuchin, S., Lee, W. Y., & Simon, G. (1996). *Mastering family therapy: Journeys of growth and transformation.* New York: Wiley.

Minuchin, S., & Nichols, M. P. (1998). Structural family therapy. In F. M. Datilio (Ed.), *Case studies in couple and family therapy: Systemic and cognitive perspectives.* New York: Guilford.

Mitchell, S. (1988). *Relational concepts in psychoanalysis.* Cambridge, MA: Harvard University Press.

Mitchell, S. A. (1993). *Hope and dread in psychoanalysis.* New York: Basic Books.

Mitchell, S., & Black, M. (1995). *Freud and beyond.* New York: Basic Books.

Mueller, W., & Kell, B. (1972). *Coping with conflict: Supervising counselors and psychotherapists.* New York: Appleton-Century-Crofts.

Najavits, L., & Strupp, H. (1994). Differences in the effectiveness of psychodynamic therapists: A process-outcome study. *Psychotherapy, 31,* 114–123.

Nathanson, D. L. (1987). (Ed.) *The many faces of shame* (pp. 246–270). New York: Guilford.

Norcross, J. C. (2002). *Psychotherapy relationships that work.* New York: Oxford University Press.

Olson, D. H., McCubbin, H. I., Barnes, H., Larsen, A., Muxen, M., & Wilson, M. (1983). *Families: What makes them work*. Newbury Park, CA: Sage.

Persons, J., Curtis, J., & Silberschatz, G. (1991). Psychodynamic and cognitive-behavioral formulations of a single case. *Psychotherapy: Theory, Research, Practice, Training, 28*, 4.

Peterson, D., Friedman, S., Geshmay, S., & Hill, C. E. (1998). Client perspectives on impasses. Unpublished manuscript, University of Maryland.

Passons, W. (1975). *Gestalt approaches in counseling*. New York: Holt, Rinehart & Winston.

Pinderhughes, H. (1989). *Understanding race, ethnicity and power*. New York: Free Press.

Pipher, M. (1994). *Reviving Ophelia*. New York: Ballantine.

Poorman, P. (2003). *Microskills*. Boston: Pearson.

Regan, A. M., & Hill, C. E. (1992). Investigation of what clients and counselors do not say in brief therapy. *Journal of Counseling Psychology, 39*, 168–174.

Rennie, D. L. (1992). Qualitative analysis of the client's experience of psychotherapy: The unfolding of reflexivity. In S. G. Toukmanian & D. L. Rennie (Eds.), *Psychotherapy process research. Paradigmatic and narrative approaches* (pp. 211–233). Newbury Park, CA: Sage.

Rhodes, R. H., Thompson, B. J., & Elliott, R. (1994). Client retrospective recall of resolved and unresolved misunderstanding events. *Journal of Counseling Psychology, 41*, 473–483.

Robbins, S. B., & Jolkovski, M. P. (1987). Managing countertransference feelings: An interactional model using awareness of feeling and theoretical framework. *Journal of Counseling Psychology, 34*, 276–282.

Robitschek, C., & McCarthy, P. (1991). Prevalence of counselor self-reference in the therapeutic dyad. *Journal of Counseling & Development, 69*, 218–221.

Rocklin R., & Levitt, D. (1987). Those who broke the cycle: Therapy with nonabusive adults who were physically abused as children. *Psychotherapy, 24*(4), 769–778.

Rogers, C. (1951). *Client-centered therapy*. Boston: Houghton-Mifflin.

Rogers, C. (1959). A theory of therapy, personality and interpersonal relationships as developed in the client-centered framework. In S. Koch (Ed.), *Psychology: A study of science: Vol. 3. Formulations of the person and the social context* (pp. 184–256). New York: McGraw-Hill.

Rogers, C. (1975). Empathy: An unappreciated way of being. *Counseling Psychologist, 21*, 95–103.

Rogers, C. (1980). *A way of being*. Boston: Houghton Mifflin.

Safran, J., & Muran, J. C. (1995). Resolving therapeutic alliance ruptures: Diversity and integration. *Psychotherapy in Practice, 1*, 81–91.

Safran, J. & Muran, J. (2003). *Negotiating the therapeutic alliance: A relational treatment guide*. New York: Guilford.

Safran, J., & Muran, J. (Eds.). (1998). *The therapeutic alliance in brief psychotherapy*. Washington, D.C.: American Psychological Association.

Safran, J. D., Muran, J. C., Samstag, L. W., & Stevens, C. (2002). Repairing alliance ruptures. In J. C. Norcross (Ed.), *Psychotherapy relationships that work: Therapist contributions and responsiveness to patients* (pp. 235–254). New York: Oxford University Press.

Safran, J. D., & Segal, Z. V. (1990). *Interpersonal process in cognitive therapy*. New York: Basic Books.

Satir, V. (1967). *Conjoint family therapy*. Palo Alto, CA: Science and Behavior Books.

Satir, V., & Bitter, J. R. (2000). The therapist and family therapy: Satir's human validation model. In A. Horne & J. L. Passmore (Eds.), *Family counseling and therapy* (3rd ed.). Pacific Grove, CA: Wadsworth–Brooks/Cole.

Schatzman, M. (1973). *Soul murder: Persecution in the family*. New York: Random House.

Seligman, M. (1975). *Helplessness*. New York: Freeman.

Seligman, M. (1995). The effectiveness of psychotherapy: The Consumer Reports study. *American Psychologist, 50*, 965–974.

Sheff, T. J. (1995). Shame and related emotions. *American Behavioral Scientist, 38*(8) [Special Issue].

Sifneos, P. (1987). *Short-term dynamic psychotherapy: Evaluation and techniques* (2nd ed.). New York: Plenum.

Simon, J. L. (2002). Analysis of the relationship between shame, guilt, and empathy in intimate relationship violence (abstract). *Dissertation Abstracts International, 63*(6-b), 3026.

Snyder, C. R. (1994). *The psychology of hope: You can get there from here.* New York: Free Press.

Solomon, M. (1973). A developmental, conceptual premise for family therapy. *Family Process, 12*(2), 179–188.

Speight, S., Myers, L., Cox, C., & Highlen, P. (1991). A redefinition of multicultural counseling. *Journal of Counseling Development, 70*(1), 29–36.

Spiegel, D., & Alpert, J. L. (2000). The relationship between shame and rage: Conceptualizing the violence at Columbine High School. *Journal for the Psychoanalysis of Culture, 5*(2), 237–246.

Springmann, R. (1986). Countertransference: Clarifications in supervision. *Contemporary Psychoanalysis, 22,* 252–277.

Steinbeck, J. (1952). *East of Eden.* New York: Viking.

Stierlin, H. (1972). *Separating parents and adolescents.* New York: Quadrangle.

Stolorow, R. D., Atwood, G. E., & Brandchaft, B. (Eds.). *The intersubjective perspective.* New York: Aronson.

Strupp, H. H. (1980a). Success and failure in time-limited psychotherapy. *Archives of General Psychiatry, 37,* 595–613.

Strupp, H. H. (1980b). Success and failure in time-limited psychotherapy: A systematic comparison of two cases. *Archives of General Psychiatry, 37,* 708–716.

Strupp, H., & Binder, J. (1984). *Psychotherapy in a new key: A guide to time-limited dynamic psychotherapy.* New York: Basic Books.

Strupp, H., & Hadley, S. (1979). Specific versus nonspecific factors in psychotherapy: A controlled study of outcome. *Archives of General Psychiatry, 36,* 1125–1136.

Sue, D., & Sue, H. (1998). *Counseling the culturally different* (3rd ed.). New York: Wiley.

Sue, S., & Zane, N. (1987). The role of culture and cultural techniques in psychotherapy: A critique and reformulation. *American Psychologist, 42,* 37–45.

Sullivan, H. S. (1968). *The interpersonal theory of psychiatry.* New York: Norton.

Sullivan, H. S. (1970). *The psychiatric interview.* New York: Norton.

Teyber, E. (1981). Structural family relations: A review. *Family Therapy, 1,* 39–48.

Teyber, E. (1983). Effects of the parental coalition on adolescent emancipation from the family. *Journal of Marital and Family Therapy, 9,* 89–99.

Teyber, E. (2001). *Helping children cope with divorce.* San Francisco: Jossey-Bass.

Teyber, E., & McClure, F. (2000). Therapist variables. In C. Snyder & R. Ingram (Eds.), *Handbook of psychological change: Psychotherapy processes and practices for the 21st century.* New York: Wiley.

Tyron, G., & Kane, A. (1993). Relationship of working alliance to mutual and unilateral termination. *Journal of Counseling Psychology, 40*(1), 33–36.

Wachtel, P. (1982). *Psychoanalysis and behavior therapy: Toward an integration.* New York: Basic Books.

Wachtel, P. (1987). *Action and insight.* New York: Guilford.

Wachtel, P. (1993). *Therapeutic communication: Principles and effective practice.* New York: Guilford.

Wachtel, P. (1997). *Psychoanalysis, behavior therapy, and the relational world.* Washington, D.C.: American Psychological Association.

Wachtel, P. (2002). Termination of therapy: An effort at integration. *Journal of Psychotherapy Integration, 12*(3), 373–383.

Walborn, F. (1996). *Process variables: Four common elements of counseling and psychotherapy.* Pacific Grove, CA: Brooks/Cole.

Wampold, B. E., Mondin, G. W., Moody, M., Stich, F., Benson, K., & Ahn, H. (1997). A meta-analysis of outcome studies comparing bona fide psychotherapies: Empirically "all must have prizes." *Psychological Bulletin, 122,* 203–215.

Watson, J. C., & Greenberg, L. (2000). Alliance ruptures and repairs in experimental therapy. *Journal of Clinical Psychology, 56*, 175–186.

Weiss, J. (1993). *How psychotherapy works*. New York: Guilford.

Weiss, J., & Sampson, H. (1986). *The psychoanalytic process: Theory, clinical observation and empirical research*. New York: Guilford.

Wenar, C., & Kerig, P. (2000). *Psychopathology from infancy through adolescence: A developmental approach* (4th ed.). New York: Random House.

Wheelis, A. (1974). *How people change*. New York: Harper & Row.

Wiger, D. (1999). *The psychotherapy documentation primer*. New York: Wiley.

Winnicott, D. W. (1965). Ego distortion in terms of true and false self. In *The maturational process and the facilitating environment*. New York: International Universities Press.

Woodward, B. (1984). *Wired*. New York: Simon & Schuster.

Yalom, I. (1995). *The theory and practice of group psychotherapy* (4th ed.). New York: Basic Books.

Yalom, I. (1981). *Existential psychotherapy*. New York: Basic Books.

Yalom, I. (2003). The gift of therapy: An open letter to a new generation of therapists and their patients. New York: Perennial Currents.

Young, J. (1999). *Cognitive therapy for personality disorders: A schema-focused approach* (3rd ed.). Sarasota, FL: Professional Resource Press.

Young, J. E., Klosko, H. S., & Weishaar, M. E. (2003). *Schema therapy: A practitioner's guide*. New York: Guilford.

NAME INDEX

SUBJECT INDEX